W9-APZ-844

EUCLIDEAN AND NON-EUCLIDEAN GEOMETRIES

EUCLIDEAN AND NON-EUCLIDEAN GEOMETRIES

M. HELENA NORONHA
California State University, Northridge

PRENTICE HALL, Upper Saddle River, New Jersey, 07458

Library of Congress Cataloging-in-Publication Data

Noronha, M. Helena.
Euclidean and non-Euclidean geometries / by M. Helena Noronha.
p. cm.
Includes index.
ISBN 013-033717-X
1. Geometry. I. Title

QA 473 .N67 2002 CIP
516'.04—dc21

2001033235

Acquisitions Editor: *George Lobell*
Editor in Chief: *Sally Yagan*
Vice President/Director of Production and Manufacturing: *David W. Riccardi*
Production Editor: *Lynn Savino Wendel*
Senior Managing Editor: *Linda Mihatov Behrens*
Executive Managing Editor: *Kathleen Schiaparelli*
Manufacturing Buyer: *Alan Fischer*
Manufacturing Manager: *Trudy Pisciotti*
Marketing Manager: *Angela Battle*
Art Director: *Jayne Conte*
Editorial Assistant: *Melanie Van Benthuysen*
Cover Designer: *Kiwi Design*
Cover Photo: *Frank Stella*, "Giufa, La Luna, I Ladri e Le Guardie" from the Cones and Pillars series (1984). Synthetic polymer paint, oil, urethane enamel, fluorescent alkyd, and printing ink on canvas, and etched magnesium, aluminum, and fiberglass, 9' 71/4" × 16" 31/4" × 24" (292.7 × 495.9 × 61 cm). The Museum of Modern Art, New York. Acquired through the James Thrall Soby Request. Photograph @2001 The Museum of Modern Art, New York.

Printed in the United States of America
10 9 8 7 6 5 4 3 2 1

ISBN 0–13–033717–X

Pearson Education Ltd., London
Pearson Education Australia PTY, Limited., Sydney
Pearson Education Singapore, Pte. Ltd.
Pearson Education North Asia, Ltd., Hong Kong
Pearson Education Canada, Ltd., Toronto
Pearson Educación de Mexico, S.A. de C.V.
Pearson Education - Japan, Tokyo
Pearson Education Malaysia, Pte. Ltd.

To my husband Darren

and my parents

José Carlos and Alice

CONTENTS

PREFACE

This is a book to be used in undergraduate geometry courses at the junior-senior level. It develops a self-contained treatment of classical Euclidean geometry through both axiomatic and analytic methods. In addition, the text integrates the study of spherical and hyperbolic geometry. Euclidean and hyperbolic geometries are constructed upon a consistent set of axioms, as well as presenting the analytic aspects of their models and their isometries. It also contains a study of the Euclidean n-space.

The text differs from the traditional textbooks on the foundations of geometry by taking a more natural route that leads to non-Euclidean geometries. The topics presented not only compare different parallel postulates, but place a certain emphasis on analytic aspects of some of the non-Euclidean geometries. I intend to show students how theories that underlie other fields of mathematics can be used to better understand the concrete models for the axiom systems of Euclidean, spherical, and hyperbolic geometries and to better visualize the abstract theorems. I use elementary calculus to compute lengths of curves in 3-space and on spheres, a topic usually found at the beginning of elementary books on differential geometry.

Another feature of this book is the inclusion in the text of a few topics in linear algebra and complex variable. I treat those topics (and only those) which will be used in the book. They are introduced as needed to advance in the study of geometry. I do not assume any prior knowledge of complex variable and only a few basic facts about matrices. The topics of linear algebra are used to do a more advanced study of rigid motions of the n-dimensional Euclidean space, while complex variables are used to thoroughly study two models of the hyperbolic plane. This text also describes the connections between the study of geometric transformations and transformations groups — for example, showing how a dihedral group is realized as the symmetry group of a regular polygon.

The prerequisites for reading this book should be quite minimal. It has been written not presupposing any knowledge of Euclidean and analytic geometry. However, basic elementary set theory is required for the axiomatic geometry part. The part containing the analytic methods is basically self-contained, assuming only that the students have had a basic single-variable calculus course.

The book has been written assuming that a standard mathematical curriculum contains at most two semesters of geometry. Although some beautiful topics, such as projective geometry, have been left out, I believe that I have chosen crucial aspects of classical and modern geometry that provide an accessible introduction to advanced geometry.

It is not unreasonable for the instructor to hope to cover the whole book in one year. Of course, it depends on the ability and experience of his/her students. But if some choices have to be made, the book is organized to permit a number of one- or two-semester course outlines so that instructors may follow their preferences. Moreover, I have attempted to discuss all topics in detail, so that the ones not covered could be undertaken as independent study. Since Chapter 1 sets the tone for the whole book and Chapter 2 contains the classical results of plane Euclidean geometry, they form the core of the book.

It is possible to teach a course only on axiomatic geometry using this text. A one-semester course on Euclidean and hyperbolic geometries using only axiomatic methods would be Chapters $1, 2$, and 8. If the instructor wishes to cover geometric transformations, without coordinatizing the plane, then Sections $3.1, 3.2, 3.3$ should be included. Students will then be ready for the models of the hyperbolic plane and its isometries in Chapter 9. Another one-semester alternative is the one on 2- and 3-dimensional Euclidean geometry. This would include Chapters 1 and 2, Sections $3.1, 3.2, 3.3$, Sections $4.1, 4.2, 4.3$, and Chapter 6.

The instructor who wants to study hyperbolic geometry mainly through the Poincaré models can move to Chapters 9 and 10 after having covered only Sections $8.1, 8.2$ and 8.3. Likewise, in a one-semester course, spherical geometry can be briefly studied in Section 9.1, which is independent of Chapter 7 where this topic is presented with some thoroughness. The diagram on page xv shows the dependencies among the various chapters, and the instructor can customize the coverage by choosing the desired topics.

This text fits very well the needs of an undergraduate in the secondary teaching option. It contains the classical Euclidean geometry required for obtaining a teaching credential, and it is an introduction to non-Euclidean geometry. The majority of mathematics majors in the secondary teaching option are not required to take courses on advanced calculus and complex variables. It is for such majors that I wrote a self-contained text. The book could also be used in an introductory graduate course for secondary or community college teachers. This text should also appeal to undergraduate mathematics majors interested in geometry. The analytic methods used in the text will complement and deepen the knowledge of certain areas of mathematics involved in such methods, including Euclidean geometry itself.

Throughout the text I follow the same guiding principles, precisely stating definitions and theorems. Proofs are presented in detail, and when some steps are missing the reader is told precisely from which exercises they will follow. A difficulty usually found at the beginning of courses that build a geometry upon a set of postulates is to have the students understand that results derived from these postulates hold in some of the non-Euclidean geometries as well, and therefore their proofs cannot rely on facts obtained from their drawings. To clarify this point, in all proofs in the first sections of Chapter 1 I precisely indicate which axioms have been used. I also clearly point out assertions that follow from results that have already been proved. Likewise, after introducing the Euclidean parallel postulate in Chapter 2, I repeatedly point out in subsequent proofs where such a postulate or an equivalent result is used.

The degree of difficulty of exercises varies. Some have the purpose of only fixing the concepts or practicing a method, while others complete the results proved in the text or extend some of the theory, sometimes proving a result that will be needed later in the book. All challenging problems contain hints that should get the students started on the solution. Moreover, bearing in mind that to write logical and coherent proofs for theorems is a difficult task for beginners, in several exercises I sketch the proofs, and students are asked to justify the steps or explain the contradictions.

Acknowledgments This text evolved from geometry courses that I taught at California State University, Northridge. I acknowledge my debt and my gratitude to all students who attended these courses in 1998 and 1999. Their reaction to my classes was the best encouragement I received to write this book. I would especially like to thank my colleagues Yuriko Baldin, Elena Marchisotto, Eliane Quelho, Patrick Shanahan, and Joel Zeitlin for having read the preliminary notes and making helpful suggestions. Particular thanks go to Adonai Seixas and Dennis Kletzing for their help with LaTeX, and to George Lobell and all Prentice Hall staff for helping me to refine the text. In addition, I wish to thank the reviewers Roger Cooke, David Ewing, Loren Johnsen, Sandy Norman, and Alvin Tinsley for numerous useful suggestions.

Finally, I want to thank my husband for spending many hours drawing the figures, for his encouragement, and tolerance.

M. Helena Noronha
maria.noronha@csun.edu

Dependencies among chapters and sections

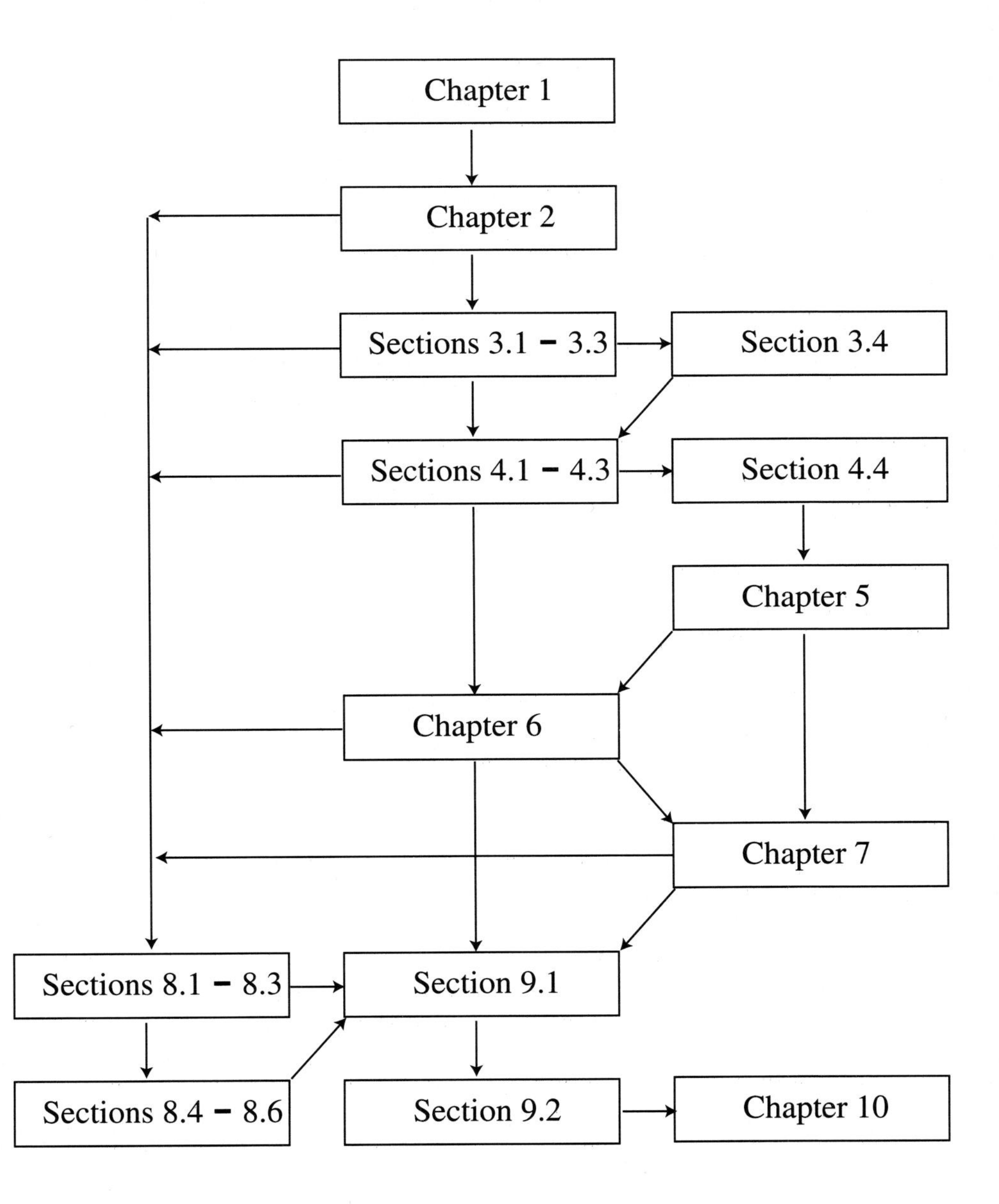

Chapter 1

NEUTRAL GEOMETRY

1.1 Introduction

In approximately 300 B.C., *Euclid* wrote his famous book entitled *The Elements.* Although the results it contained were already known and had been proved, his contribution was to present them using a logical structure that became a model for many theories in mathematics and philosophy. Such a structure is called an *axiom system.* It begins with some undefined terms, definitions, and *axioms* or *postulates* (statements that we assume to be true) and builds a theory upon such terms, definitions and postulates. This theory contains results called *theorems* or *propositions* — statements that are proved using only the axioms and the earlier results. Euclid's first five postulates are as follows:

1. A line may be drawn between any two points.
2. Any line segment may be extended indefinitely.
3. A circle may be drawn with any given point as center and any given radius.
4. All right angles are congruent.
5. If two lines are cut by a transversal in such way that the sum of the two interior angles on one side of the transversal is less than two right angles, then the lines meet on that side of the transversal.

From the very beginning, the fifth postulate was attacked, since it was not as self-evident as the first four. For almost two thousand years mathematicians tried to prove the fifth postulate from the others. All attempts resulted in statements that turned out to be equivalent to the fifth postulate. Among them we find *Playfair's postulate* (John Playfair, 1748–1819), also called the *parallel postulate.* It states that *through a point not on a given line, exactly one line can be drawn in the plane parallel to the given line.* We will show this equivalence in Chapter 2.

At the beginning of the nineteenth century several noted mathematicians (Gauss, Bolyai, Lobachevsky) realized Euclid was correct; that is, the fifth postulate is independent of the first four. The reason is that other geometries exist for which the first four postulates hold but not the fifth. These have been called *non-Euclidean* geometries.

The discovery of non-Euclidean geometries led mathematicians to reexamine the foundations of Euclidean geometry. At the beginning of the twentieth century, a lot of criticism was made of Euclid's work. For instance, quite a few of his proofs relied on diagrams; some undefined terms were never considered by him; and mainly, in order to make his proofs rigorous, a larger set of axioms is needed. Several axiom systems have been composed to furnish rigorous proofs for the results of Euclid.

In this chapter we will present a set of axioms that does not include any statement equivalent to Euclid's fifth postulate. The set of geometric theorems we obtain is called *neutral geometry*, since they do not require the postulate that characterizes Euclidean geometry. Such a postulate will be assumed in Chapter 2; the aim is to clarify the role of the parallel postulate in the classical results of Euclidean geometry.

1.2 Axioms of Incidence and Betweenness

For the first set of axioms we assume two undefined objects, *point* and *line*, and an undefined relation between them, namely, *lines contain points.* This relation is also called *incidence.* If a line l is incident with point P, then we will say that *point P lies on l.* The *axioms of incidence* are:

$\mathbf{I_1}$ Given two distinct points P, Q there exists a unique line incident with P and Q. This line will be denoted by $\overleftrightarrow{PQ}$.

$\mathbf{I_2}$ For every line l there exist at least two distinct points incident with l.

I$_3$ There exist three distinct points with the property that no line is incident with all three of them.

Definition 1.2.1 *Three points A, B, and C are called* collinear *if there exists a line which is incident with all three of them.*

Definition 1.2.2 *Two lines l and m are are said to be* parallel *if they do not have a point in common. Two distinct lines with a common point are called* concurrent *lines.*

Observe that the undefined relation "lines contain points" suggests that we consider a line as a specified set of points. Thus, in order to make some of the statements easier to write, in this text we will be using the standard notation of set theory. For instance, if a point P lies on line l, we will write $P \in l$. Likewise, if a point lies on lines l and m, we write $P \in l \cap m$.

Some results can be proved from the postulates above. It is common in mathematics to reserve the word "theorem" for a result that has a special importance and significance in the subject and is central to the development of a theory, while other statements of less impact are called "propositions." Such a distinction is not so clear for many results, and the use of the words "theorem" or "proposition" for such results really depends on the author. We call the next set of results a proposition.

Some of the results below will be proved *by contradiction.* That is to say, we assume that the result we wish to prove does *not* hold. Then we make an argument and ultimately deduce a statement that we know is not true. Such a negation is called a contradiction.

In the first set of problems the reader will practice this method of doing proofs. The method of *proof by contradiction* will be used several times in this text.

Proposition 1.2.3 *The axioms of incidence imply:*
(a) If l and m are two distinct lines, then they either are parallel or have a unique point in common.
(b) There exist three distinct lines with no common point.
(c) For every line there exists at least one point not lying on it.
(d) For every point there exists at least one line not passing through it.
(e) For every point P there exist at least two lines through P.

Proof *(a)* Let us suppose that l and m are not parallel. Then these two lines do not satisfy the definition of parallel lines. It follows that they have at least one common point. Now we suppose that they have more than one common point; that is, we suppose that $P, Q \in l \cap m$, $P \neq Q$. Then Axiom $\mathbf{I_1}$ implies that $l = m$, and this contradicts what we assumed about l and m, namely that they are distinct lines.

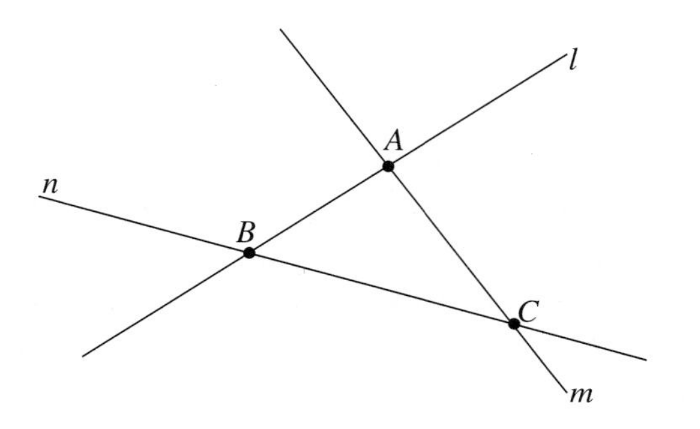

Figure 1.1

(b) It follows from $\mathbf{I_3}$ that there are three points, say A, B, and C that are noncollinear (see Figure 1.1). Using Axiom $\mathbf{I_1}$, we conclude that there exist three distinct lines l, m, and n, which are the unique lines incident with A and B, A and C, and B and C, respectively. Let us suppose that there exists a point P such that $P \in l \cap m \cap n$; then $P \in l \cap m$ and then $P = A$ by part (a) of this proposition. The point P is also in the intersection of m and n and hence $P = C$, again by part (a). This is clearly a contradiction, since we have assumed that $A \neq C$.

For the proofs of *(c)*, *(d)*, and *(e)* see Exercise 1. □

The next set of axioms assumes an undefined relation among three points called *betweenness*. The *axioms of betweenness* are also called *axioms of order*. They are:

> $\mathbf{B_1}$ If B is between A and C, then A, B, and C are three distinct collinear points and B is between C and A.
>
> $\mathbf{B_2}$ Given two distinct points B and D, let l be a line incident with B and D. Then there exist points A, C, and E

such that B is between A and D, C is between B and D, and D is between B and E.

$\mathbf{B_3}$ Given three distinct points lying on the same line, one and only one of the points is between the other two.

Figure 1.2 Axiom $\mathbf{B_2}$

Definition 1.2.4 *Given two distinct points A and B, the* line segment $\overline{AB}$ *is the set whose elements are A, B, and all points that lie on the line through A and B and are between A and B. The points A and B are called the* endpoints *of the line segment $\overline{AB}$.*

Definition 1.2.5 *A* ray $\overrightarrow{AB}$ *is the set of points lying on the line segment $\overline{AB}$ and all points E on line through A and B such that B is between A and E. The point A is called vertex of the ray $\overrightarrow{AB}$, and we say that ray $\overrightarrow{AB}$* emanates from *A.*

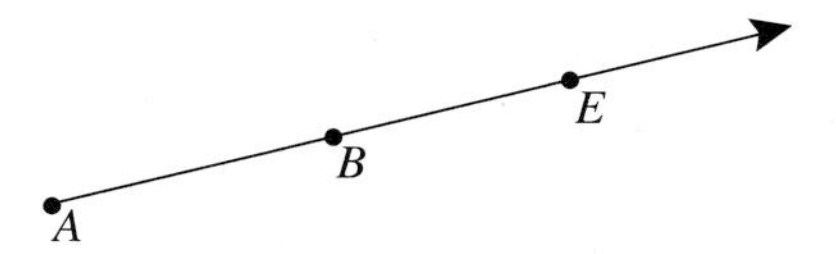

Figure 1.3

It follows from Axiom $\mathbf{B_2}$ that given two points A and B on a line l there exist infinitely many points of l between A and B. $\mathbf{B_2}$ also implies that there are infinitely many points lying on $\overrightarrow{AB}$ that are not on $\overline{AB}$.

The next result is a consequence of the first three axioms of betweenness. Before stating it we recall that a general precedure in set theory for showing that two sets S_1 and S_2 are the same is to show $S_1 \subset S_2$ and $S_2 \subset S_1$. We will use this method to prove the next proposition.

Proposition 1.2.6 *For any two points A and B we have:*
(a) $\overrightarrow{AB} \cup \overrightarrow{BA} = \overleftrightarrow{AB}$.
(b) $\overrightarrow{AB} \cap \overrightarrow{BA} = \overline{AB}$.

Proof *(a)* We will show that $\overrightarrow{AB} \cup \overrightarrow{BA}$ and $\overleftrightarrow{AB}$ are the same set of points; that is, we will show that

$$(\overrightarrow{AB} \cup \overrightarrow{BA}) \subset \overleftrightarrow{AB} \quad \text{and} \quad \overleftrightarrow{AB} \subset (\overrightarrow{AB} \cup \overrightarrow{BA}).$$

Since $\overrightarrow{AB} \subset \overleftrightarrow{AB}$ and $\overrightarrow{BA} \subset \overleftrightarrow{AB}$ we have that $(\overrightarrow{AB} \cup \overrightarrow{BA}) \subset \overleftrightarrow{AB}$. Now, for the second part, let P be a point lying on $\overleftrightarrow{AB}$. From $\mathbf{B_3}$ we get that one and only one of the following cases holds:

(1) P is between A and B and hence $P \in \overline{AB}$.

(2) A is between P and B and hence $P \in \overrightarrow{BA}$.

(3) B is between A and P and hence $P \in \overrightarrow{AB}$.

In each case $P \in l$, since $\overline{AB}$, $\overrightarrow{BA}$, and $\overrightarrow{AB}$ are all subsets of $\overleftrightarrow{AB}$. Therefore $\overleftrightarrow{AB} \subset (\overrightarrow{AB} \cup \overrightarrow{BA})$.

The proof of *(b)* is sketched in Exercise 2. □

Definition 1.2.7 *Let l be a line and A and B two distinct points not lying on l. We say that A* and *B* are on the same side of *l if the line segment $\overline{AB}$ does not intersect l. In the case that $\overline{AB}$ does intersect l, we say that A and B are on opposite sides of l.*

Observe that if point C is between A and B then A and B are on opposite sides of any line $l \neq \overleftrightarrow{AB}$ that intersects line $\overleftrightarrow{AB}$ at C. The last betweenness axiom is:

> $\mathbf{B_4}$ Given a line l and three points A, B, and C not lying on l, we have:
> (a) If A and B are on the same side of l and B and C are on the same side of l, then A and C are on the same side of l.
> (b) If A and B are on opposite sides of l and B and C are on opposite sides of l, then A and C are on the same side of line l.

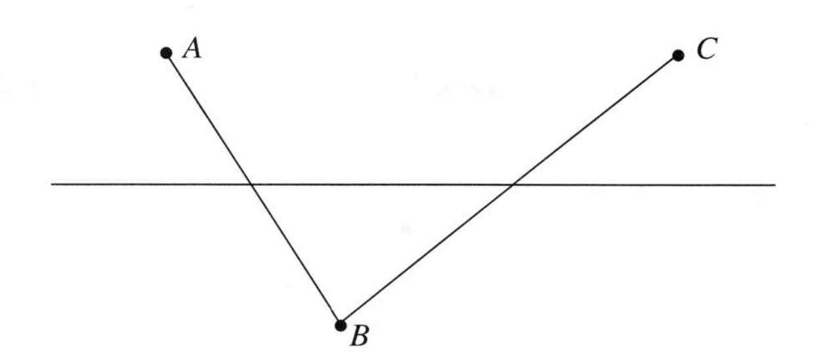

Figure 1.4 Axiom $\mathbf{B_4}$

Axiom $\mathbf{B_4}$ implies the following: "If A and B are on the same side of l and B and C are on opposite sides of l, then A and C are on opposite sides of l."

Definition 1.2.8 *Let l be a line and A a point not lying on l. The set of all points that are on the same side of l as A is called the (open)* half-plane bounded by l *containing A. We denote it by $H_{l,A}$* (see Figure 1.5).

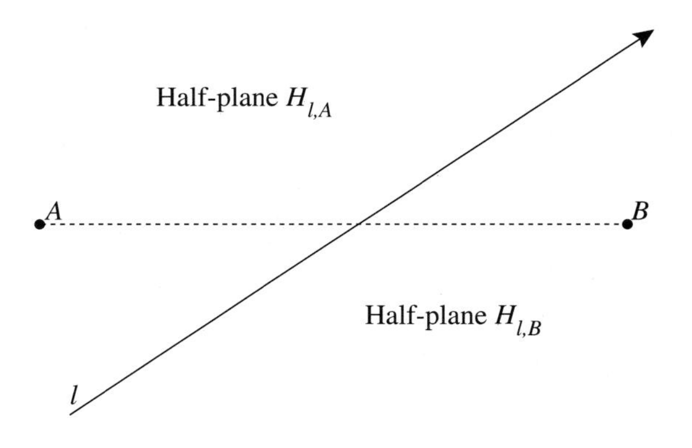

Figure 1.5

Notice that $\mathbf{B_4}$(a) implies that if A and B are on the same side of l then $H_{l,A} = H_{l,B}$.

Theorem 1.2.9 (Plane separation property) *Every line determines exactly two disjoint half-planes.*

Proof The axioms of incidence imply that given a line l there exist points A and C such that $A \notin l$ and $C \in l$. From $\mathbf{B_2}$ we get that there exists a point B such that C is between A and B. Therefore A and B are on opposite sides of l. The theorem will be proved if we show the following:

(i) $H_{l,A} \cap H_{l,B} = \phi$.

(ii) if $P \notin l$ then either $P \in H_{l,A}$ or $P \in H_{l,B}$.

(i) Suppose that there is point P such that $P \in H_{l,A} \cap H_{l,B}$. From the definition of half-plane we conclude that P and A are on the same side of l and that P and B are on the same side of l. In this case Axiom $\mathbf{B_4}$(a) implies that A and B are on the same side of l, which contradicts that A and B are on opposite sides of l. Therefore $H_{l,A} \cap H_{l,B}$ is the nullset.

(ii) If P is a point not lying on l and $P \notin H_{l,A}$, then P and A are on opposite sides of l. Since A and B are on opposite sides of l, Axiom $\mathbf{B_4}$ implies that P is on the same side of l as B, and then $P \in H_{l,B}$. □

Theorem 1.2.10 (Line separation property) *Let A, B, and C be three points such that A is between C and B and let l denote the line through them. Then for every $P \neq A$ lying on l one, and only one, of the following occurs:*

(i) $P \in \overrightarrow{AB}$.

(ii) $P \in \overrightarrow{AC}$.

Proof Let m be any line intersecting line l at point A. It follows from the plane separation property that if $P \in l$ and $P \neq A$, then either $P \in H_{m,B}$ or $P \in H_{m,C}$ but not both. Let us then suppose that $P \in H_{m,B}$, that is, P and B are on the same side of m. If A is between P and B we have that $\overline{PB}$ intersects line m at A, contradicting that P and B are on the same side of line m. Since A is not between P and B, Axiom $\mathbf{B_3}$ implies that only one of the two following cases holds for points A, B, P:

$$(1) \quad P \text{ is between } A \text{ and } B \text{ and hence } P \in \overrightarrow{AB}.$$

$$(2) \quad B \text{ is between } A \text{ and } P \text{ and hence } P \in \overrightarrow{AB}.$$

Now we suppose that $P \in H_{m,C}$, and a similar argument shows that in this case $P \in \overrightarrow{AC}$. □

Proposition 1.2.11 *Let A, B, C be three points such that A is between B and C. Then*
(a) $\overline{BC} = \overline{BA} \cup \overline{AC}$.
(b) $\overline{BA} \cap \overline{AC} = \{A\}$.
(c) $\overrightarrow{BA} = \overrightarrow{BC}$.

Proof The proof of part *(a)* is sketched in Exercise 6. We leave parts *(b)* and *(c)* to the reader. □

Theorem 1.2.12 (Pasch's theorem) *Let A, B, and C be distinct noncollinear points and l be any line intersecting $\overline{AB}$ at a point between A and B. If C does not lie on l, then l intersects only one of the other two sides.*

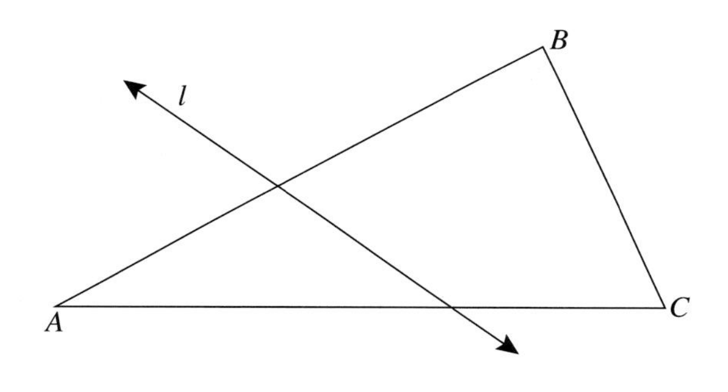

Figure 1.6

Proof Since $\overline{AB}$ intersects l, we conclude that A and B are on opposite sides of l. Since $C \notin l$, from Proposition 1.2.9 we obtain that either $C \in H_{l,A}$ or $C \in H_{l,B}$ but not both. The first case implies that l intersects $\overline{BC}$, while the second gives that l intersects $\overline{AC}$. □

Definition 1.2.13 *An* angle *with vertex A is a point A together with two rays $\overrightarrow{AB}$ and $\overrightarrow{AC}$ emanating from A. The rays $\overrightarrow{AB}$ and $\overrightarrow{AC}$ are called* sides.

Definition 1.2.14 *We say that two rays $\overrightarrow{AB}$ and $\overrightarrow{AC}$ are* opposite *if they are distinct, part of the same line, and have a common vertex A. In this case the angle formed by them is called a* straight angle (see Figure 1.7).

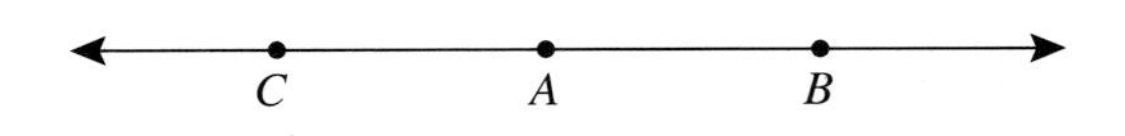

Figure 1.7 Opposite rays

Definition 1.2.15 *A point D is said to be an interior point of a nonstraight angle $\angle CAB$ if D is on the same side of line $\overleftrightarrow{AC}$ as B and D is on the same side of line $\overleftrightarrow{AB}$ as C.*

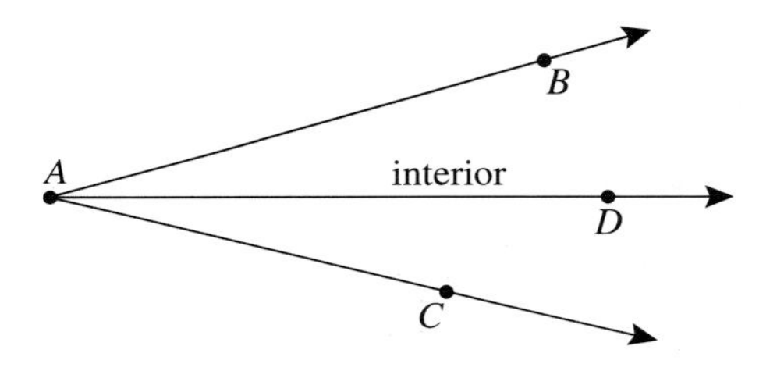

Figure 1.8

The next result contains two statements and claims that the first holds "if and only if" the second does. Observe that the "if and only if" means that the two statements are equivalent, that is to say each implies the other. Therefore the proof of an "if and only if " result has two parts: assuming that the first statement is true and we prove the second, and then the converse, that is, from the second statement we prove the first.

Proposition 1.2.16 *Let $\angle CAB$ be a nonstraight angle and D a point lying on line $\overleftrightarrow{CB}$. Then D is an interior point of $\angle CAB$ if and only if D is between B and C.*

Proof Let us assume that D is an interior point of $\angle CAB$. From $\mathbf{B_3}$ we have that either B is between D and C, or C is between B and D, or D is between B and C. If B is between D and C, then D and C are on opposite sides of any line that intersects line $\overleftrightarrow{DC}$ at B. In particular,

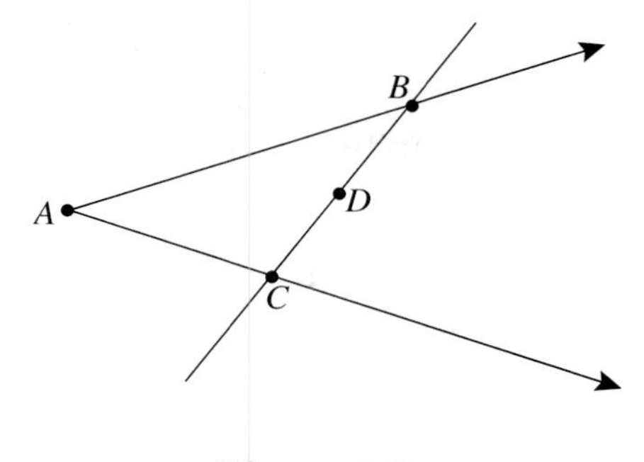

Figure 1.9

D and C are on opposite sides of $\overleftrightarrow{AB}$, contradicting the definition of interior point. If C is between D and B, using a similar argument we obtain the same contradiction. Therefore the only possible case is D between B and C.

Conversely, let us now assume that D is between B and C. If D is not an interior point, then either D and C are on opposite sides of line $\overleftrightarrow{AB}$ or D and B are on opposite sides of line $\overleftrightarrow{AC}$. Suppose that D and C are on opposite sides of line $\overleftrightarrow{AB}$; then $\overline{CD}$ intersects $\overleftrightarrow{AB}$ at some point that we call E. We claim that $E = B$. In fact, suppose $E \neq B$. Since both B and E lie on line $\overleftrightarrow{AB}$ and on line $\overleftrightarrow{CD}$, we would have that lines $\overleftrightarrow{AB}$ are $\overleftrightarrow{CD}$ are the same line, which contradicts the fact that $\angle CAB$ is a nonstraight angle. Then $E = B$, and then B is between C and D, which in turn contradicts the assumption that D is between B and C. The case that D and B are on opposite sides of line $\overleftrightarrow{AC}$ leads to a similar contradiction. Then we conclude that D is an interior point of $\angle CAB$. □

Proposition 1.2.17 *Let D be an interior point of angle $\angle CAB$. Then every point $E \neq A$ on ray $\overrightarrow{AD}$ is an interior point of $\angle CAB$.*

Proof Let $E \in \overrightarrow{AD}$. If E is not an interior point of $\angle CAB$, then either B and E are on opposite sides of line $\overleftrightarrow{AC}$ or C and E are on opposite sides of line $\overleftrightarrow{AB}$. If B and E are on opposite sides of $\overleftrightarrow{AC}$, since B and D are on the same side of line $\overleftrightarrow{AC}$, we get that D and E are on opposite sides of line $\overleftrightarrow{AC}$ by $\mathbf{B_4}$. Then $\overline{DE}$ intersects line $\overleftrightarrow{AC}$ at a point between

D and E. But $\overline{DE} \subset \overrightarrow{AD}$, and then $\overline{DE}$ intersects line $\overleftrightarrow{AC}$ at A. This is a contradiction, for A is not between D and E, since it is the vertex of ray $\overrightarrow{AD}$. The second case leads to the same contradiction. □

Definition 1.2.18 *A ray $\overrightarrow{AD}$ is said to be between rays $\overrightarrow{AC}$ and $\overrightarrow{AB}$ if $\overrightarrow{AC}$ and $\overrightarrow{AB}$ are not opposite rays and D is an interior point of angle $\angle CAB$.*

The next two results will not be called propositions or theorems. In general, results that are stepping stones for proving basic theorems are called "lemmas."

Lemma 1.2.19 *Let D be an interior point of $\angle CAB$. If A is between C and E, then B is an interior point of $\angle DAE$.*

Proof Let l be the line through C, A, and E (see Figure 1.10). We have that B and D are on the same side of l. We want to show that E and B are on the same side of the line through points A and D. If not, $\overline{EB}$ intersects line $\overleftrightarrow{AD}$ at a point, say F, and then F is an interior point of $\angle BAE$ by Proposition 1.2.16. We claim that $D \in \overrightarrow{AF}$. In fact, since F is on the same side of l as B and B is on the same side of l as D, we get that F and D are on the same side of l by $\mathbf{B_4}$, and then $D \in \overrightarrow{AF}$. Then D is also an interior point of $\angle BAE$. Since D is also an interior point of $\angle BAC$, we obtain that D is in the two half-planes bounded by line $\overleftrightarrow{AB}$, contradicting Theorem 1.2.9. □

Lemma 1.2.20 *Given three rays $\overrightarrow{AB}, \overrightarrow{AC}$, and $\overrightarrow{AD}$, if $\overrightarrow{AD}$ does not intersect $\overline{BC}$, then one of following occurs:*

(i) *Points B and C are on the same side of line $\overleftrightarrow{AD}$.*
(ii) *D is not an interior point of $\angle CAB$.*

Proof We prove this lemma by showing that if (i) does not happen, then we conclude (ii). Let us then suppose that $\overline{BC}$ intersects line $\overleftrightarrow{AD}$ at a point F. Since $\overline{BC}$ does not intersect ray $\overrightarrow{AD}$, we have that $\overrightarrow{AD}$ and $\overrightarrow{AF}$ are opposite rays, by the line separation property. Then A is between D and F. Moreover, since ray $\overrightarrow{AF}$ intersects $\overline{BC}$, F is an

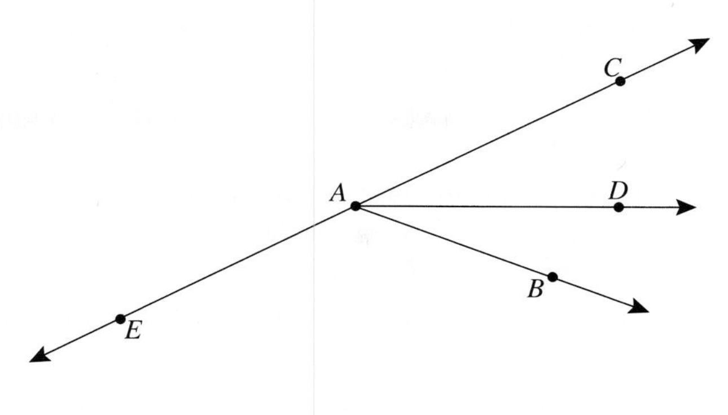

Figure 1.10

interior point of $\angle CAB$, by Proposition 1.2.16. It follows that F is on the same side of line $\overleftrightarrow{AC}$ as B. Now, if we suppose that D is an interior point of $\angle CAB$, then D and B are also on the same side of line $\overleftrightarrow{AC}$. Therefore D and F are on the same side of line $\overleftrightarrow{AC}$ by Axiom $\mathbf{B_4}$(a). This gives that $\overline{DF} \cap \overleftrightarrow{AC} = \phi$, which contradicts that A is between D and F. It follows that D is not an interior point of $\angle CAB$. $\square$

Theorem 1.2.21 (Crossbar theorem) *If $\overrightarrow{AD}$ is between rays $\overrightarrow{AC}$ and $\overrightarrow{AB}$, then $\overrightarrow{AD}$ intersects $\overline{BC}$.*

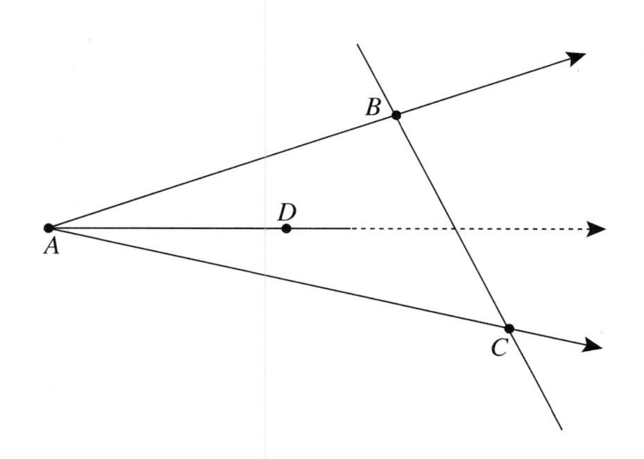

Figure 1.11 Crossbar theorem

Proof Let us suppose that $\overrightarrow{AD}$ does not intersect $\overline{BC}$. Since D is an interior point of $\angle CAB$, Lemma 1.2.20 implies that B and C are on the same side of line $\overleftrightarrow{AD}$. We also have that D and B are on the same side of the line through points A and C, because D is an interior point of $\angle CAB$. Therefore B is an interior point of $\angle CAD$. In this case, Lemma 1.2.19 implies that D is an interior point of $\angle BAE$, where E is such that A is between C and E. Since D is also an interior point of $\angle BAC$, we obtain that D is in the two half-planes bounded by line $\overleftrightarrow{AB}$, contradicting Theorem 1.2.9. $\square$

Exercises

1. Prove parts (c),(d), and (e) of Proposition 1.2.3 by answering the following questions.
   (c) Suppose all points are coincident with a given line. Then all points are collinear. Which axiom has been contradicted?
   (d) Suppose that all lines pass through a point P. This implies that all lines are concurrent at P. Does this fact contradict a previous result?
   (e) Given a point P, we know that there exists a line l not incident with P; why? Further, if A and B are two distinct points on line l, then $\overleftrightarrow{AP}$ and $\overleftrightarrow{BP}$ are two distinct lines through P. Justify the last assertion.

2. Prove Proposition 1.2.6(b) by justifying the steps below.
   (1) If $P \in \overline{AB}$, then $P \in \overrightarrow{AB}$ and $P \in \overrightarrow{BA}$.
   (2) $\overline{AB} \subset \overrightarrow{AB} \cap \overrightarrow{BA}$.
   (3) Suppose $P \in \overrightarrow{AB} \cap \overrightarrow{BA}$ and $P \notin \overline{AB}$. Then B is between A and P and A is between B and P.
   (4) We have a contradiction (why?) and then $\overrightarrow{AB} \cap \overrightarrow{BA} \subset \overline{AB}$.
   (5) $\overline{AB} = \overrightarrow{AB} \cap \overrightarrow{BA}$.

3. Prove that if A and B are on the same side of line l and $C \in \overline{AB}$, then C and A are on the same side of l.
   Here is the proof. Justify each step.
   (1) Suppose that C and A are on opposite sides of l. Then there exists $P \in l$ such that $\overline{CA} \cap l = \{P\}$.

(2) Since A and B are on the same side of l, C and B are on opposite sides of l.
(3) There exists $Q \in l$ such that $\overline{CB} \cap l = \{Q\}$.
(4) $\overleftrightarrow{AC}=\overleftrightarrow{BC}=\overleftrightarrow{AB}$.
(5) It follows that $P = Q$.
(6) Then point P is between C and A and between C and B.
(7) Therefore point $P \in \overleftrightarrow{CA}$ and $P \in \overleftrightarrow{CB}$ and this is the required contradiction.

4. Let A and B be two points such that point A lies on line l and $B \notin l$. Show that if $C \in \overline{AB}$, $C \neq A, B$, then B and C are on the same side of l.
Hint: Follow the idea used in Exercise 3.

5. Let A, B, and C be three points such that A is between B and C. Show that if P is between B and A, then P is between B and C. Here is the proof. Justify each step.
(1) Let l be any line intersecting line $\overleftrightarrow{AB}$ at P. Then A and B are on opposite sides of l.
(2) Point C is on either $\overrightarrow{PB}$ or $\overrightarrow{PA}$.
(3) If $C \in \overrightarrow{PB}$ then C and B are on the same side of l (see the proof of the line separation property).
(4) Then points C and A are on opposite sides of l.
(5) It follows that P is between A and C and hence $P \in \overrightarrow{AC}$.
(6) We have a contradiction (why?) and then $C \in \overrightarrow{PA}$.
(7) Then C and B are on opposite sides of l.
(8) Therefore P is between B and C.

6. (a) Prove Proposition 1.2.11(a) by justifying each step below.
(1) If $P = B$ or $P = C$, then $P \in \overline{BA} \cup \overline{AC}$.
(2) Let $P \in \overline{BC}$ such that $P \neq B, C$. Then $A \in \overrightarrow{PB}$ or $A \in \overrightarrow{PC}$.
(3) If $A \in \overrightarrow{PB}$ and B is between A and P, we have a contradiction (use Exercise 5).
(4) If $A \in \overrightarrow{PC}$ and C is between A and P, we have a contradiction.
(5) It follows that either A is between B and P or A is between C and P.

(6) Then $P \in \overline{BA} \cup \overline{AC}$.
(7) Now, if $P \in \overline{BA} \cup \overline{AC}$ and $P \neq A, B, C$, then either P is between B and A or P is between C and A. This implies that P is between B and C (use Exercise 5).
(b) Write a paragraph containing the steps above.

7. (a) Show that if C and D are on opposite sides of $l = \overleftrightarrow{AB}$, then either B is an interior point of $\angle CAD$ or A is an interior point of $\angle CBD$.
Here is the proof. Justify each step.
(1) There exists $E \in l$ such that $\overline{CD} \cap l = \{E\}$.
(2) If $E = A$, then E is an interior point of $\angle CBD$.
(3) If $E = B$, then E is an interior point of $\angle CAD$.
(4) If $E \neq A, B$, then E is an interior point of $\angle CAD$.
(5) If $B \in \overrightarrow{AE}$ then B is also an interior point of $\angle CAD$.
(6) If $B \notin \overrightarrow{AE}$, then $A \in \overrightarrow{BE}$.
(7) It follows from this that E is also an interior point of $\angle CBD$ and so is A.
(b) Write a paragraph containing the steps above.

8. Let $\overrightarrow{AB}, \overrightarrow{AC}$, and $\overrightarrow{AD}$ be three distinct rays such that C and D are on the same side of line $\overleftrightarrow{AB}$. Show that either $\overrightarrow{AD}$ is between $\overrightarrow{AB}$ and $\overrightarrow{AC}$ or $\overrightarrow{AC}$ is between $\overrightarrow{AB}$ and $\overrightarrow{AD}$.
Hint: Use Lemma 1.2.20.

9. Consider four rays $\overrightarrow{AB}, \overrightarrow{AC}, \overrightarrow{AD}$, and $\overrightarrow{AE}$ such that $\overrightarrow{AC}$ is between $\overrightarrow{AB}$ and $\overrightarrow{AD}$, and $\overrightarrow{AD}$ is between $\overrightarrow{AC}$ and $\overrightarrow{AE}$. Show that $\overrightarrow{AC}$ is between $\overrightarrow{AB}$ and $\overrightarrow{AE}$.

1.3 Convex Sets

Definition 1.3.1 *A subset $\mathcal{S}$ of the plane is said to be* convex *if given any two points $A, B \in \mathcal{S}$, the line segment $\overline{AB}$ is contained in $\mathcal{S}$* (see Figure 1.12).

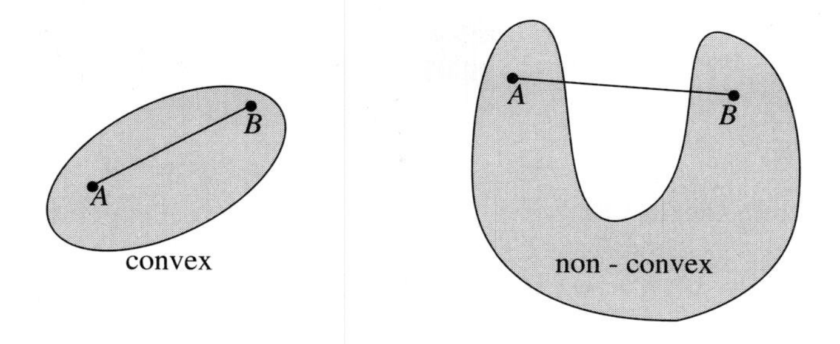

Figure 1.12

Proposition 1.3.2 *(a) A half-plane is a convex set.*
(b) Let H be a half-plane bounded by line l. Then the closed half-plane given by $\overline{H} = H \cup l$ is convex.
(c) The intersection of two convex sets is a convex set.
(d) The intersection of n convex sets is convex.
(e) The intersection of n half-planes (closed or not) is a convex set.

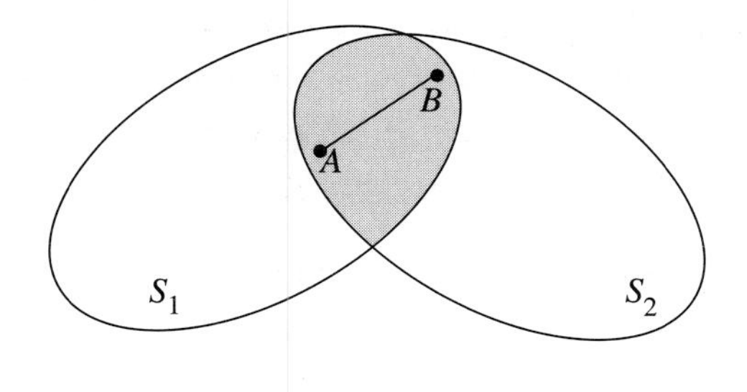

Figure 1.13

Proof Part (a) follows from Exercise 3 and (b) from Exercise 4 of Section 1.2.
(c) Let $\mathcal{S}_1$ and $\mathcal{S}_2$ be two convex sets and suppose that $\mathcal{S}_1 \cap \mathcal{S}_2$ contains at least two distinct points A and B. Since $A, B \in \mathcal{S}_1$ and $\mathcal{S}_1$ is convex, we have that $\overline{AB} \subset \mathcal{S}_1$. Likewise, we obtain that $\overline{AB} \subset \mathcal{S}_2$, since $A, B \in \mathcal{S}_2$ and $\mathcal{S}_2$ is convex. Therefore $\overline{AB} \subset \mathcal{S}_1 \cap \mathcal{S}_2$.
(d) We prove this part by induction on n. Let $S_1, \ldots, S_n$ be a collection of n convex sets. From part (b) of this proposition we conclude that the

result holds for $n = 2$. Suppose by induction that the result holds for $n - 1$. Then the set given by

$$\mathcal{U}_1 = S_1 \cap \cdots \cap S_{n-1}$$

is convex. Let $\mathcal{U}_2$ denote S_n. Then

$$\mathcal{U}_1 \cap \mathcal{U}_2 = S_1 \cap \cdots \cap S_n$$

is convex, by part (b).
(e) It follows from *(a)*, *(b)*, and *(d)*. □

Definition 1.3.3 *Let $P_1, \ldots, P_n, n \geq 3$, be n points of the plane such that no three consecutive points are collinear and such that any pair of segments $\overline{P_1P_2}, \ldots, \overline{P_{n-2}P_{n-1}}$, and $\overline{P_{n-1}P_1}$ either have no point in common or have only an endpoint in common. A* polygon *is defined as the union of the line segments $\overline{P_1P_2}, \ldots, \overline{P_{n-2}P_{n-1}}$, and $\overline{P_{n-1}P_1}$, which are called its* sides. *The points $P_1, \ldots, P_n$ are called vertices. A polygon of n sides is also called an n*-gon.

Definition 1.3.4 *A polygon of n sides is said to be* convex *if for all pairs of consecutive vertices P_i, P_{i+1}, the points P_j, $\forall\, j \neq i, i+1$, are on the same side of line $\overleftrightarrow{P_iP_{i+1}}$.*

It follows immediately from the definition that *triangles* (polygons of three sides) are convex.

Notice that the word *convex* was used in two different contexts. One describes a property of sets of points and another a property of polygons. Actually these concepts are related to each other, as shown by Proposition 1.3.6 below.

Definition 1.3.5 *Let $\mathcal{P}_n$ be a convex polygon of vertices $P_1, \ldots, P_n$. For $i = 1, \ldots, n-1$, let l_i denote line $\overleftrightarrow{P_iP_{i+1}}$ and $l_n = \overleftrightarrow{P_nP_1}$. Let H_i be the half-plane determined by l_i and the points P_j, for $j \neq i, i+1$. We define the* interior *of $\mathcal{P}_n$ as*

$$\operatorname{Int}(\mathcal{P}_n) = \bigcap_{i=1}^{n} H_i.$$

Proposition 1.3.6 *Let $\mathcal{P}_n$ be a convex polygon of vertices $P_1, \ldots, P_n$ and $\overline{H}_i$ the closed half-plane given by $\overline{H}_i = H_i \cup l_i$, where H_i is as in Definition 1.3.5. Then*

$$\mathcal{P}_n \cup \mathrm{Int}(\mathcal{P}_n) = \bigcap_{i=1}^{n} \overline{H}_i$$

and hence $\mathcal{P}_n \cup \mathrm{Int}(\mathcal{P}_n)$ is a convex set.

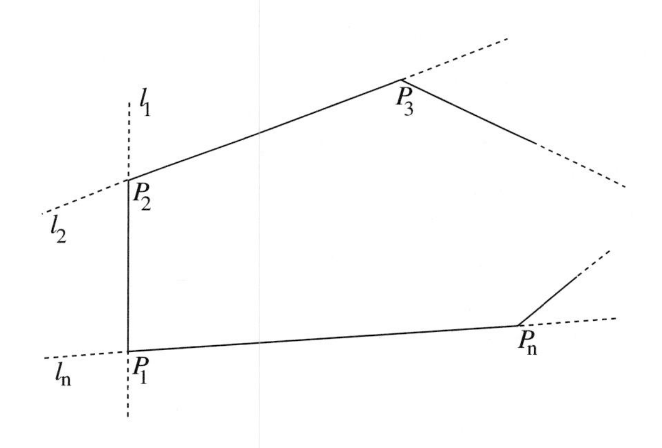

Figure 1.14

Proof We will show first that $\mathcal{P}_n \cup \mathrm{Int}(\mathcal{P}_n) \subset \bigcap_{i=1}^{n} \overline{H}_i$. The definition of $\mathrm{Int}(\mathcal{P}_n)$ implies immediately that $\mathrm{Int}(\mathcal{P}_n) \subset \bigcap_{i=1}^{n} \overline{H}_i$. Let us now consider a point $P \in \mathcal{P}_n$. If P is a vertex, without loss of generality, we assume that $P = P_1$. Then $P \in l_1$ and $P \in l_n$. Therefore $P \in \overline{H}_1$ and $P \in \overline{H}_n$. Moreover, from the definition of convex polygon we obtain that $P \in H_i$ for all $i = 2, \ldots, n-1$. It follows that $P \in \bigcap_{i=1}^{n} \overline{H}_i$. If P is not a vertex, then without loss of generality we assume that P is between P_1 and P_2. Then $P \in \overline{H}_1$ and $P \in H_i$ for all $i = 2, \ldots, n$, implying again that $P \in \bigcap_{i=1}^{n} \overline{H}_i$.

Now we show that $\bigcap_{i=1}^{n} \overline{H}_i \subset \mathcal{P}_n \cup \mathrm{Int}(\mathcal{P}_n)$. Let $P \in \bigcap_{i=1}^{n} \overline{H}_i$. There are two cases here:

(i) $P \in H_i$ for all $i = 1, \ldots, n$; then $P \in \mathrm{Int}(\mathcal{P}_n)$.

(ii) $P \in l_i$, for some i, say l_1 and $P \in H_i$ for all $i \geq 2$. If $P = P_1$ or $P = P_2$, then obviously $P \in \mathcal{P}_n$. If $P \neq P_1, P_2$, since $P \in H_2$ and $P \in H_n$ we conclude that P is between P_1 and P_2 and then P is on the side $\overline{P_1P_2}$ of the polygon. $\square$

We finish this section with a proposition that will be used in the next chapter.

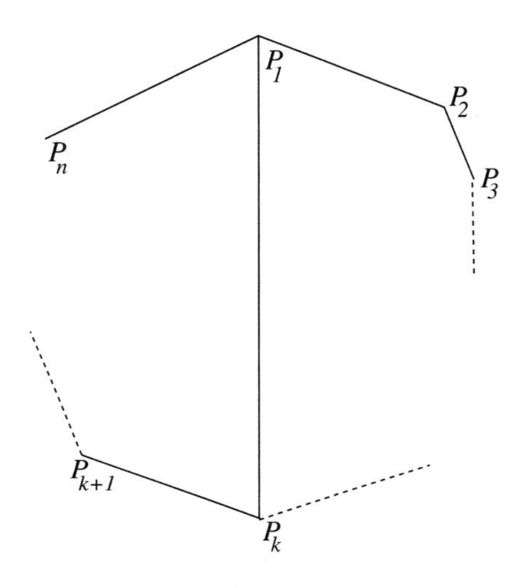

Figure 1.15

Proposition 1.3.7 *Any diagonal (a line segment connecting nonconsecutive vertices) of a convex polygon $\mathcal{P}_n$ divides it into two convex polygons.*

Proof Let us consider the diagonal that connects vertex P_1 to vertex P_k (see Figure 1.15). We want to show that points $P_2, \ldots, P_{k-1}$ are on the same side of line $\overleftrightarrow{P_1P_k}$. Since $\mathcal{P}_n$ is convex, we conclude that P_1 and P_k, $k > 3$, are on the same side of $\overleftrightarrow{P_2P_3}$, which implies that P_2 and P_3 are on the same side of line $\overleftrightarrow{P_1P_k}$. If $k > 4$, the same reasoning implies that P_3 and P_4 are on the same side of line $\overleftrightarrow{P_1P_k}$. Repeating this argument several times, we conclude that points $P_2, \ldots, P_{k-1}$ are all on the same side of line $\overleftrightarrow{P_1P_k}$, by Axiom $\mathbf{B_4}$.

Similarly we show that $P_{k+1}, \ldots, P_n$ are on the same side of line $\overleftrightarrow{P_1 P_k}$. It follows that a polygon of vertices $P_1, \ldots, P_k$ and a polygon of vertices $P_k, P_{k+1}, \ldots, P_n, P_1$ are both convex. Observe that P_1 in this proof is an arbitrary vertex, since we can always relabel the vertices. □

Exercises

1. Show that the intersection of any collection (not necessarily countable) of convex sets is convex.

2. *Definition:* Let S be a set of points of the plane. Let $\{C_\lambda\}$ be the collection of convex sets such that $S \subset C_\lambda$, $\forall \lambda$. The *convex hull* of a set of points S is defined as

$$\mathcal{C}(S) = \bigcap_\lambda C_\lambda.$$

   Show the following: A set S is convex if and only if $\mathcal{C}(S) = S$. *Hint*: Observe that $S \subset \mathcal{C}(S)$. Now if S is convex, then S is one of the sets in the collection of all convex sets that contain S.

3. (a) Given a nonstraight angle $\angle A$, let $\text{Int}(\angle A)$ denote the set of interior points of $\angle A$. Show that $\text{Int}(\angle A)$ is a convex set.
   (b) Given a triangle $\triangle ABC$, define its interior as

$$\text{Int}(\Delta) = \text{Int}(\angle A) \cap \text{Int}(\angle B) \cap \text{Int}\angle C).$$

   Show that $\text{Int}(\Delta)$ is a convex set.
   (c) Show that the convex hull of three noncollinear points A, B, and C is

$$\triangle ABC \cup \text{Int}(\triangle ABC).$$

   Hint: Use Pasch's theorem.

4. A polygon of four sides is called a quadrilateral. A *parallelogram* is a quadrilateral with the property that opposite sides are parallel. Show that a parallelogram is a convex polygon.

5. Show that the diagonals of a convex quadrilateral intersect each other.
   Hint: Use the crossbar theorem.

6. Let $\mathcal{P}_n$ be a convex polygon of vertices $P_1, \dots, P_n$. Show that the convex hull of set $\{P_1, \dots, P_n\}$ is $\mathcal{P}_n \cup \mathrm{Int}(\mathcal{P}_n)$.
   Hint: Let $\mathcal{C}$ denote the convex hull of $\{P_1, \dots, P_n\}$. Recall that Proposition 1.3.6 implies that $\mathcal{P}_n \cup \mathrm{Int}(\mathcal{P}_n)$ is convex and then $\mathcal{C} \subset \mathcal{P}_n \cup \mathrm{Int}(\mathcal{P}_n)$. In order to show that $\mathcal{P}_n \cup \mathrm{Int}(\mathcal{P}_n) \subset \mathcal{C}$ follow the steps below:
   (1) $\mathcal{P}_n \subset \mathcal{C}$.
   (2) Any diagonal of $\mathcal{P}_n$ is contained in $\mathcal{C}$.
   (3) If P is any point in $\mathrm{Int}(\mathcal{P}_n)$, show that $P \in \mathrm{Int}(\angle P_{i-1}P_iP_{i+1})$, for all i.
   (4) Choose i such that P is in the interior of triangle $\triangle P_{i-1}P_iP_{i+1}$. Let Q be any point between P_i and P_{i+1}; apply Pasch's theorem to triangle $\triangle P_{i-1}P_iP_{i+1}$ and line $\overleftrightarrow{PQ}$.

1.4 Measuring Segments and Angles

In this section we introduce axioms that assign real numbers to segments and angles. In stating such axioms we shall proceed as if the real number system were already constructed, not entering into any discussion of the axioms used for its construction. Many textbooks on real analysis contain in their introductions a list of axioms for the real numbers. The *axioms for segment measurement* are:

> $\mathbf{S_1}$ To every pair of points A and B there corresponds a real number $x \geq 0$ such that $x = 0$ if and only if $A = B$. The number x will be called the *length* of $\overline{AB}$ and will be denoted by AB.
>
> $\mathbf{S_2}$ There is a one-to-one correspondence between the points of a line l and the set of real numbers $\mathbf{R}$ such that, if to point A corresponds number a and to point B corresponds b, then $AB = |b - a|$. The real number corresponding to a point of a line will be called *coordinate* of the point, relative to this correspondence.
>
> $\mathbf{S_3}$ If C is between A and B, then $AC + CB = AB$.

Remark Axiom $\mathbf{S_2}$ has the following consequences.
(i) The length of a line segment is in fact unique.

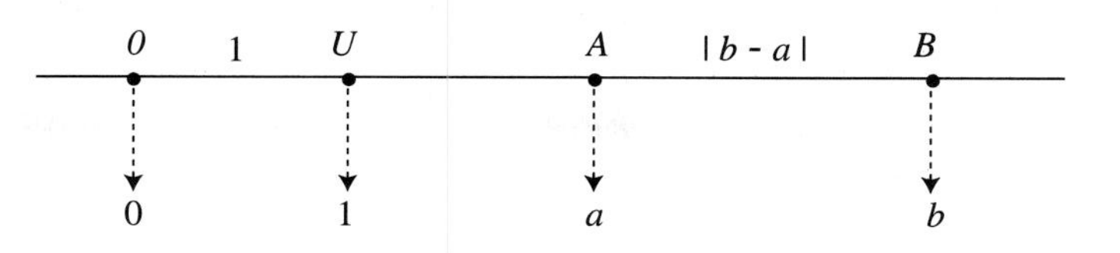

Figure 1.16 Axiom $\mathbf{S_2}$

(ii) There exists a line segment $\overline{OU}$ of length 1. The line segment $\overline{OU}$ is called the *unit of length.*
(iii) Given a point R on a line and a real number r, a one-to-one correspondence can be established such that $R \mapsto r$. In fact, suppose that a one-to-one correspondence $X \mapsto x$ has assigned number z to point R. Then the correspondence $X \mapsto x' = x + (r - z)$ is still one-to-one and maps R to r. Moreover, lengths are preserved, since

$$\begin{aligned} AB &= |b - a| = |b + (r - z) - a - (r - z)| \\ &= |b + (r - z) - (a + (r - z))| = |b' - a'|. \end{aligned}$$

In particular, there exists a one-to-one correspondence that maps a particular point to number 0.
(iv) Given a line segment $\overline{AB}$ and a ray $\overrightarrow{CD}$, there exists only one point $E \in \overrightarrow{CD}$ such that $\overline{AB} \simeq \overline{CE}$ (Exercise 1 of this section).

Definition 1.4.1 *Two line segments $\overline{AB}$ and $\overline{CD}$ are said to be* congruent, *and are denoted by $\overline{AB} \simeq \overline{CD}$, if $AB = CD$.*

Proposition 1.4.2 *Let C be a point lying on ray $\overrightarrow{AB}$ such that $AC < AB$. Then C is between A and B.*

Proof From $\mathbf{B_3}$ and the fact that B and C are both on the ray emanating from A we have that either B is between A and C or C is between A and B. If the former happens, we obtain that $AB + BC = AC$ by $\mathbf{S_3}$, which in turn implies that $AB < AC$, contradicting our hypothesis. □

Proposition 1.4.3 *Let $A, B,$ and C be three collinear points and let $a, b,$ and c be their respective coordinates. Then C is between A and B if and only if c is between a and b.*

Proof First, we point out that a, b, and c are three distinct real numbers, since the correspondence between the line and the set of real numbers is one-to-one and A, B, and C are three distinct points.

Let us assume that C is between A and B. From $\mathbf{S_3}$ and the fact that $AB = AC + CB$ we obtain

$$|b - a| = |c - a| + |b - c|.$$

Suppose that $a < b$. If $c > b$ and hence $c > a$, we would have

$$b - a = c - a + c - b = 2c - b - a$$

implying $b = c$. Therefore $c < b$. If $c < a$ we would get

$$b - a = b - c + a - c = b + a - 2c$$

implying $a = c$. Therefore $a < c < b$. If we suppose $a > b$, using similar arguments, we conclude that $b < c < a$.

Conversely, let us assume now that $a < c < b$. Then $|c - a| = c - a$ and $|b - c| = b - c$. Therefore

$$|c - a| + |b - c| = |a - b|$$

which in turn implies $AC + CB = AB$. In particular, $AC < AB$ and $CB < AB$. We claim that A is not between B and C. If so, then $BA + AC = BC$, implying $CB > AB$. A similar argument implies that B is not between A and C. Therefore the only possible case is C between A and B by Axiom $\mathbf{B_3}$. If $b < c < a$, the proof is similar. □

Definition 1.4.4 *A point M is called the* midpoint *of line segment $\overline{AB}$ if M is between A and B and $AM = MB$.*

Proposition 1.4.5 *Any line segment has exactly one midpoint.*

Proof Given the line segment $\overline{AB}$, let a and b be the respective coordinates of A and B. Let $m = (a + b)/2$. Axiom $\mathbf{S_2}$ implies that there exists M on line through A and B whose coordinate is m. Then M is between A and B because m is between a and b. Computing

$$AM = |a - m| = \frac{|a - b|}{2} \quad \text{and} \quad MB = |b - m| = \frac{|a - b|}{2},$$

we conclude that M is a midpoint. This shows the existence of the midpoint. In order to show its uniqueness, we suppose that N is also a midpoint and let n denote its coordinate. Since $AN = NB$, we have $|n-a| = |b-n|$, which implies either $n-a = (b-n)$ or $n-a = -(b-n)$. In the latter case we conclude $a = b$, which contradicts that $A \neq B$. The first case implies $n = (a+b)/2$. Therefore M and N have the same coordinates, and then $M = N$ by the uniqueness part of Axiom $\mathbf{S_2}$. □

The *measurement axioms for angles* are:

$\mathbf{A_1}$ To every angle $\angle ABC$ corresponds a unique real number x such that $0 \leq x \leq 180$. Further,
(i) $x = 0$ if and only if $\overrightarrow{BA} = \overrightarrow{BC}$.
(ii) $x = 180$ if and only if $\overrightarrow{BA}$ and $\overrightarrow{BC}$ are opposite rays.
This real number x is called the *measure* of angle $\angle ABC$ and will be denoted by $m(\angle ABC)$. The unit used here is called *degree* and is denoted by $^\circ$.

$\mathbf{A_2}$ Let $\overrightarrow{AB}$ be a ray and H one half-plane determined by the line through points A and B. Then for every x between 0 and 180 there is only one ray $\overrightarrow{AC}$, with $C \in H$ such that $m(\angle CAB) = x$.

Definition 1.4.6 *We say that a ray* $\overrightarrow{AD}$ divides *an angle* $\angle BAC$ *if* $\overrightarrow{AD}$ *intersects line segment* $\overline{BC}$.

Notice that the crossbar theorem implies that if a ray $\overrightarrow{AD}$ is between rays $\overrightarrow{AB}$ and $\overrightarrow{AC}$, then ray $\overrightarrow{AD}$ divides $\angle BAC$. Observe also that given points A, B, C such that A is between B and C and any point D in one half-plane determined by $\overleftrightarrow{AB}$, ray $\overrightarrow{AD}$ divides the straight angle $\angle BAC$. The last measurement axiom for angles is:

$\mathbf{A_3}$ If ray $\overrightarrow{AD}$ divides angle $\angle BAC$, then

$$m(\angle BAC) = m(\angle BAD) + m(\angle DAC).$$

It follows from $\mathbf{A_2}$, $\mathbf{A_3}$, and the crossbar theorem that if $0 < m(\angle ABC) < m(\angle DEF)$, then there exists a point $G \in \overline{DF}$ such that $\angle ABC \simeq \angle GEF$ (see Exercise 2 of this section).

Definition 1.4.7 *Two angles $\angle ABC$ and $\angle DEF$ are said to be* congruent, *and are denoted by $\angle ABC \simeq \angle DEF$, if $m(\angle ABC) = m(\angle DEF)$.*

Definition 1.4.8 *An angle is said to be an* acute *angle if its measure is less than $90°$. An angle whose measure is $90°$ is called a* right angle. *An* obtuse *angle is an angle of measure greater than $90°$ and less than $180°$.*

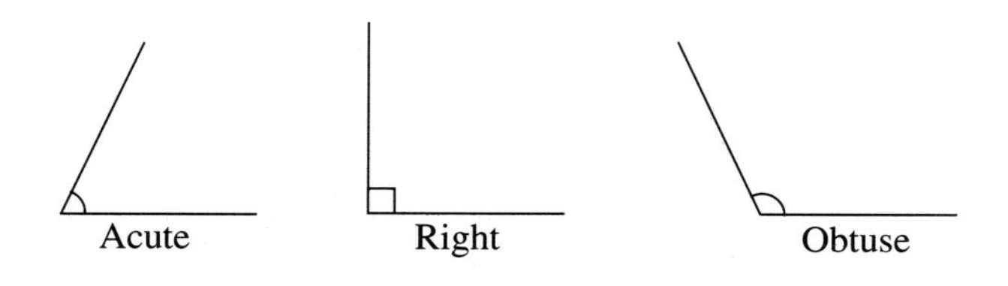

Figure 1.17

The definition above implies that all right angles are congruent, one of the first four Euclid's postulates.

Definition 1.4.9 *Two angles $\angle DAC$ and $\angle DAB$ are called* supplementary angles *if they have a common side $\overrightarrow{AD}$ and the other two sides $\overrightarrow{AB}$ and $\overrightarrow{AC}$ are opposite rays. In this case, each angle is said to be the* supplement *of the other* (see Figure 1.18).

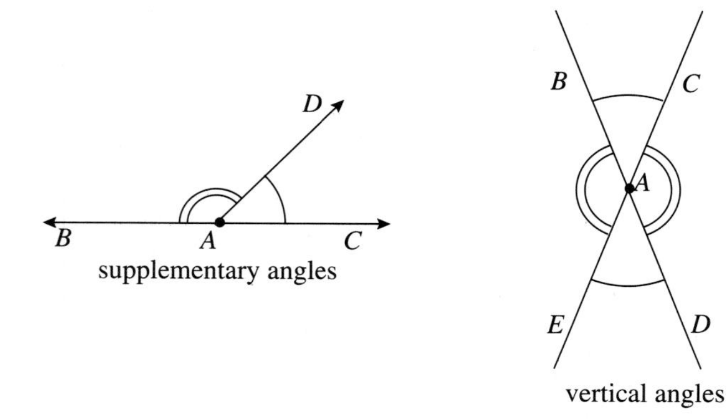

Figure 1.18

Definition 1.4.10 *Let l and m be two lines concurrent at point A. Let B and D be on l such that A is between B and D. Let C and E be on m such that A is between C and E. The pairs of angles $\{\angle BAC, \angle DAE\}$ and $\{\angle BAE, \angle CAD\}$ are called* vertical angles *(see Figure 1.18).*

The axioms above imply the following results.

Proposition 1.4.11 *(a) The sum of the measures of supplementary angles is* $180°$.
(b) Supplements of congruent angles are congruent.
(c) Vertical angles are congruent.

Definition 1.4.12 *Two concurrent lines are said to be* perpendicular *if one of the angles formed by them is a right angle.*

It follows from Proposition 1.4.11 that if two lines are perpendicular, then the four angles formed by them are right angles. Proposition 1.4.11 also implies that two concurrent and nonperpendicular lines determine only one pair of acute vertical angles.

Definition 1.4.13 *A line l is said to be a* perpendicular bisector *of line segment $\overline{AB}$ if l passes through the midpoint of $\overline{AB}$ and is perpendicular to line $\overleftrightarrow{AB}$.*

Theorem 1.4.14 *For every line l and for every point $A \in l$ there exists a unique line m through A such that m is perpendicular to l.*

Proof Let B be another point on line l and let us consider ray $\overrightarrow{AB}$. From Axiom $\mathbf{A_2}$ we obtain that there is only one ray $\overrightarrow{AC}$ such that $m(\angle CAB) = 90°$. The unique line m through A and C is perpendicular to l. Now let us suppose that there exists line m' perpendicular to l and through A. Let $C' \in m'$ such that C' is on the same side of l as C. Then $m(\angle C'AB) = 90°$ by the definition of perpendicular lines. Therefore $\overrightarrow{AC} = \overrightarrow{AC'}$ and A, C, and C' are collinear. □

The next result will be called a "corollary." This term is used for a statement that is an immediate consequence of a theorem or has a straightforward proof from a theorem.

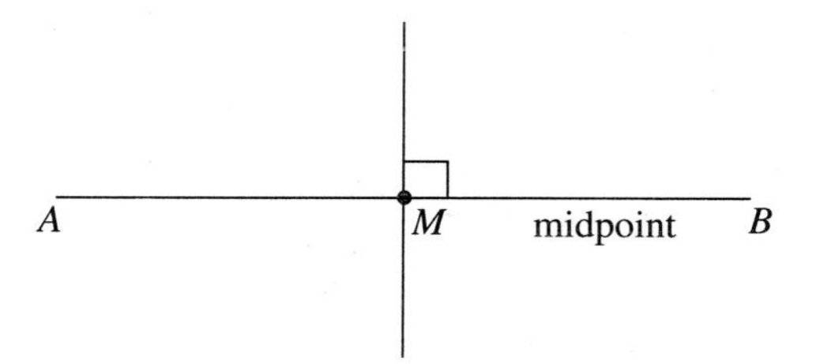

Figure 1.19

Corollary 1.4.15 *Every segment has a unique perpendicular bisector.*

Exercises

1. Given a line segment $\overline{AB}$ and a ray $\overrightarrow{CD}$, show that there exists only one point $E \in \overrightarrow{CD}$ such that $\overline{AB} \simeq \overline{CE}$.

2. Let $\angle ABC$ and $\angle DEF$ be two angles such that $0 < m(\angle ABC) < m(\angle DEF)$. Show that there is a point $G \in \overline{DF}$ such that $\angle ABC \simeq \angle GEF$.

1.5 Congruence of Triangles

Recall that a triangle is a polygon of three sides, that is, a geometric figure formed by three line segments connecting three noncollinear points.

Definition 1.5.1 *Two triangles $\triangle ABC$ and $\triangle DEF$ are said to be* congruent *if there is a one-to-one correspondence between their vertices such that corresponding sides and corresponding angles are congruent.*

Notice that if two triangles are congruent, then six congruence relations hold — three between line segments and three between corresponding angles. However, in order to establish congruence between two triangles we do not need to verify all of them, provided we assume the *congruence axiom* below.

> *SAS* If two sides and the included angle of one triangle are congruent to two sides and the included angle of another triangle, then these two triangles are congruent.

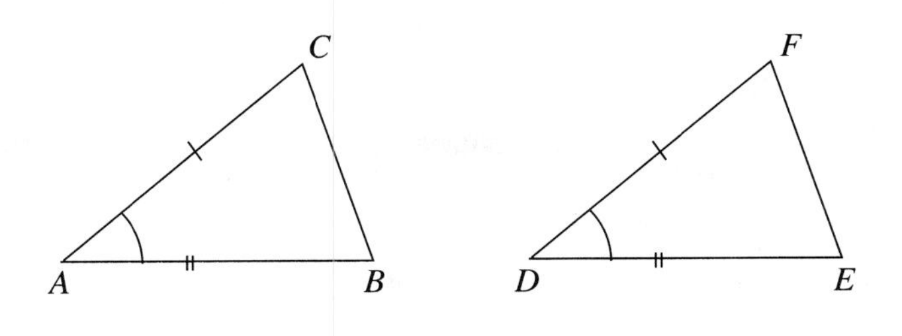

Figure 1.20 *SAS*

Proposition 1.5.2 *Given $\triangle ABC$ and a line segment $\overline{DE}$ such that $\overline{AB} \simeq \overline{DE}$, then there exists a unique point F on a given side of line $\overleftrightarrow{DE}$ such that $\triangle ABC \simeq \triangle DEF$.*

Proof From $\mathbf{A_2}$ we know that there exists a unique ray r emanating from D such that if F lies on r, then $\angle EDF \simeq \angle BAC$. Moreover, F can be chosen such that $\overline{AC} \simeq \overline{DF}$ by $\mathbf{S_2}$ (Exercise 1 of Section 1.4). Then $\triangle ABC \simeq \triangle DEF$ by SAS. □

Proposition 1.5.3 *(ASA) If $\overline{AB} \simeq \overline{DE}$, $\angle CAB \simeq \angle FDE$, and $\angle CBA \simeq \angle FED$, then $\triangle ABC \simeq \triangle DEF$.*

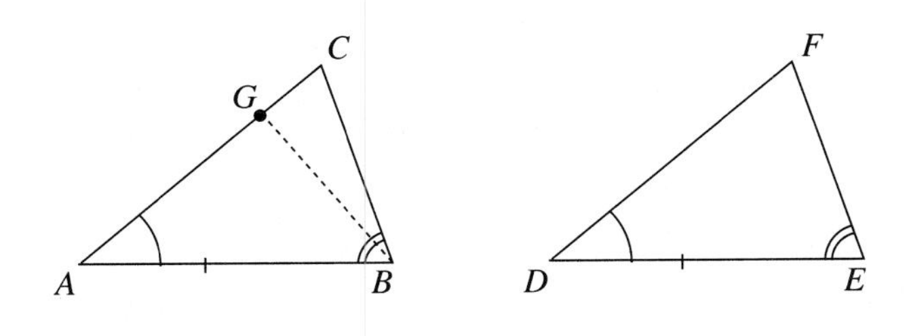

Figure 1.21 *ASA*

Proof It follows from Exercise 1 of Section 1.4 that we can consider a point $G \in \overline{AC}$ such that $\overline{AG} \simeq \overline{DF}$. Then $\triangle ABG \simeq \triangle DEF$ by SAS and hence $\angle GBA \simeq \angle FED$. But $\angle CBA \simeq \angle FED$ and therefore $\angle CBA \simeq \angle GBA$. This implies $\overrightarrow{BC}=\overrightarrow{BG}$. We claim that $G = C$. In fact, suppose $G \neq C$; then C and G lie on line $\overleftrightarrow{AC}$ and on line $\overleftrightarrow{BC}$, implying that $\overleftrightarrow{AC}=\overleftrightarrow{BC}$. But this contradicts that A, B, and C are noncollinear. □

Definition 1.5.4 *A triangle is called* isosceles *if two of its sides are congruent.*

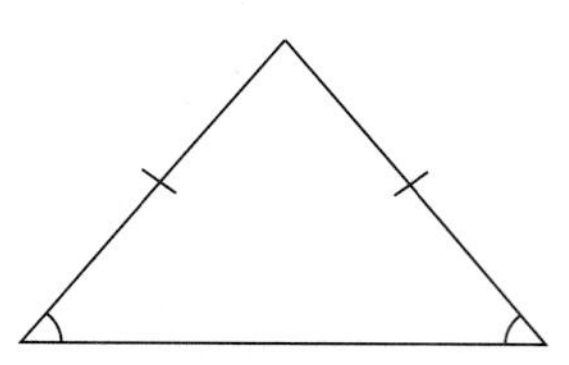

Figure 1.22 Isosceles triangles

Proposition 1.5.5 *Given* $\triangle ABC$ *we have:*
(i) *If* $\overline{AB} \simeq \overline{AC}$, *then* $\angle B \simeq \angle C$.
(ii) *If* $\angle B \simeq \angle C$, *then* $\overline{AB} \simeq \overline{AC}$.

Proof Consider the following pair of triangles: $\triangle ABC$ and $\triangle ACB$. The hypothesis in (i) implies that they are congruent by SAS and then $\angle B \simeq \angle C$. In (ii) we conclude that the triangles are congruent by ASA, which implies that $\overline{AB} \simeq \overline{AC}$. □

Proposition 1.5.6 *(SSS) If* $\overline{AB} \simeq \overline{DE}$, $\overline{AC} \simeq \overline{DF}$, *and* $\overline{BC} \simeq \overline{EF}$, *then* $\triangle ABC \simeq \triangle DEF$.

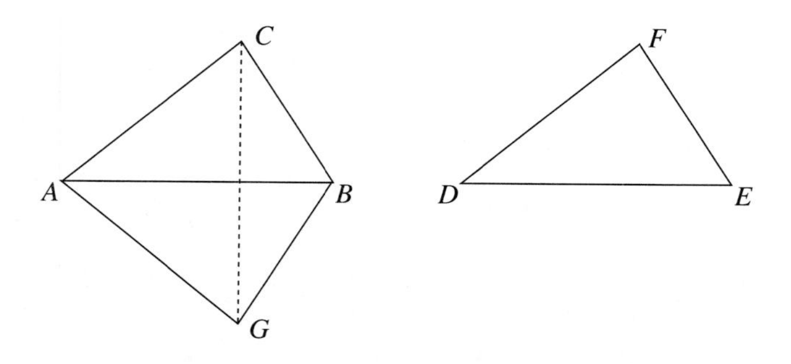

Figure 1.23 SSS

Proof Let H be the half-plane determined by the line through A and B such that $C \notin H$. There exists G in H such that $\angle GAB \simeq \angle D$ (see

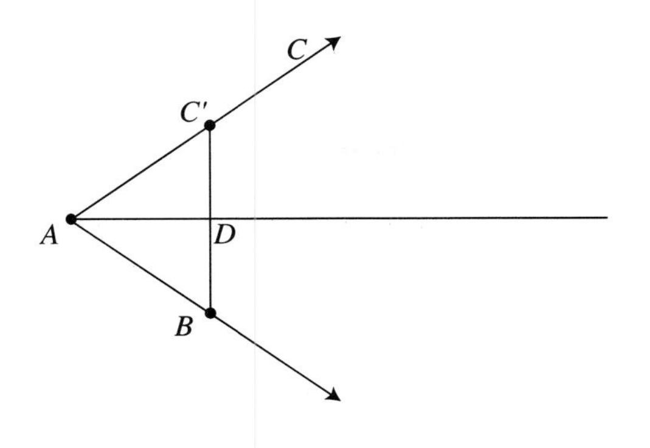

Figure 1.24

Figure 1.23). Moreover, the point G can be chosen so that $\overline{AG} \simeq \overline{DF}$. Then $\triangle ABG \simeq \triangle DEF$ and then $\overline{BG} \simeq \overline{EF}$. Therefore, if we show that $\triangle ABG \simeq \triangle ABC$, the proof will be complete. For that, consider line segment $\overline{CG}$. Our assumptions imply that $\triangle ACG$ and $\triangle BCG$ are isosceles, implying $\angle ACG \simeq \angle AGC$ and $\angle BCG \simeq \angle BGC$. Then Axiom $\mathbf{A_3}$ implies that $\angle C \simeq \angle G$, and the congruence between $\triangle ABC$ and $\triangle ABG$ follows by SAS. □

Definition 1.5.7 *We say that ray $\overrightarrow{AD}$ bisects $\angle BAC$ if $\angle BAD \simeq \angle CAD$. Ray $\overrightarrow{AD}$ is called a* bisector *of angle $\angle BAC$.*

Proposition 1.5.8 *Every angle has a unique bisector.*

Proof Given $\angle CAB$, let us consider a point C' on ray $\overrightarrow{AC}$ such that $AB = AC'$ (see Figure 1.24). Let D be the midpoint of line segment $\overline{BC'}$. Then $\triangle ABD \simeq \triangle AC'D$ by SSS. It follows that

$$\angle BAD \simeq \angle C'AD \simeq \angle CAD.$$

Now to show the uniqueness, let us suppose that ray $\overrightarrow{AD'}$ is a bisector of $\angle BAC'$. Then $\angle BAD \simeq \angle C'AD$, and hence $\triangle ABD' \simeq \triangle AC'D'$ by SAS. This implies that $BD' = D'C'$, and then D' is the midpoint of $\overline{BC'}$. Therefore $D' = D$, showing the uniqueness of the bisector. □

Theorem 1.5.9 *For every line l and for every point P not lying on l there exists a line m through P such that m is perpendicular to l.*

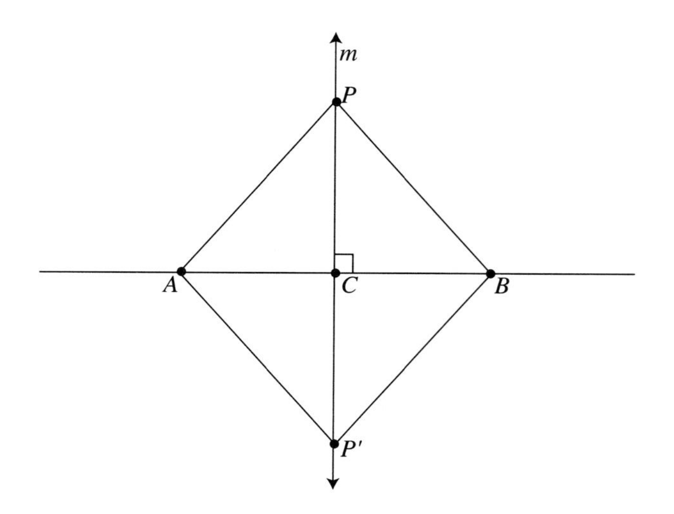

Figure 1.25

Proof Let A and B be two distinct points lying on l. Let $\mathcal{H}$ be the half-plane determined by line l that does not contain P (see Figure 1.25). There is a point P' in $\mathcal{H}$ such that $\angle PAB \simeq \angle P'AB$ and $\overline{PA} \simeq \overline{P'A}$. Let m be the line through P and P'. We will show that m is perpendicular to l. In fact, since P and P' are on opposite sides of l, line segment $\overline{PP'}$ intersects line l at a point, say C. If $C = A$, then $m(\angle PAB) = 90°$, since it is congruent to its supplement. In this case we conclude that $m \perp l$. If $C \neq A$, we obtain that $\triangle PAC \simeq \triangle P'CA$ by SAS. It follows then that $\angle PCA \simeq \angle P'CA$. Since they are supplementary angles, we conclude that $m(\angle PCA) = 90°$ and hence $m \perp l$. □

Line m above is called the *perpendicular dropped from P to l*, and the point of intersection C is called the *foot* of the perpendicular line. We will show in this section that the perpendicular line is unique. This is a consequence of the result below on alternate interior angles. Alternate interior angles appear when three lines l, m and t have no common point and, say, line t intersects both l ad m. In this case line t is said to be *transversal* to l and m.

Definition 1.5.10 *Let us consider a pair of lines l and m cut by a transversal line t* (see Figure 1.26). *Let $A = l \cap t$ and $D = m \cap t$.*

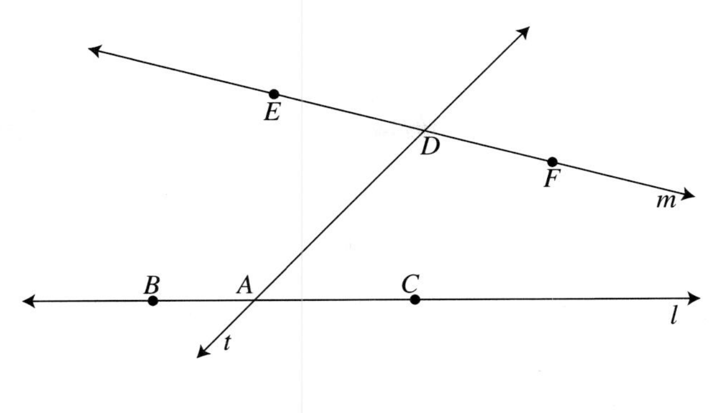

Figure 1.26

Let us choose points B and C on l, E and F on m, such that A is between B and C, E and B are on the same side of t, and D is between E and F. The pairs of angles $\{\angle EDA, \angle CAD\}$ and $\{\angle FDA, \angle BAD\}$ are called alternate interior angles.

Theorem 1.5.11 (Alternate interior angle theorem) *If two lines cut by a transversal have a pair of congruent alternate interior angles, then the two lines are parallel.*

Proof Let us use the notation of Definition 1.5.10. Let us suppose that l and m are not parallel. Then, there exists a point G such that $G = l \cap m$ and G is on the same side of t as C (see Figure 1.27). Let H be on ray $\overrightarrow{DE}$ such that $\overline{DH} \simeq \overline{AG}$. Since $\angle DAG \simeq \angle HDA$, we obtain that $\triangle DAG \simeq \triangle ADH$ by SAS. In particular, $\angle ADG \simeq \angle HAD$. But our assumption implies that $\angle ADG \simeq \angle BAD$, since they are the supplements of $\angle HDA$ and $\angle DAG$, respectively, and then $\angle BAD \simeq \angle HAD$. Therefore $H \in l$. Since H also lies on m, we get that $l = m$, which is the desired contradiction. □

Corollary 1.5.12 *(a) Let l and l' be two distinct lines. If $l \perp m$ and $l' \perp m$, then $l \parallel l'$.*
(b) If l is any line and $P \notin l$, then the perpendicular dropped from P to l is unique.
(c) If l is any line and $P \notin l$, then there exists at least one line l' incident with P that is parallel to l.

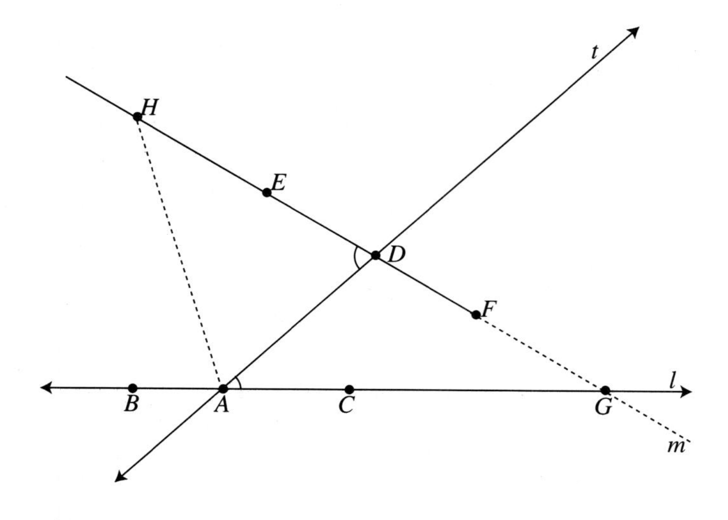

Figure 1.27

Proof *(a)* We have lines l and l' cut by the transversal m, and the alternate interior angles are right angles (see Figure 1.28). Since all right angles are congruent, the result follows from Theorem 1.5.11.
(b) Suppose that lines m_1 and m_2 are perpendicular to l and pass through P. If $m_1 \neq m_2$, then part (a) implies that $m_1 \parallel m_2$. But this is contradiction, since P is a common point.
(c) Let m be the perpendicular dropped from P to l. Let $l' \perp m$ and incident with P. Then $l \parallel l'$ by part (a). □

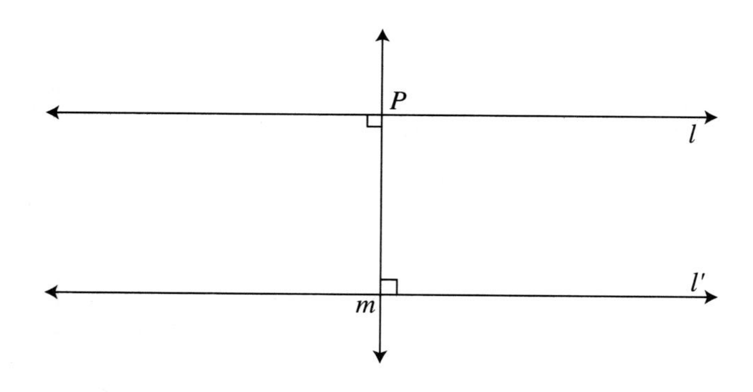

Figure 1.28

Another important consequence of the axioms and results on congruence of triangles is the exterior angle theorem. Before stating it, we make another definition.

Definition 1.5.13 *An angle supplementary to an angle of a triangle is called an* exterior angle *of the triangle. Given an exterior angle of a triangle, its* remote *angles are the angles of the triangle not adjacent to it.*

Theorem 1.5.14 (Exterior angle theorem) *The measure of an exterior angle of a triangle is greater than the measure of either of its remote angles.*

Proof Let $\angle DBC$ be an exterior angle of $\triangle ABC$. We will show that $m(\angle DBC) > m(\angle C)$. Observe that lines $\overleftrightarrow{AC}$ and $\overleftrightarrow{DA}$ are cut by the transversal $\overleftrightarrow{BC}$. Therefore, $m(\angle DBC) \neq m(\angle C)$, otherwise we would have $\overleftrightarrow{AC} \parallel \overleftrightarrow{DA}$, by the alternate interior angle theorem. Let us then suppose that $m(\angle DBC) < m(\angle C)$. Using the result of Exercise 2 of Section 1.4, we conclude that there exists a point $E \in \overline{AB}$ such that $\angle ECB \simeq \angle DBC$ (see Figure 1.29). Then, line $\overleftrightarrow{BC}$ cuts lines $\overleftrightarrow{DE}$ and $\overleftrightarrow{CE}$, forming congruent alternate interior angles. Therefore they must be parallel, which is a contradiction, since they meet at point E. It follows that $m(\angle DBC) > m(\angle C)$. The case $m(\angle DBC) > m(\angle A)$ is proved in a similar manner. $\square$

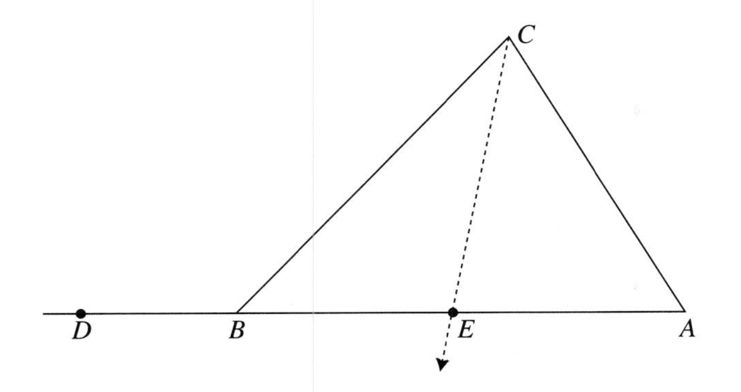

Figure 1.29 Exterior angle theorem

Corollary 1.5.15 *(a) The sum of the measures of any two angles of a triangle is less than* $180°$.
(b) In a triangle $\triangle ABC$, $AB > BC$ *if and only if* $m(\angle C) > m(\angle A)$.

Proof The proof of *(a)* is left as an exercise.
(b) Let us suppose that $AB > BC$. Then there exists $D \in \overline{AB}$ such that $BD = BC$. Then $\triangle BDC$ is isosceles, which in turn implies that $\angle BDC \simeq DCB$. Therefore $m(\angle BDC) < m(\angle C)$. Since $m(\angle BDC) > m(\angle A)$, because $\angle BDC$ is an exterior angle of $\triangle ADC$, we conclude $m(\angle C) > m(\angle A)$.

Conversely, if $m(\angle C) > m(\angle A)$, then $AB \neq BC$. Let us suppose that $AB < BC$. Then the first part of this proof implies that $m(\angle C) < m(\angle A)$, contradicting the hypothesis. □

Theorem 1.5.16 (Triangle Inequality) *If* A, B, *and* C *are three noncollinear points, then* $AC < AB + BC$.

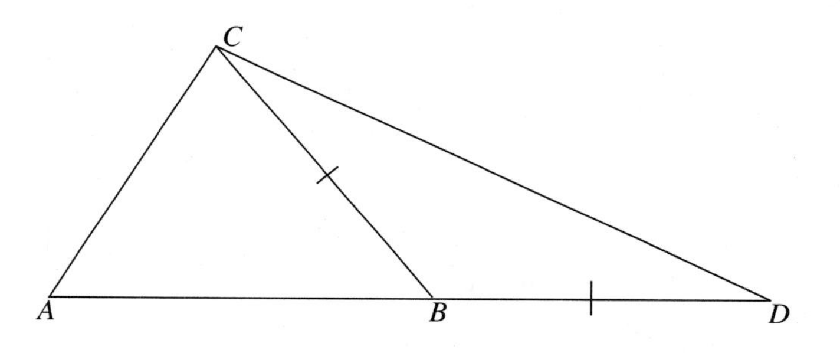

Figure 1.30

Proof Let $D \in \overrightarrow{AB}$ such that $CB = BD$ and B is between A and D. Then $AD = AB + BD = AB + BC$. In addition $\triangle CBD$ is isosceles and hence $\angle BCD \simeq \angle BDC$. This implies that $m(\angle D) < m(\angle C)$. Now we consider $\triangle ACD$. Applying Corollary 1.5.15(b) to the last inequality, we conclude that $AC < AD$. Therefore $AC < AB + BC$. □

Exercises

1. (a) Show that if opposite sides of a quadrilateral are congruent, then so are its opposite angles.
   (b) Show that if opposite sides of a quadrilateral are congruent,

then such a quadrilateral is a parallelogram.
(c) Show that if the diagonals of a quadrilateral bisect each other, then such a quadrilateral is a parallelogram.

2. A *rhombus* is a parallelogram with all sides congruent.
(a) Show that if all sides of a quadrilateral are congruent, then such a quadrilateral is a rhombus.
(b) Show that the diagonals of a rhombus bisect opposite angles.
(c) Show that the diagonals of a rhombus bisect each other.
(d) Show that the diagonals of a rhombus are perpendicular to each other.

3. Given $\triangle ABC$, show that the line that bisects angle $\angle A$ is the perpendicular bisector of $\overline{BC}$ if and only if $AB = AC$.

4. Show by means of an example that the following statement is false: *If one triangle has two sides and one angle congruent to two sides and an angle of a second triangle, then these triangles are congruent.* Such an example is called a *counterexample* to the statement. *Hint*: Start with an isosceles triangle. Then consider a suitable ray emanating from one vertex and obtain two noncongruent triangles.

5. *Definition:* A *kite* is a quadrilateral with the property that two distinct pairs of consecutive sides are congruent. Prove the following:
(a) The diagonals of a kite are perpendicular to each other.
(b) A line containing one diagonal bisects the other.
(c) A line containing one diagonal bisects nonconsecutive angles.

6. Prove that there exists a triangle that is not isosceles.

7. (*AAS*) Let $\triangle ABC$ and $\triangle DEF$ such that $\overline{AC} \simeq \overline{DF}$, $\angle A \simeq \angle D$, and $\angle B \simeq \angle E$. Show that $\triangle ABC \simeq \triangle DEF$.

8. *Definition:* A triangle with one right angle is called a *right triangle*. The side opposite the right angle is called the *hypotenuse* and the other two are called the *legs* of the triangle. Show that if the hypotenuse and a leg of a right triangle are congruent to the hypotenuse and a leg of another right triangle, then these triangles are congruent.

9. Show that a leg of right triangle is less than its hypotenuse.

10. *Definition:* A line segment whose endpoints are a vertex of a triangle and the foot of the perpendicular dropped from this vertex to the line containing its opposite side is called an *altitude* of a triangle. The foot of the perpendicular line is also called the *foot of the altitude.* Show the following:
    If $\angle A$ of $\triangle ABC$ is obtuse then A is between C and the foot of the altitude dropped from B.

1.6 The Circle

Definition 1.6.1 *Given two points O and A. The set of all points P such that $\overline{OP} \simeq \overline{OA}$ is called the* circle of center O, *and each segment $\overline{OP}$ is called a* radius *of the circle.*

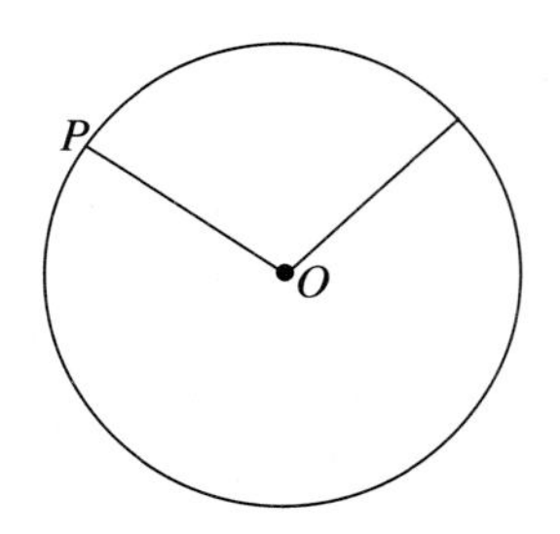

Figure 1.31

It follows that all radii are congruent to each other and hence they have the same length. Such a length will be called the *radius* of the circle.

Definition 1.6.2 *(a) Let γ be a circle with center O. A line segment connecting any two distinct points on γ is called a* chord *of γ.*
(b) A chord passing through the center of the circle is called a diameter *of the circle.*

Observe again that all diameters have the same length. Let d denote such a length. If r denotes the radius, then $d = 2r$. The number d is called the *diameter* of the circle.

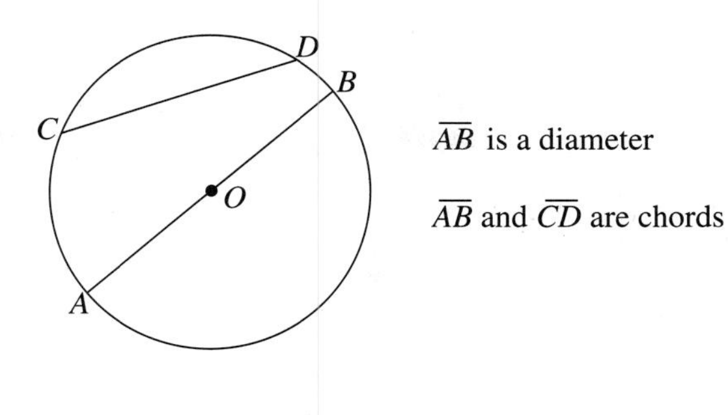

Figure 1.32

Definition 1.6.3 *Given two points A and B on a circle, the line $\overleftrightarrow{AB}$ divides the circle into two disjoint parts; each part is contained in one of the two disjoint half-planes determined by $\overleftrightarrow{AB}$; each part is called an (open)* arc *of γ determined by A and B. A* closed *arc is an open arc together with its endpoints A and B. An arc determined by two points that are endpoints of a diameter is called a* semicircle.

Proposition 1.6.4 *Let γ be a circle with center O.*
(a) Let A and B be two points on γ and M be the midpoint of chord $\overline{AB}$. If $M \neq O$, then line $\overleftrightarrow{OM}$ is perpendicular to line $\overleftrightarrow{AB}$.
(b) Let $\overline{AB}$ be a chord of γ. Then the perpendicular bisector of $\overline{AB}$ passes through the center O.

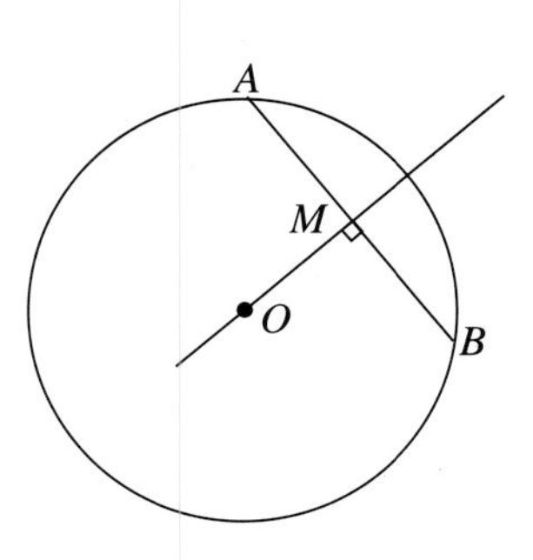

Figure 1.33

Proof *(a)* If $M \neq O$, then $\overline{AB}$ is not a diameter and hence points O, A, and B are noncollinear (see Figure 1.33). Moreover, we have that

$\triangle AOM \simeq \triangle BOM$ by SSS. Then $\angle AMO \simeq \angle BMO$. Since they are supplementary angles, each of them is a right angle.
(b) If $\overline{AB}$ is not a diameter, then its midpoint $M \neq O$. Then part (a) implies that line $\overleftrightarrow{MO}$ is the perpendicular bisector of $\overline{AB}$. If $\overline{AB}$ is a diameter, then its midpoint is the center O, and hence O lies on the perpendicular bisector of $\overline{AB}$. □

Notice that the points on chord $\overline{AB}$ fall within the circle. In fact, if $M = O$, then each point $P \neq O$ between A and B is either between A and O or between B and O. If $M \neq O$, then from Exercise 1 of this section we get that either $OM < OP < OA$ or $OM < OP < OB$.

Definition 1.6.5 *A line l is said to be* tangent *to a circle γ if it intersects the circle in exactly one point. This point is called a* point of tangency. *A line l is said to be a* secant *of a circle γ if it intersects the circle in two points.*

The next result states a geometric property of a line through a point of circle that is equivalent (if and only if) to being tangent to the circle. The usual symbol for "equivalent" is $\Leftrightarrow$. For this reason, it is common that we indicate the two parts of the proof with the symbols $\Rightarrow$ and $\Leftarrow$. For instance, if the result is $a \quad \Leftrightarrow \quad b$, we write:
($\Rightarrow$): this part assumes a to prove b.
($\Leftarrow$): this part assumes b to prove a.

Proposition 1.6.6 *Let γ be a circle with center O. Let $A \in \gamma$ and l a line through A. Then l is tangent to γ if and only if l is perpendicular to line $\overleftrightarrow{AO}$.*

Proof ($\Rightarrow$) Let m be a line perpendicular to l and passing through O. Let $P = m \cap l$. We want to show that $P = A$. Let us suppose $P \neq A$. Now consider a point A' on line m such that $PA = PA'$ (see Figure 1.34). Since $\angle APO$ is a right angle, we conclude that $\triangle APO \simeq \triangle A'PO$ by SAS. This implies $OA = OA'$, which in turn gives that $A' \in \gamma$. Since $\overrightarrow{AP} \subset l$, we obtain that A and A' are in $l \cap \gamma$, and this contradicts that l is tangent to γ.
($\Leftarrow$) Now we want to show that if l is perpendicular to line $\overleftrightarrow{AO}$, then $l \cap \gamma = A$. Let $P \neq A$ be an arbitrary point of l. Then $\triangle OAP$ is a

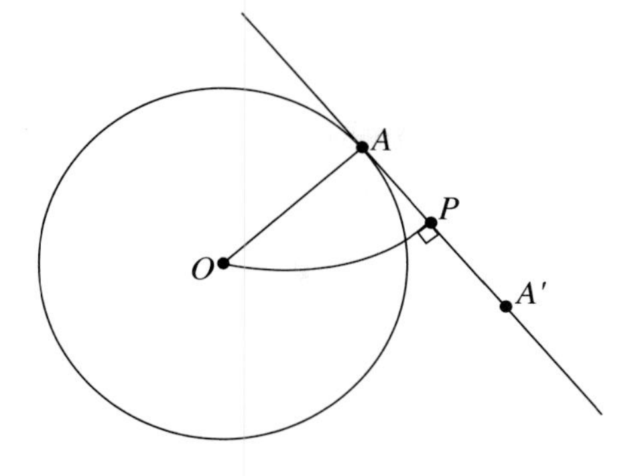

Figure 1.34

right triangle with hypotenuse OP and hence $OP > OA$, by Exercise 8 of Section 1.5. Therefore $P \notin \gamma$, since $\overline{OP}$ and $\overline{OA}$ are not congruent.$\square$

Definition 1.6.7 *We say that a circle is* inscribed in a triangle *if all three sides of the triangle are tangent to the circle.*

Proposition 1.6.8 *Every triangle has one and only one inscribed circle.*

The proof of this result depends on the following lemma.

Lemma 1.6.9 *The bisectors of the angles of any triangle all meet at a single point.*

Proof Let us consider $\triangle ABC$. Let rays $\overrightarrow{AG}$ and $\overrightarrow{BH}$ bisect $\angle A$ and $\angle B$, respectively (see Figure 1.35). Since $\overrightarrow{AG}$ is between rays $\overrightarrow{AB}$ and $\overrightarrow{AC}$, ray $\overrightarrow{AG}$ intersects $\overline{BC}$ at point D between B and C, by the crossbar theorem. We also have that ray $\overrightarrow{BH}$ is between rays $\overrightarrow{BA}$ and $\overrightarrow{BD}$; then ray $\overrightarrow{BH}$ intersects $\overline{AD}$ at a point O between A and D. We will show that $\overrightarrow{CO}$ bisects $\angle C$ and then O is the desired point. For that let L, M, N be the feet of the perpendicular lines dropped from O to $\overline{AB}$, $\overline{AC}$, and $\overline{BC}$, respectively (see Figure 1.36). Using Exercise 7 of Section 1.5, we conclude that $\triangle ALO \simeq \triangle AMO$ and hence $OL = OM$. Similarly we obtain that $\triangle BLO \simeq \triangle BNO$, implying $OL = ON$. Then

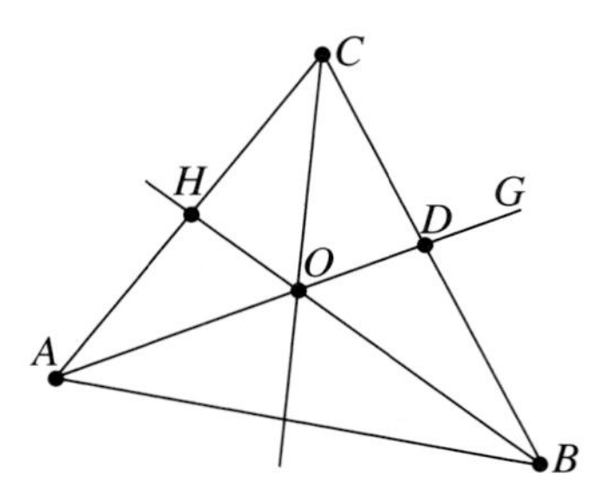

Figure 1.35

we have $OM = ON$, and since $\triangle NOC$ and $\triangle MOC$ are both right triangles, we conclude that they are congruent, using the result of Exercise 8 of Section 1.5. This implies that $\angle NCO \simeq \angle MCO$ and therefore ray $\overrightarrow{CO}$ bisects $\angle C$. $\square$

This concurrence point O is called the *incenter* of the triangle.

Proof of Proposition 1.6.8 With the same notation used in the previous lemma, let γ denote the circle centered at the incenter O and radius $OL = OM = ON$ (see Figure 1.36). Since $\overleftrightarrow{AB}$ is perpendicular to $\overline{OL}$, $\overleftrightarrow{AB}$ is tangent to γ at L. Similarly, we have that $\overleftrightarrow{AC}$ and $\overleftrightarrow{BC}$ are tangent to γ at N and M, respectively.

To complete the proof, let us suppose that γ' is a circle centered at O' and inscribed in $\triangle ABC$. Let $L' \in \overline{AB}$ and $M' \in \overline{AC}$ be points of tangency. Then $\triangle L'O'A$ and $\triangle M'O'A$ are both right triangles and Exercise 8 implies that they are congruent, since $O'L' = O'M'$. This implies that $\angle O'AL' \simeq \angle O'AN'$, and then ray $\overrightarrow{AO'}$ bisects $\angle A$. From the uniqueness of the bisector we conclude that $\overrightarrow{AO'} = \overrightarrow{AO}$. Similarly one obtains $\overrightarrow{BO'} = \overrightarrow{BO}$. Since lines $\overleftrightarrow{AO}$ and $\overleftrightarrow{BO}$ have only one intersection point, we conclude that $O = O'$. Now, since the perpendicular dropped from O to $\overleftrightarrow{AB}$ is unique, we conclude that $L = L'$. It follows that γ and γ' are both centered at the same point and have the same radii; therefore $\gamma = \gamma'$. $\square$

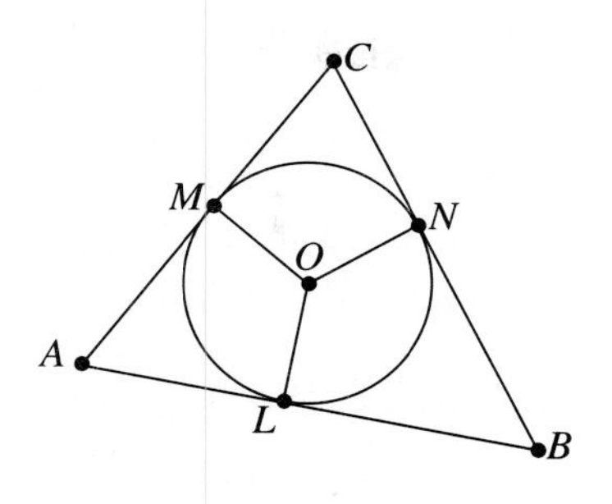

Figure 1.36

Definition 1.6.10 *Let γ be a circle of center O and radius r. A point P is said to be* inside *the circle if $OP < r$. If $OP > r$, we say that P is* outside a circle *the circle. The set of points inside circle γ is called the* interior *of γ.*

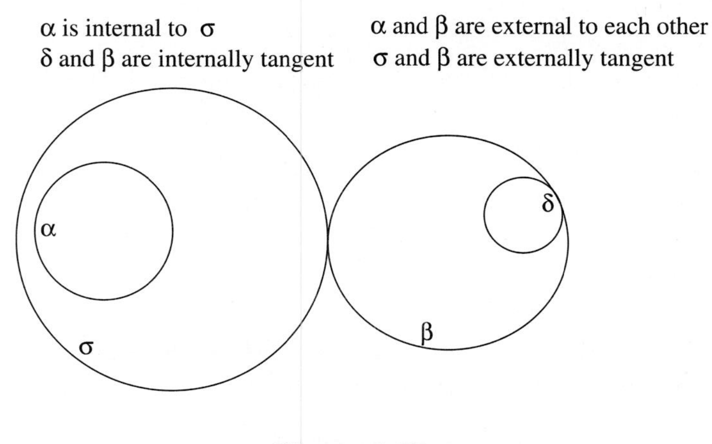

Figure 1.37

Definition 1.6.11 *Let γ and γ' be two distinct circles* (see Figure 1.37).
(a) We say that γ and γ' are external *to one another if every point on γ is outside γ' and every point on γ' is outside γ.*
(b) We say that γ' is internal *to γ if every point on γ' is inside γ.*

(c) We say that γ' *is* internally tangent *to* γ *if they have only one common point and all other points of* γ' *are inside of* γ.
(d) We say that γ' *is* externally tangent *to* γ *if they have only one common point and neither has points inside the other.*

Proposition 1.6.12 *If two circles are externally tangent, then their centers and the point of contact are collinear.*

Proof Let γ and γ' be circles of centers O and O' and radii r and r', respectively (see Figure 1.38). Consider point B on ray $\overrightarrow{OO'}$ such that $OB = r$; similarly, consider point $B' \in \overrightarrow{O'O}$ such that $O'B' = r'$. Suppose that $B \neq B'$. Since the circles are externally tangent, let A be their point of contact. Then we have:

$$OA + O'A = r + r' = OB + O'B' < OO'$$

which contradicts the triangle inequality. Therefore $A = B = B'$. □

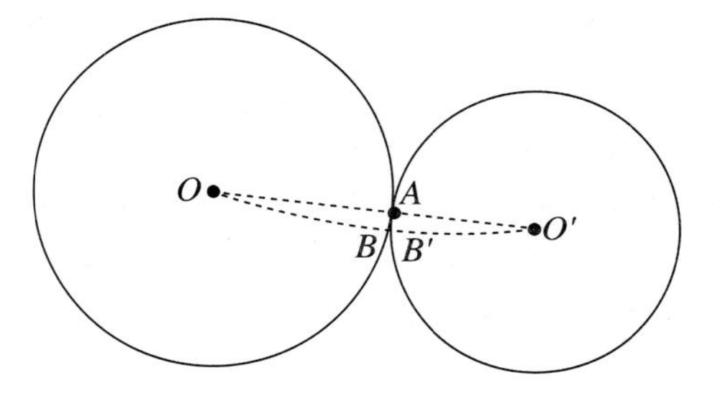

Figure 1.38

Proposition 1.6.13 *If two circles are internally tangent, then their centers and the point of contact are collinear.*

Proof Let γ and γ' be circles of centers O and O' and radii r and r', respectively. Let us assume that γ' and γ are internally tangent, with A denoting their point of contact (see Figure 1.39). Observe that Axiom $\mathbf{S_2}$ implies that point O' is inside circle γ; in fact, if O' is not inside γ, then any circle centered at O' has a point on ray $\overrightarrow{O'O}$ and a point on the opposite ray. The latter point is outside γ, contradicting the assumption that γ' and γ are internally tangent.

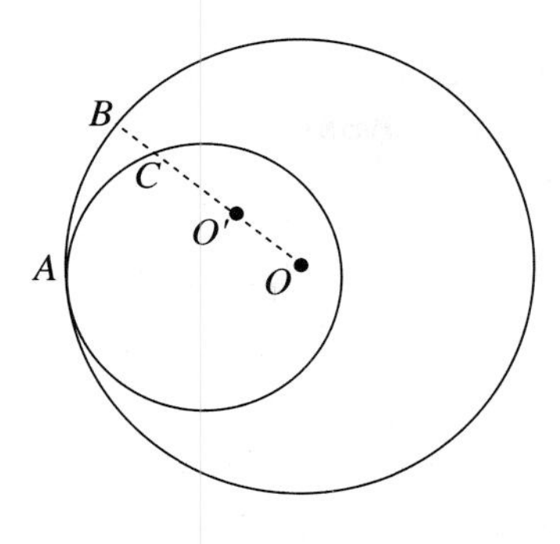

Figure 1.39

Now, let $B \in \overrightarrow{OO'}$ such that $OB = r$ and $C = \gamma' \cap \overrightarrow{OO'}$. Notice that C is between O' and B, because C is inside γ. Since O' is between O and B, we have

$$OB = OO' + O'B.$$

But $OA = OB$, and the triangle inequality implies

$$OA < OO' + O'A.$$

Therefore $O'B < O'A$. But if $C \neq A$, we also have $O'A = O'C$, because both lie on γ' and therefore $O'B < O'C$. But this contradicts the fact that C is between O' and B. □

Remark: We point out that the proof above does not depend on the uniqueness of the point of contact; it depends only on the fact that some points of γ' lie on γ and the remaining points of γ' are inside γ. The same observation applies to the proof of Proposition 1.6.12.

Exercises

1. (a) Let D be between B and C and A such that $\overleftrightarrow{AB} \perp \overleftrightarrow{BC}$. Show that $AB < AD < AC$.
   (b) Let l be a line with a point inside a circle γ. Show that the foot of the perpendicular from the center to l is also inside γ.
   (c) Given any triangle $\triangle ABC$ and any point D between B and C, show that either $AD < AC$ or $AD < AB$.

2. Show that the interior of a circle is a convex set.
   Hint: Use the previous exercise.

3. Let γ be a circle with center A and radius r. Let l_1 and l_2 be two lines that are tangent to γ at P and Q, respectively. Let $B = l_1 \cap l_2$.
   (a) Show that $\overline{PB} \simeq \overline{QB}$.
   (b) Show that $\overline{AB} \perp \overline{PQ}$.

4. Let γ_1 and γ_2 be a pair of circles with centers O and O', respectively, and intersecting each other in two points P, Q. Let M be the midpoint of $\overline{OO'}$.
   (a) Show that $\overline{PM} \simeq \overline{QM}$.
   (b) Let l and l' be lines tangent to γ and γ' and intersecting at point P. The angle formed by them is called *angle subtended* by the circles at P. Show that the circles subtend the same angle at P as they do at Q.

5. Let σ be a semicircle of endpoints A, B, and center O. Let $O' \in \overleftrightarrow{AB}$ such that A is between O and O'. Show that if $P, Q \in \sigma$ and P is between A and Q, then $O'A < O'P < O'Q < O'B$.

6. Let γ and γ' be two distinct circles, with centers O and O' and radii r, r', respectively. Show the following:
   (a) If $OO' > r + r'$, then the two circles are external to one another.
   (b) If $OO' = r + r'$, then the two circles are externally tangent.
   (c) If $OO' < |r - r'|$, then the lesser circle is internal to the greater.
   (d) If $OO' = |r - r'|$, then the two circles are internally tangent.

1.7 Principles of Continuity

Some problems in geometry are solved by finding points determined by the intersection of either two lines, or a line with a circle or two circles. The incidence axioms guarantee the existence of concurrent lines. In the previous section we studied some properties of circles that are either tangent or intersect each other. However, under certain relative positions of lines and circles, the existence of points of intersection of a circle with a line and of a circle with a circle cannot be proved using only the postulates assumed in this chapter, even if we add Euclid's fifth postulate, which will be introduced in the next chapter.

The principles of continuity prove the existence of such points. They are consequences of Axiom $\mathbf{S_2}$ combined with the *completeness axiom* of the real numbers. In order to make clear the statement of the completeness axiom, as well as its application in the proofs of this section, we introduce some terminology.

Definition 1.7.1 *Let S be a set of real numbers. We say that S is* bounded above *if there exists $\alpha \in \mathbf{R}$ such that*

$$x \leq \alpha, \quad \forall\, x \in S.$$

The number α is called an upper bound *of S. Similarly, we say that S* is bounded below *if there exists $\beta \in \mathbf{R}$ such that*

$$\beta \leq x, \quad \forall\, x \in S,$$

and we call β a lower bound *of S.*

It is clear that if α is an upper bound of a set S and $\delta \geq \alpha$, then δ is also an upper bound of S. Now let us suppose that α is an upper bound of S with the property that if $\delta < \alpha$, then δ is not an upper bound of S. Then α is called the *least upper bound* or the *supremum* of S, and we write

$$\alpha = \sup S.$$

Similarly, the *greatest lower bound* or the *infimum* of a set S bounded below, denoted by

$$\beta = \inf S,$$

means that if $\delta > \beta$, then δ is not a lower bound of S. Next we state the axiom for the real numbers that is used to prove the principles of continuity.

> **Completeness axiom** Let S be a nonempty set of real numbers.
> (i) If S is bounded above, then $\sup S$ exists.
> (ii) If S is bounded below, then $\inf S$ exists.

The completeness axiom implies the *Archimedean property* of the real numbers that we state and prove below.

Proposition 1.7.2 (Archimedean property) *Let $x, y \in \mathbf{R}, x > 0$. Then there exists a positive integer n such that $nx > y$. In particular, for any positive $\epsilon > 0$ there exists a positive integer n such that $n\epsilon > 1$, that is,*

$$\frac{1}{n} < \epsilon.$$

Proof Let $\mathbf{N}$ denote the set of positive integers and consider

$$S = \{nx, \mid n \in \mathbf{N}\}.$$

If $y \geq nx$, for all $n \in \mathbf{N}$, then y is an upper bound of S. Then S has a least upper bound. Let $s = \sup S$. Since $x > 0$, $s - x < s$, then there is a positive integer k such that $s - x < kx$ (because $s - x \geq kx$ for all $k \in \mathbf{N}$ would imply that $s - x$ is also an upper bound of S). Then we have

$$s - x < k\,x \Rightarrow s < (k+1)\,x \in S,$$

which is a contradiction, since s is an upper bound of S. □

The first geometric consequence of the completeness axiom that we present in this text is the Saccheri-Legendre theorem. This result is proved using the Archimedean property of the real numbers. Several other results whose proofs depend upon the completeness axiom and the Archimedean property will appear in subsequent chapters.

Theorem 1.7.3 (Saccheri-Legendre theorem) *The sum of the measures of the angles of any triangle is less than or equal to* $180°$.

To prove this result we use the following lemma:

Lemma 1.7.4 *For any triangle $\triangle ABC$ there exists another triangle $\triangle A_1B_1C_1$ such that*

$$m(\angle A) + m(\angle B) + m(\angle C) = m(\angle A_1) + m(\angle B_1) + m(\angle C_1)$$

and $m(\angle A_1) \leq (1/2)\, m(\angle A)$.

Proof Let D be the midpoint of $\overline{BC}$ (see Figure 1.40). Consider point E on ray $\overrightarrow{AD}$ such that $AD = DE$. Then $\triangle BDA \simeq \triangle CDE$, by SAS. Notice that the angle sum of $\triangle ABC$ is given by

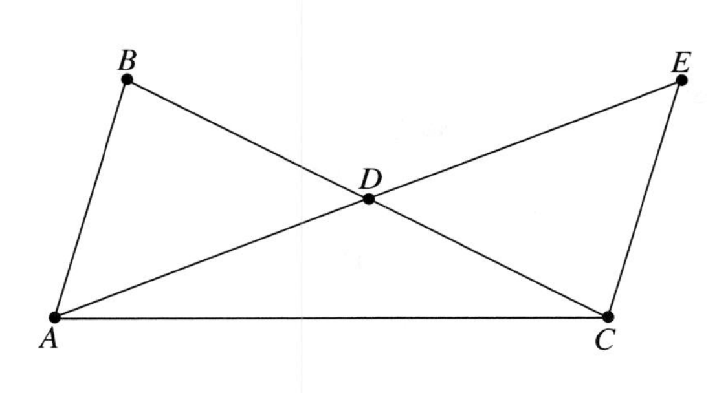

Figure 1.40

$$m(\angle B) + m(\angle A) + m(\angle C)$$
$$= m(\angle B) + m(\angle BAD) + m(\angle DAC) + m(\angle DCA).$$

But the congruence of the triangles implies that $m(\angle E) = m(\angle BAD)$ and $m(\angle B) = m(\angle DCE)$. Substituting in the equality above, we obtain

$$m(\angle B) + m(\angle A) + m(\angle C)$$
$$= m(\angle DCE) + m(\angle E) + m(\angle DAC) + m(\angle DCA).$$

The right-hand side of the equation above is the angle sum of $\triangle ACE$. Furthermore,

$$m(\angle A) = m(\angle BAD) + m(\angle DAC) = m(\angle E) + m(\angle DAC)$$

and hence either $2m(\angle E) \leq m(\angle A)$ or $2m(\angle DAC) \leq m(\angle A)$. Therefore $\triangle ACE$ implies the result of the lemma. □

Proof of the Saccheri-Legendre theorem Given triangle $\triangle ABC$, let us suppose that

$$m(\angle A) + m(\angle B) + m(\angle C) = 180 + \epsilon$$

where ϵ is a positive real number. Using Lemma 1.7.4, we find $\triangle A_1B_1C_1$ such that $m(\angle A_1) \leq (1/2)\ m(\angle A)$ and the angle sum is $180 + \epsilon$. Now Lemma 1.7.4 can be used for $\triangle A_1B_1C_1$ to obtain $\triangle A_2B_2C_2$, whose angle sum is $180 + \epsilon$, and

$$m(\angle A_2) \leq \frac{1}{2}\ m(\angle A_1) \leq \frac{1}{4}\ m(\angle A).$$

Repeating this argument n times, we get a triangle with angle sum as the original triangle, and the measure of one of its angles satisfies

$$m(\angle A_n) \leq \frac{1}{2^n}\, m(\angle A).$$

Since for any positive real number ϵ we find a positive integer number N such that

$$\frac{1}{2^N} < \frac{\epsilon}{m(\angle A)},$$

(justify) we conclude that there exists a $\triangle A_N B_N C_N$ such that

$$m(\angle A_N) + m(\angle B_N) + m(\angle C_N) = 180 + \epsilon$$

and $m(\angle A_N) < \epsilon$. This implies then that $m(\angle B_N) + m(\angle C_N) > 180$, which contradicts Corollary 1.5.15(a). □

In the rest of this section we shall prove the two principles of continuity that are needed to justify the geometric constructions at the end of this chapter.

Theorem 1.7.5 (Elementary continuity principle) *If a line l has one point inside and one point outside a circle γ, then it has two points common with the circle γ.*

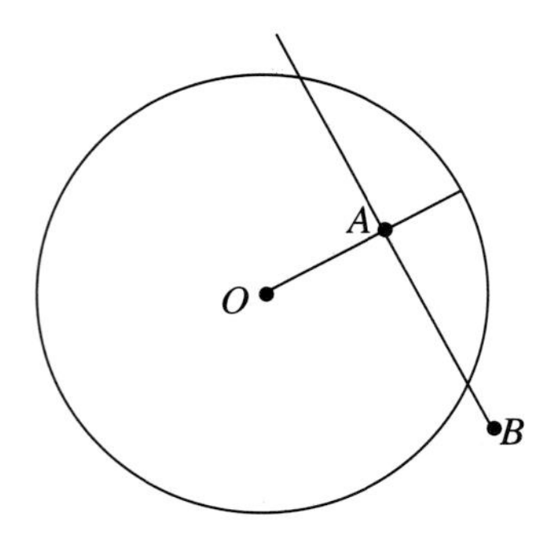

Figure 1.41

Proof Let O denote the center of the circle and A the foot of the perpendicular dropped from O to the line. From Exercise 1(b) of Section 1.6 we get that $OA < r$, where r is the radius of the circle. Let B denote the point outside γ, that is, $OB > r$. Let a, b be coordinates of A and B, respectively. Applying Axiom $\mathbf{S_2}$, we find a one-to-one correspondence f, from segment $\overline{AB}$ onto the interval of real numbers $[a, b]$. Moreover, such a correspondence preserves betweenness, by Proposition 1.4.3. We define

$$\Sigma_1 = \{P \in \overline{AB} \mid OP < r\}, \quad \Gamma_1 = \{x \in [a,b] \mid f(P) = x,\ P \in \Sigma_1\},$$

$$\Sigma_2 = \{P \in \overline{AB} \mid OP > r\}, \quad \Gamma_2 = \{x \in [a,b] \mid f(P) = x,\ P \in \Sigma_2\}.$$

Then $A \in \Sigma_1$ and $B \in \Sigma_2$. Notice that b is an upper bound of Γ_1 and hence there exists $s = \sup \Gamma_1$. Let $S \in \overline{AB}$ such that $f(S) = s$. We claim that $OS \geq r$. In fact, suppose that $OS < r$. Let $\epsilon = r - OS > 0$. Since $S \neq B$, there exists $S' \in \overline{AB}$, between S and B, and such that $SS' < \epsilon$. Let s' be the coordinate of S'. From Proposition 1.4.3 we obtain that $s < s' < b$. On the other hand, the triangle inequality implies

$$OS' < OS + SS' < OS + \epsilon = OS + r - OS = r$$

and hence $S' \in \Sigma_1$. Then $s' \in \Gamma_1$ and hence $s' < s < b$, for s is an upper bound of Γ_1. We have then obtained a contradiction. Therefore $OS \geq r$.

Now consider Σ_2 and Γ_2, which is bounded below by a. Let $i = \inf \Gamma_2$ and $I \in \overline{AB}$ such that $f(I) = i$. We will show that $OI \leq r$. Suppose that $OI > r$. Then let $\epsilon = OI - r > 0$. Since $I \neq A$, there exists $I' \in \overline{AB}$, between I and A, and such that $II' < \epsilon$. Let i' be the coordinate of I'. Using again Proposition 1.4.3, we obtain that $a < i' < i$. The triangle inequality implies

$$OI < OI' + II' < OI' + \epsilon = OI' + OI - r$$

yielding $OI' > r$. Therefore $I' \in \Sigma_2$ and then $i < i'$, since i is a lower bound. This is a contradiction, and we conclude that $OI \leq r$.

We will show now that $S = I$, which in turn implies that $OS = r$ and then $S \in \gamma$. For that, it is enough to show that $s = i$.

Observe that Exercise 1 of Section 1.6 implies that if $X \in \Sigma_1$ and $Y \in \Sigma_2$, then X is between A and Y (if not, we would have $OY < OX < r$). It follows that if x and y are their respective coordinates, then $x < y$. Therefore every number in Γ_2 is an upper bound of Γ_1 and hence

$$s \leq y, \quad \forall y \in \Gamma_2,$$

since s is the least upper bound. Therefore s is a lower bound of Γ_2 implying that $s \leq i$, by the definition of i. We will then suppose that $s < i$ and obtain a contradiction. If $s < i$, let t be a real number such that $s < t < i$ and $T \in \overline{AB}$ such that $f(T) = t$. Applying again Proposition 1.4.3, we obtain that T is between S and I, and from Exercise 1 of Section 1.6 we get that $OT > OS \geq r$. Then $T \in \Sigma_2$ and hence $t \in \Gamma_2$, implying that $i \leq t$, which is the desired contradiction.

The argument above shows that there exists only one point S on ray $\overrightarrow{AB}$ common with the circle γ. On the opposite ray, emanating from A, there exists a point B' outside the circle (why?). Then, the same proof implies that γ intersects $\overrightarrow{AB'}$ at only one point R. Now, if $X \in \gamma \cap l$, the line separation property implies that either $X \in \overrightarrow{AB}$ or $X \in \overrightarrow{AB'}$ and, therefore, either $X = S$ or $X = R$, showing that l has exactly two points in common with γ. $\square$

Now we study points of intersection of two circles. The circular continuity principle is also proved using the *completeness axiom*. But, in order to apply it, we need to establish a one-to-one correspondence between arcs of a circle and sets of real numbers that preserve betweenness. We then define betweenness for points on arcs of circles and prove some preliminary results.

Definition 1.7.6 *Let $\overline{AB}$ be a diameter of a circle γ with center O. A* semicircle *σ with* endpoints *A, B, and* center *O is the intersection of γ with one of the half-planes determined by line $\overleftrightarrow{AB}$. The union of a semicircle σ with its endpoints will be denoted by $\overline{\sigma}$.*

Definition 1.7.7 *Let σ be a semicircle and $P, Q, R \in \overline{\sigma}$. We say R is* between *P and Q if ray $\overrightarrow{OR}$ is between $\overrightarrow{OP}$ and $\overrightarrow{OQ}$, where O is the center of σ.*

Lemma 1.7.8 *Let σ be a semicircle with endpoints A, B, and center O. Consider a point M on σ such that $\overleftrightarrow{AB} \perp \overleftrightarrow{MO}$. Let $\Lambda = \overline{AM} \cup \overline{MB}$. Then*

(a) For any point $P \in \bar{\sigma}$ the ray $\overrightarrow{OP}$ intersects Λ at a point P'.

(b) The mapping $P \mapsto P'$ is one-to-one and onto Λ.

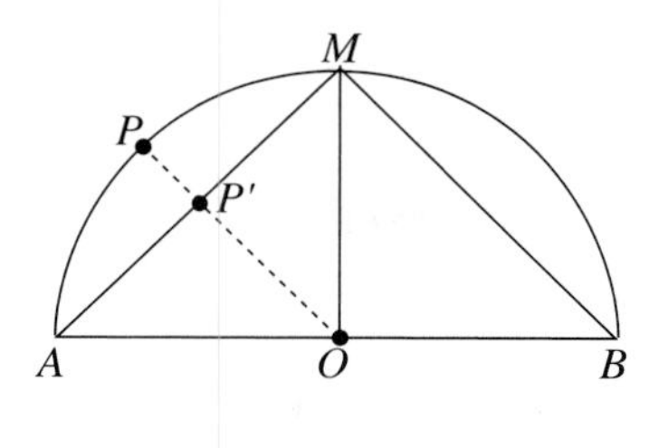

Figure 1.42

Proof (a) If P is either A or M or B, then their images are themselves. Let us consider $P \neq A, M, B$ (see Figure 1.42). The ray $\overrightarrow{OP}$ is either between rays $\overrightarrow{OA}$ and $\overrightarrow{OM}$ or between rays $\overrightarrow{OM}$ and $\overrightarrow{OB}$. Therefore ray $\overrightarrow{OP}$ intersects Λ by the crossbar theorem.
(b) If $P' = Q'$ then $\overrightarrow{OP} = \overrightarrow{OQ}$, and since $OP = OQ$ we have that $P = Q$ by Axiom $\mathbf{S_2}$. Now, given any point X on Λ, consider on ray $\overrightarrow{OX}$ the point P such that OP equals the radius of σ. Then $P \in \sigma$ and f is onto. □

Proposition 1.7.9 *There is a one-to-one correspondence between the points of $\overline{\sigma}$ and a closed and bounded interval of real numbers, $[a, b]$, which preserves betweenness; that is, a point R on $\overline{\sigma}$ is between P and Q if and only if the corresponding number r is between the corresponding numbers p and q.*

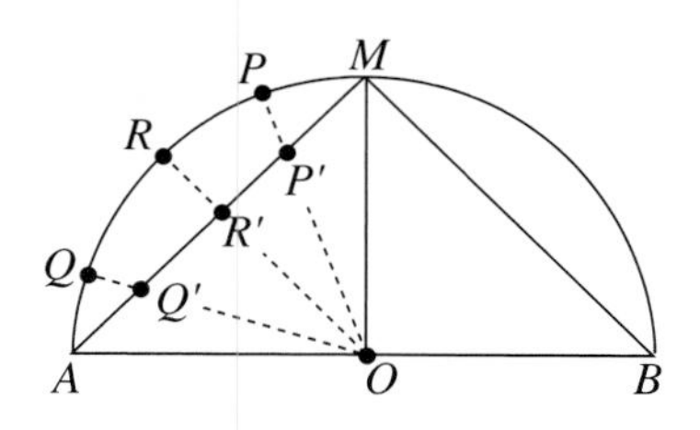

Figure 1.43

Proof First, we remark that the map f defined in Lemma 1.7.8 and restricted to $\overline{AM}$ preserves betweenness. In fact, if R is between P and

Q, then point R is an interior point of $\angle POQ$. But $\angle POQ = \angle P'OQ'$, and if R is an interior point of $\angle P'OQ'$, so is R', since R and R' are both on ray $\overrightarrow{OR}$. Hence R' is between P' and Q' by Proposition 1.2.16. Similarly, f restricted to $\overline{MB}$ preserves betweenness. Now, since Λ is the union of two line segments, Axiom $\mathbf{S_2}$ implies that there exists a one-to-one map g from Λ onto a closed and bounded interval $[a, b]$ (consider a one-to-one correspondence from $\overline{AM}$ onto an interval $[a, m]$ and a one-to-one correspondence from $\overline{MB}$ onto $[m, b]$). In addition, g preserves betweenness by Proposition 1.4.3. Therefore $g \circ f$ is a one-to-one correspondence between $\overline{\sigma}$ and $[a, b]$ preserving betweenness. □

Theorem 1.7.10 (Circular continuity principle) *If a circle γ has one point X inside and one point Y outside another circle γ', then the two circles intersect in two points.*

We must first prove two preliminary lemmas.

Lemma 1.7.11 *The hypotheses of Theorem 1.7.10 imply that there exists a semicircle $\sigma \subset \gamma$ such that one endpoint is inside and the other outside circle γ'.*

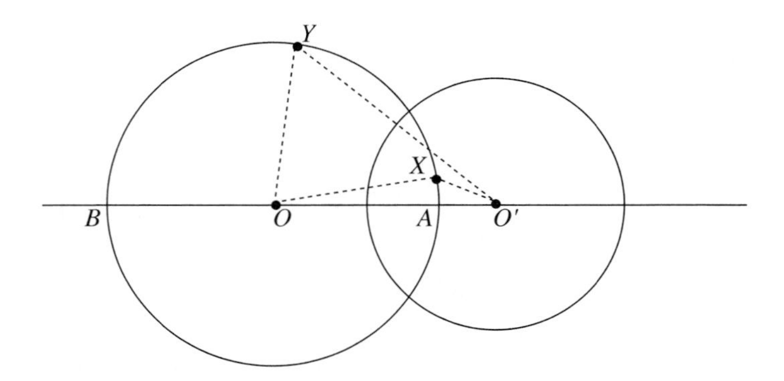

Figure 1.44 Case (i)

Proof Let O and O' be the centers of the two circles γ and γ' and r and r' be their radii, respectively. Let B be a point on ray $\overrightarrow{O'O}$ such that $OB = r$. Then $B \in \gamma$. We claim that B is outside γ', that is, $O'B > r'$. In fact, the triangle inequality implies that

$$O'Y < O'O + OY = O'O + r.$$

Since $O'Y > r'$, by hypothesis, we have

$$r' < O'Y < OO' + r = OO' + BO = O'B.$$

Now we consider $A \in \gamma$ such that $\overline{AB}$ is a diameter of γ. We will show that A is inside γ'. We consider the two possible cases:
(*i*) A is between O and O': In this case we have $OO' = OA + O'A = r + O'A$. On the other hand, since X is on γ and inside γ', we have $OX + O'X < r + r'$. Combining these two equalities with the triangle inequality, we obtain

$$OO' = r + O'A < OX + O'X < r + r'$$

which implies $O'A < r'$.

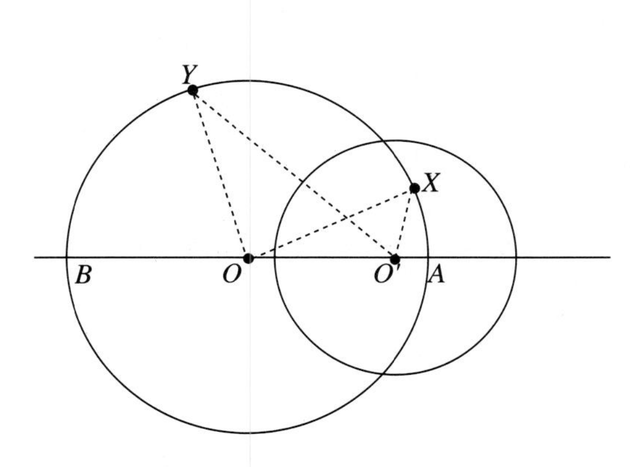

Figure 1.45 Case (*ii*)

(*ii*) O' is between O and A: Here we have $O'A = OA - OO' = r - OO'$ and hence $r = O'A + OO'$. From the triangle inequality we get $OX < OO' + O'X$ and, since X is on γ, $OX = r$, and inside γ', $O'X < r$. Then we obtain $r < OO' + r'$. Therefore

$$O'A + OO' = r < OO' + r'$$

which implies $O'A < r'$. □

Lemma 1.7.12 *Let P be on a semicircle σ and A be one endpoint of σ. Then for each positive real number ϵ there exists a point Q between P and A such that $PQ < \epsilon$.*

Proof Let t be the bisector of $\angle AOP$ and l be a line through P that intersects t at point $Z \neq P$ (see Figure 1.46). Let m be a real number given by

$$m = \min\{\frac{\epsilon}{2}, PZ\}.$$

On ray $\overrightarrow{PZ}$, we choose a point X such that $PX < m$. Let $Y \in \overrightarrow{PZ}$ such that $\angle YOX \simeq \angle XOP$. Since $\overrightarrow{OX}$ is between $\overrightarrow{OP}$ and $\overrightarrow{OZ}$, we have

$$m(\angle POX) < m(\angle POZ) = \frac{1}{2}m(\angle POA)$$

and hence

$$\begin{aligned} 2m(\angle POX) &= m(\angle POX) + m(\angle XOY) \\ &= m(\angle POY) < m(\angle POA). \end{aligned}$$

This implies that ray $\overrightarrow{OY}$ intersects σ at a point Q between P and A. Moreover, line $\overleftrightarrow{OX}$ is the bisector of $\angle POY$, and then it is the perpendicular bisector of $\overline{PQ}$. Let M be the midpoint of $\overline{PQ}$. Observe that $\triangle MPX$ is a right triangle of hypotenuse $\overline{PX}$. Therefore $PM < PX$ and hence $PQ = 2PM < \epsilon$. □

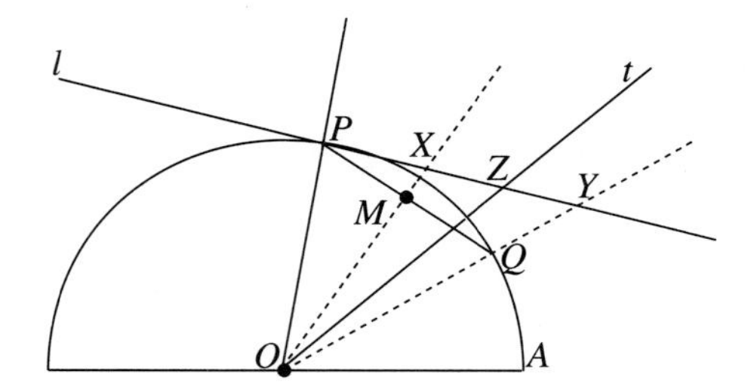

Figure 1.46

Proof of Theorem 1.7.10 This proof is similar to the proof of Theorem 1.7.5. We will sketch it, and the reader is asked to fill in the details.

Applying Lemma 1.7.11, we suppose that $\sigma \subset \gamma$ is a semicircle such that one endpoint A is inside and the other endpoint B is outside γ'. We want to show that there is a point on σ that lies on γ'. We then consider

$$\begin{aligned} \Sigma_1 &= \{P \in \sigma \mid O'P < r'\}, \\ \Sigma_2 &= \{P \in \sigma \mid O'P > r'\}. \end{aligned}$$

Then $A \in \Sigma_1$ and $B \in \Sigma_2$. Let f be a one-to-one correspondence between $\overline{\sigma}$ and the interval of real numbers $[a, b]$. We consider

$$\begin{aligned} \Gamma_1 &= \{x \in [a,b] \mid f(P) = x,\ P \in \Sigma_1\}, \\ \Gamma_2 &= \{x \in [a,b] \mid f(P) = x,\ P \in \Sigma_2\}. \end{aligned}$$

Let $s = \sup\Gamma_1$ and $S \in \sigma$ such that $f(S) = s$. Applying Lemma 1.7.12 we show (as in the proof of Theorem 1.7.5) that $O'S \geq r'$. Let $i = \inf S_2$ and $I \in \sigma$ such that $f(I) = i$. Similarly, one shows that $O'I \leq r'$. It follows from Exercise 5 and the definitions of sup and inf that $s \leq t$. Supposing $s < i$ and applying Proposition 1.7.9 and Exercise 5, we obtain a contradiction. Then we conclude that $s = i$, which in turn implies that $S = I$, giving that $O'S = r'$, that is, $S \in \gamma'$. Observe that the intersection of σ with γ' is only one point.

Now let δ be a semicircle such that $\gamma = \sigma \cup \delta$. There exists a point $Y' \in \delta$ that is outside γ'. Then there exists only one point on δ that is common with γ'. □

Proposition 1.7.13 *Let γ and γ' be two circles. Then only one of the following occurs:*
(a) $\gamma \cap \gamma' = \phi$.
(b) The circles are tangent to each other.
(c) The circles intersect one another in two points.
(d) $\gamma = \gamma'$.

Proof Suppose that $\gamma \cap \gamma' \neq \phi$ and $\gamma \neq \gamma'$. Then if γ has a point inside and a point outside γ', we have *(c)* by the circular continuity principle.

Let us then suppose that γ' has a point inside γ and no point of γ' is outside γ. We will show that in this case γ' are γ internally tangent. If not, there exist points $A, B \in \gamma \cap \gamma'$ and both lie on line $\overleftrightarrow{OO'}$ by Proposition 1.6.13 (see the remark at the end of the previous section). Since O' is inside γ, we can suppose, without loss of generality, that O' is between O and A. Then

$$OA = OO' + O'A.$$

But $OA = OB$, $O'A = O'B$. Substituting above, we have

$$OB = OO' + O'B$$

implying that O' is between O and B. Therefore, A and B are both on the ray opposite to $\overrightarrow{O'O}$, and this contradicts that O' is between points A and B.

To complete the proof, we suppose that no point of γ' is inside γ. Using arguments similar to those in the last paragraph we conclude that γ and γ' are externally tangent; we leave the details to the reader. □

Exercises

1. Show the converse of the triangle inequality, i.e., if a, b, and c are lengths of segments such that the sum of any two is greater than the third, then there exists a triangle whose sides have those lengths.
   Hint: Assume $a \geq b \geq c$. Then consider points A, B, C, and D on a line l such that $AB = a$, $BC = b$, $CD = c$, B is between A and C, and C is between B and D. Use the circular continuity principle for the circle centered at B of radius a and the circle centered at C of radius c.

2. Show that the converse of the triangle inequality implies the circular continuity principle.

3. A triangle is said to be *equilateral* if all its sides are congruent. Prove that there exists an equilateral triangle.

4. Use the circular continuity principle to show the elementary continuity principle.

Hint: Consider a circle γ centered at O and line through a point P inside γ. Let m be the line perpendicular to l through O. Let A be the foot on the perpendicular m and consider point B on m such that $OA = AB$. Use the circular continuity principle for γ and the circle centered at B having the same radius as γ.

5. Let a, b be two real numbers such that $b < a$. Show that there exists a right triangle whose hypotenuse has length a and one leg has length b.
   Hint: On a line l consider points O, A, and B such that $OB = b$, $OA = a$, and B is between O and A. Let m be a line perpendicular to l through B. Use the elementary continuity principle for line m and the circle centered at O and radius a.

6. Let γ and γ' be two distinct circles, with centers O and O' and radii r, r', respectively.
   (a) Show that if
   $$|r - r'| < OO' < r + r'$$
   then the two circles have two common points.
   Hint: Use the circular continuity principle.
   (b) Show that if γ and γ' have two common points, then
   $$|r - r'| < OO' < r + r'.$$

7. Use Axiom $\mathbf{S_2}$ and the Archimedean property to show the following:
   Given any line segment $\overline{AB}$ and any ray $\overrightarrow{CD}$, then for every point $E \in \overrightarrow{CD}$, $E \neq C$, there is a positive integer n such that when $\overline{AB}$ is laid off n times on $\overrightarrow{CD}$ starting at C, a point F is reached such that $CF = nAB$ and either $F = E$ or E is between C and F.
   In some classical axiom systems, the statement above is called *Archimedes' axiom.*

8. Use Axiom $\mathbf{S_2}$ and the elementary continuity principle to show the following:
   If S_1 and S_2 are two subsets of a line l such that $l = S_1 \cup S_2$, $S_1 \cap S_2 = \phi$, and no point of either subset is between two points of the other, then there exists a unique point D on l such that one of

the sets is a ray with vertex D and the other set is its complement. In some axiom systems, the statement above is called *Dedekind's axiom.*

1.8 The Basic Geometric Constructions

Euclid's first three postulates state that lines and circles can be drawn, and line segments extended. His first proposition states:

Given any line segment, there is an equilateral triangle having the given segment as one of its sides.

Euclid proved this result using a geometric construction; he considered two circles centered at the endpoints of the segment and having this segment as their radii. The third vertex of the triangle is one of the points determined by the intersection of these two circles. This proof is an example that the concept of a construction was part of Euclid's axiom system and in fact was used as a method of proving propositions.

As pointed out in Section 1.7, Euclid's first four postulates do not imply that the two circles intersect each other. Therefore, there is a gap in the very beginning of his work. Probably Euclid thought that the existence of the intersection points was obvious. This does not mean that the proposition is invalid (actually, it can be proved using the circular continuity principle or his fifth postulate), but rather that the proof is incorrect. Many times, to prove a theorem reasoning only from diagrams can lead not only to a wrong proof but to a false result as well. The diagrams may show a particular configuration of points or just represent a special case.

In geometry, diagrams and figures are very helpful for understanding proofs, illustrating theorems and building geometric intuition for other results. However, every conclusion must be theoretically justified by the axioms or by the results that they imply.

With this in mind, we turn now to the actual construction of geometric figures with a straightedge and compass. A *straightedge* is a ruler without marks, and it is used for the constructions specified in Euclid's postulates 1 and 2. The device used to draw circles is called a *compass.* These two instruments are usually called *Euclidean tools.* Constructions have been studied since ancient Greece. They constituted an area of much interest for many mathematicians during the nineteenth century, and some of the construction problems were solved as applications of theories developed in fields of mathematics other than geometry.

A geometric construction consists essentially of drawing some lines and arcs of circles; their intersections produce other points, and then other lines and circles are drawn, and so on. As we made clear before, we will use the theoretical results to justify each step of these constructions, proving then that the constructions made are correct. The geometric constructions that are justified using only the results of neutral geometry are called *basic constructions*.

Construction 1 *To copy a circle*

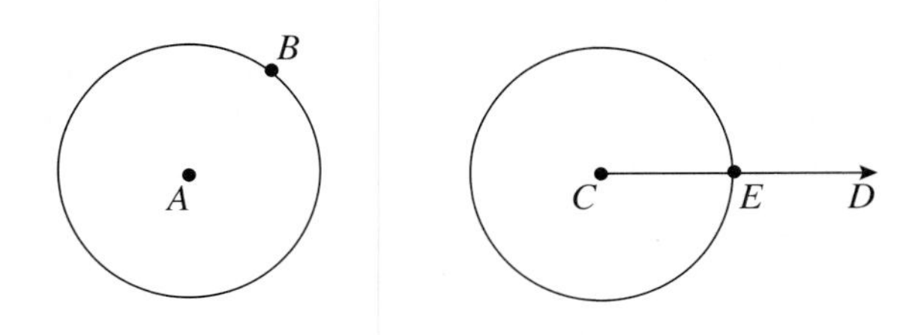

Construction 1

Given: a circle of center A and radius $\overline{AB}$, and another point C.

We want to construct a circle centered at C and radius r.

Open the compass so that the spike is on A and the pencil point is on B lying on the circle; lifting the compass and keeping the same opening, draw the circle centered at C using the radius AB.

Justification: Because the compass is rigid, $AB = CE$, where E is any point on the circle centered at C (see Exercise 1 of this section).

Construction 1′ *To copy a line segment*

Given: a line segment $\overline{AB}$ and a ray $\overrightarrow{CD}$.

We want to find point $E \in \overrightarrow{CD}$ such that $AB = CE$.

The method here is the same used in Construction 1.

Justification: The circle centered at C and radius AB intersects ray $\overrightarrow{CD}$ at a point E, by Exercise 1 of Section 1.4.

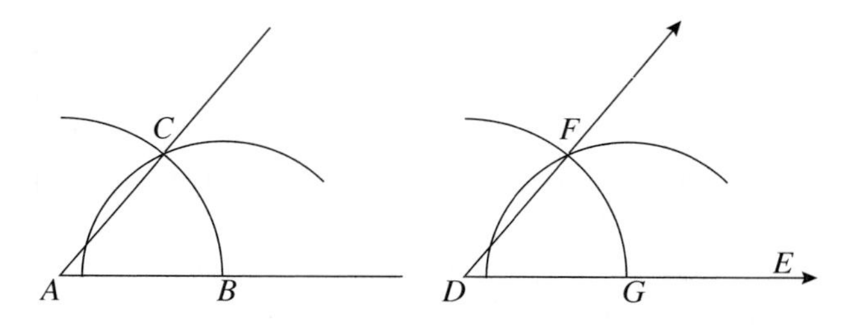

Construction 2

Construction 2 *To copy an angle*
Given: an angle $\angle A$ and ray $\overrightarrow{DE}$.

We want to construct a ray $\overrightarrow{DF}$ such that $m(\angle FDE) = m(\angle A)$.
1. Draw an arc centered at point A; this arc intersects $\angle A$ at two points that we call B and C; lifting the compass and keeping the same opening, draw an arc centered at D that will intersect ray $\overrightarrow{DE}$ at point G.
2. Copy the circle centered at B and radius BC so that point G is the center of the copy. Observe that it is not necessary to copy the whole circle, just an arc of it that intersects the arc of step 1. Let F denote this intersection point. Draw ray $\overrightarrow{DF}$.
Justification: By the construction above, $AB = AC = DG$ and $BC = GF$; then $\triangle ABC \simeq \triangle DGF$ by SSS, which implies that $m(\angle FDG) = m(\angle CAB)$, i.e., $m(\angle FDE) = m(\angle A)$.

Construction 3 *To bisect an angle*
Given: an angle $\angle BAC$.

We want to find a ray $\overrightarrow{AD}$ such that $\angle BAD \simeq \angle CAD$.
1. Draw an arc centered at point A that crosses the sides of the angle; let P and Q denote the intersection points.
2. Center the compass at point P and open to point Q; draw a circle of radius $r = PQ$; then lift the compass and using the same opening draw a circle centered at Q. The two circles intersect each other and one of the intersection points is interior to the angle $\angle PAQ$. Let D denote such a point.
Justification: Since $r = PQ$, the proof of the converse of the triangle inequality (Exercise 1 of Section 1.7) implies that the circles intersect

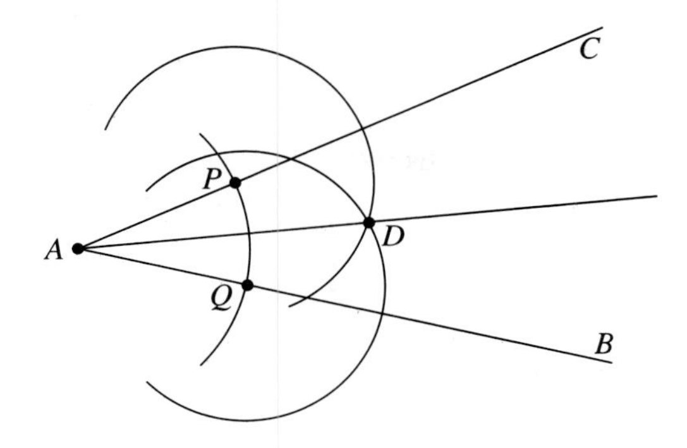

Construction 3

each other and we have an equilateral triangle of side of length r. Point D is one vertex of such a triangle.

3. Draw ray $\overrightarrow{AD}$.

Justification: From the construction we obtain $AP = AQ$ and $PD = QD$ and therefore $\triangle APD \simeq \triangle AQD$, which implies that $m(\angle PAD) = m(\angle QAD)$.

Construction 4 *To construct a perpendicular to a line through a point on the line*

Given: a line l and a point $A \in l$.

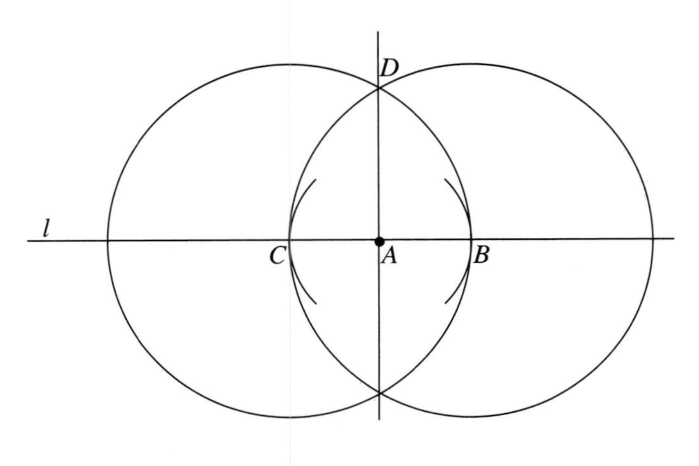

Construction 4

We want to find a point D such that line $\overleftrightarrow{AD} \perp l$.
Pick two points B and C on l such that A is between B and C; then $\angle BAC$ is a straight angle; repeat Construction 3 and draw line $\overleftrightarrow{AD}$.
Justification: Repeating Construction 3, we obtain a point D so that $\angle BAD \simeq \angle CAD$. Since these angles are supplementary, they must both be right angles, and then $\overleftrightarrow{AD}$ is perpendicular to line l.

Construction 5 *To construct a perpendicular to a line from a point not on the line.*
Given: a line l and a point $A \notin l$.

We want to find a point D such that line $\overleftrightarrow{AD} \perp l$.
1. Select a point $B \in l$ and draw a circle centered at A and radius AB. Then, either $\overleftrightarrow{AB} \perp l$ or the circle intersects line l at another point C.
Justification: If the circle intersects l only at point B, then l is tangent to the circle and hence $\overleftrightarrow{AB} \perp l$, by Proposition 1.6.6.
2. Bisect angle $\angle BAC$ using Construction 3 and draw line $\overleftrightarrow{AD}$.
Justification: We have that $\triangle ABC$ is isosceles, for $AB = AC$ by the construction; since ray $\overrightarrow{AD}$ bisects $\angle A$, we conclude that $\overleftrightarrow{AD}$ is the perpendicular bisector of $\overline{BC}$, by Exercise 3 of Section 1.5.
Remark In the proof of Proposition 1.6.6 we used the existence of the perpendicular to a line from a point not on the line, and we have

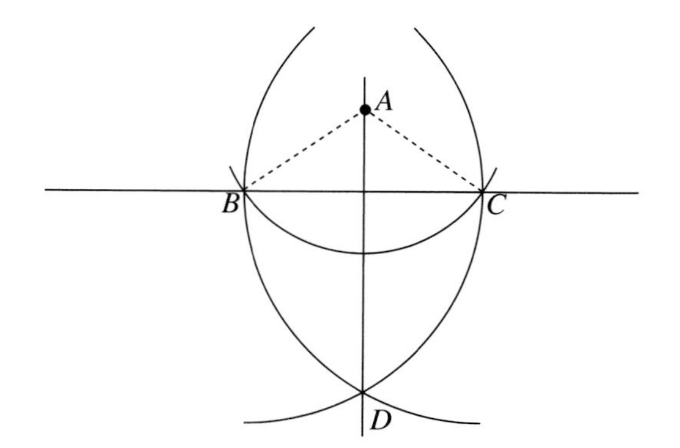

Construction 5

used Proposition 1.6.6 to justify the construction of such a line. We remark that such a perpendicular line can be constructed with compass and straightedge following the proof of Proposition 1.5.9 and using Constructions $1'$ and 2 to copy segments and angles. We described a different method in Construction 5 because it involves less geometric constructions.

Construction 6 *To construct the perpendicular bisector of a segment.*
Given: a line segment $\overline{BC}$.

We want to find a line $\overleftrightarrow{AD}$ through the midpoint M of $\overline{BC}$ which is perpendicular to line $\overleftrightarrow{BC}$.
1. Draw a circle centered at B of radius $r = BC$; then, using the same opening, draw a circle centered at C. The two circles intersect at two points A and D.

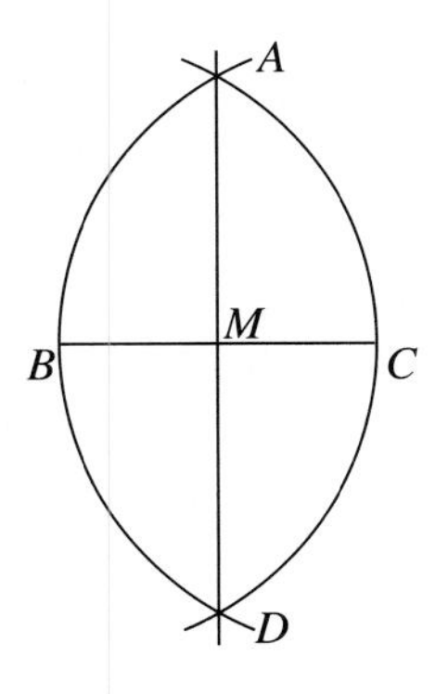

Construction 6

Justification: Since $r = BC$, we apply the converse of the triangle inequality. We have then triangles $\triangle ABC$ and $\triangle DBC$.
2. Draw line $\overleftrightarrow{AD}$.
Justification: We have that $\triangle ABD \simeq \triangle ACD$ by SSS, which implies $\angle BAD = \angle CAD$. Therefore line $\overleftrightarrow{AD}$ bisects $\angle A$, and since $\triangle ABC$ is isosceles, we conclude that $\overleftrightarrow{AD}$ is the perpendicular bisector of $\overline{BC}$, using again Exercise 3 of Section 1.5.

Exercises

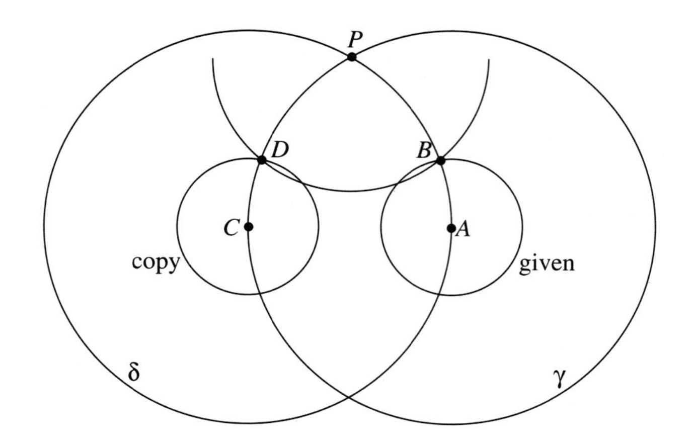

1. The compass we use today is rigid and we can easily construct a circle with a specified radius and a given center (Construction 1). But this modern compass is not the one assumed by Euclid; in his time, a compass was used to draw a circle centered at some point O and passing through a point P; but when lifted, it would collapse and then the length OP could not be transferred. This instrument is called a a *collapsing compass* or a *divider.* But a circle can still be copied using only a straightedge and a *collapsing* compass, using the procedure described below.
   (a) Justify the following procedure:
   Given: a circle with center A and radius r, and a point C.
   Open the collapsing compass so that the spike is on point A and the pencil point is on C, and draw a circle γ; repeat the procedure but now centering the collapsing compass at point C and obtaining circle δ. The two circles intersect one another at points P and Q. Moreover, the circle γ intersects the given circle at a point, say B. Draw a circle centered at P and radius PB. Such a circle intersects circle δ at point D. The circle centered at C and radius CD has radius $r = AB$.
   Hint: Show first that $\triangle PDA \simeq PBC$ by SSS; then show that $\triangle PDC \simeq PBA$ by SAS, which implies that $CD = AB$.

(b) Conclude that any construction that can be done with a modern compass can be done with a collapsing compass.

2. Draw any line segment and call it $\overline{AB}$. Construct an equilateral triangle such that $\overline{AB}$ is one of its sides.

3. Draw any line segment and call it $\overline{AB}$. Construct an isosceles right triangle such that $\overline{AB}$ is one of its legs.

4. (a) Construct a triangle given two sides of length a and b and their included angle $\angle C$.

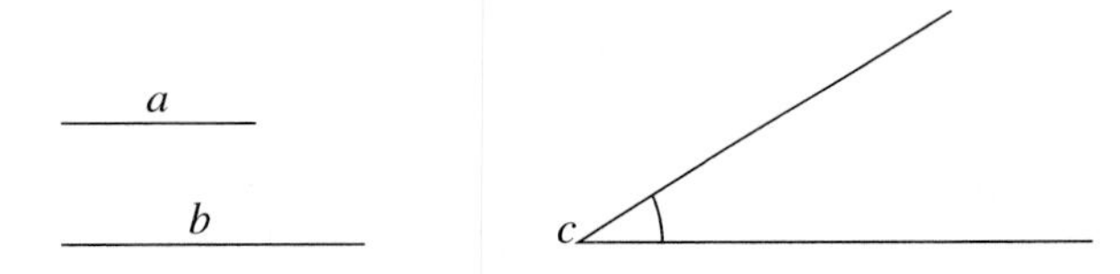

(b) Construct a triangle given two angles $\angle A$ and $\angle B$ and their included side of length c.

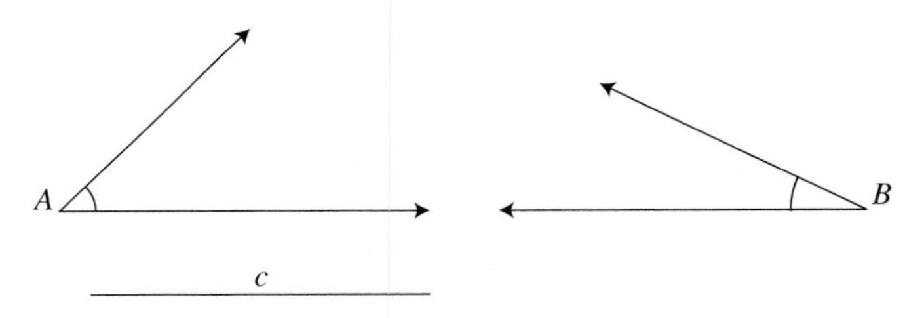

5. Draw any triangle $\triangle ABC$ and construct its three altitudes.

6. *Definition:* A *median* of a triangle is a line segment whose endpoints are a vertex of the triangle and the midpoint of the side opposite that vertex.
   Draw any triangle $\triangle ABC$ and construct its three medians.

7. Draw any triangle $\triangle ABC$ and construct its incenter in the inscribed circle.

8. Draw three line segments; let a, b, and c be their lengths. Construct a triangle whose sides have lengths a, b, and c. Which assumption is necessary?
   Hint: See Exercise 1 of Section 1.7.

9. Draw two line segments; let a denote the greater length and b the smaller. Construct a right triangle with hypotenuse of length a and a leg of length b.
   Hint: See Exercise 5 of Section 1.7.

10. Draw two external circles. Construct a third circle that is externally tangent to both circles.

11. Draw two circles such that one is internal to the other. Construct a third circle that is externally tangent to the lesser circle and internally tangent to the greater one.

12. Draw any circle and call it γ. Let A be any point on the circle and P any point outside γ. Construct another circle σ such that:
   (a) σ goes through P and intersects γ at point A. Is your answer unique? Justify.
   (b) σ goes through P, intersects γ at point A, and is centered at point on γ. Which assumption on P is necessary?
   (c) σ goes through P and intersects γ orthogonally at point A, that is, the lines tangent to γ and σ at A are perpendicular.

13. Draw any circle and let P be a point on the circle. Construct the line tangent to the circle at a point P.

14. Construct a line tangent to the circle from a point P outside the circle. (see the next figure)
   Hint: Let O and r denote the center and the radius of the circle, respectively. Then, using Exercise 9 of this section, construct a right triangle $\triangle ABC$ whose hypotenuse $\overline{BC}$ has length OP and whose leg BA has length r. Copy $\angle ABC$ onto line segment $\overline{OP}$.

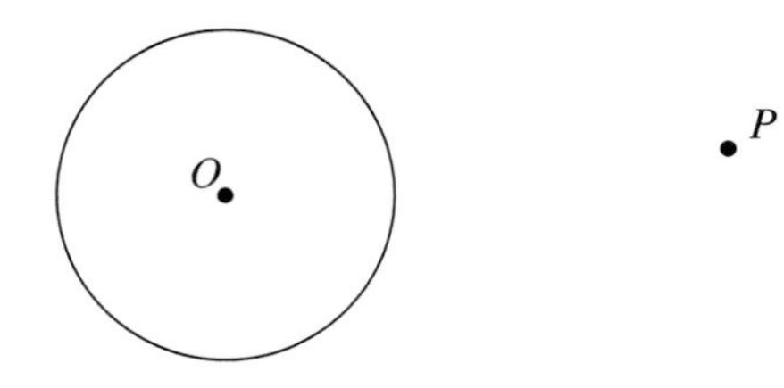

Chapter 2

EUCLIDEAN PLANE GEOMETRY

2.1 Euclid's Fifth Postulate

The axioms assumed in Chapter 1 implied that given a line l and a point P not lying on l, there exists at least one line through P parallel to l. The uniqueness of such a parallel line cannot be proved using the axioms and results of Chapter 1. The postulate that assumes uniqueness is known as the *Euclidean parallel postulate* (also as Playfair's postulate) because it is equivalent to Euclid's fifth postulate. In this section we show this equivalence.

> **Euclid's fifth postulate** If two lines are cut by a transversal in such a way that the sum of the two interior angles on one side of the transversal is less than two right angles, then the lines meet on that side of the transversal.
>
> **Euclidean parallel postulate** For every line l and every point P not lying on l, there exists a unique line through P that is parallel to l.

Theorem 2.1.1 *Euclid's fifth postulate* $\Leftrightarrow$ *Euclidean parallel postulate.*

Proof ($\Leftarrow$) Let us assume the Euclidean parallel postulate. Let l and m denote a pair of lines cut by a transversal t. Consider points $B = l \cap t$ and $B' = m \cap t$. Let A and A' be points on lines l and m, respectively, and on the same side of t. The assumption of Euclid's fifth postulate is that

$$m(\angle A'B'B) + m(\angle ABB') < 180^\circ.$$

Let $C \in l$ such that C is on the opposite side of t from A and A'. Since

$$m(\angle CBB') + m(\angle ABB') = 180^\circ$$

we obtain that $m(\angle A'B'B) < m(\angle CBB')$. Therefore there exists a ray $\overrightarrow{B'C'}$ such that $\angle C'B'B \simeq \angle CBB'$. Let m' be the line through B' and

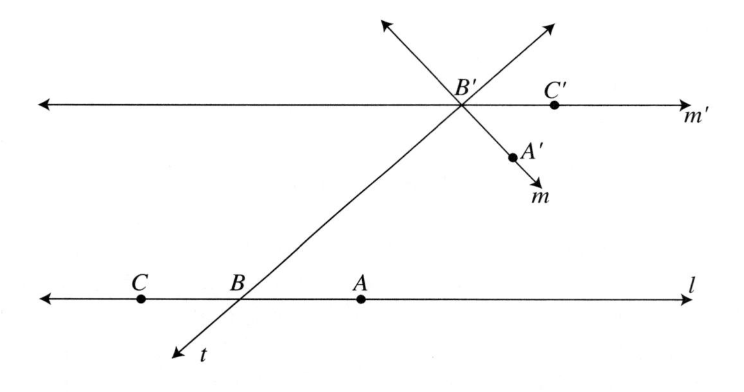

Figure 2.1

C'. Then $m' \parallel l$, because they have congruent alternate interior angles with respect to the transversal t. Since $m' \neq m$, the Euclidean parallel postulate implies that m intersects l. To finish this part of the proof, we must show that m intersects l on the same side of t as A'. Let us suppose that the lines meet at a point D on the opposite side. Then $\angle A'B'B$ is an exterior angle for $\triangle B'BD$ that is smaller than the remote angle $\angle DBB' = \angle CBB'$.

($\Rightarrow$) Now we assume Euclid's fifth postulate. Let l be a line and $P \notin l$. Let m and t be lines through P such that $m \parallel l$ and $t \perp l$. Let n be any other line through P (see Figure 2.2). Since $n \neq m$, n and t form an acute angle. Let α denote such an angle. Then we have two lines l and n cut by the transversal t, and the sum of two interior angles on one side of t is $m(\alpha) + 90^\circ < 180^\circ$. Euclid's fifth postulate implies that n meets l, and therefore n is not parallel to l. $\square$

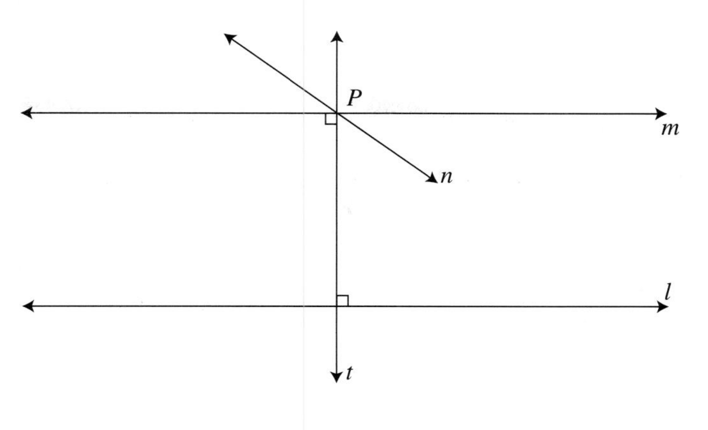

Figure 2.2

Proposition 2.1.2 *Euclidean parallel postulate* $\Leftrightarrow$ *"if l is parallel to lines m_1 and m_2, then either $m_1 = m_2$ or $m_1 \parallel m_2$."*

Proof ($\Rightarrow$) Let us suppose that $m_1 \neq m_2$ and m_1 is not parallel to m_2. Then m_1 intersects m_2 at P. This clearly contradicts the parallel postulate, since $P \notin l$ and m_1 and m_2 are two distinct lines parallel to l passing through P.
($\Leftarrow$) Let m_1 and m_2 be two lines through point P and parallel to line l. Then either $m_1 = m_2$ or $m_1 \parallel m_2$. Since P is a common point, then m_1 and m_2 cannot be parallel. Therefore $m_1 = m_2$, and the parallel is unique. □

Proposition 2.1.3 *Euclidean parallel postulate* $\Leftrightarrow$ *"if a line intersects one of the two parallel lines, then it also intersects the other."*

Proof ($\Rightarrow$) Let $m_1 \parallel m_2$. If line l intercepts m_1 and not m_2, then $l \parallel m_2$. We have then $m_2 \parallel l$ and $m_2 \parallel l$. The previous proposition implies that $l \parallel m_1$, and this contradicts that they intersect each other.
($\Leftarrow$) Let m_1 be a line through point P and $m_1 \parallel l$. If m_2 is another line through P such that $m_2 \parallel l$, then m_2 intersects m_1, and hence m_2 must intersect line l. This contradicts that $m_2 \parallel l$. □

Proposition 2.1.4 *Euclidean parallel postulate* $\Leftrightarrow$ *"if two lines are cut by a transversal, then alternate interior angles are congruent if and only if the two lines are parallel."*

Proof ($\Rightarrow$) It has already been proved that if alternate interior angles are congruent, then the lines are parallel. Let us suppose now that the lines are parallel. We keep the same notation of Theorem 2.1.1, that is, l and m denote the lines cut by a transversal t, $B = l \cap t$ and $B' = m \cap t$. If the alternate interior angles $\angle CBB'$ and $\angle C'B'B$ are not congruent, then there is a ray $\vec{B'D}$ such that D and C' are on the same side of t

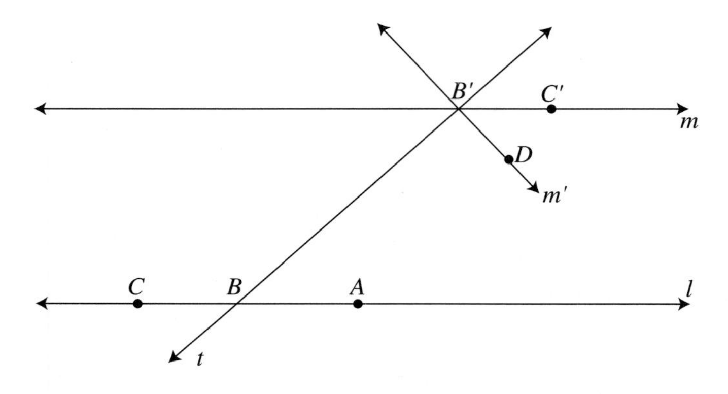

Figure 2.3

and $\angle DB'B \simeq \angle CBB'$. Let m' be line through B' and D. Then $m' \parallel l$, for they have alternate interior angles congruent and this contradicts the parallel postulate.
($\Leftarrow$) Let us consider line l and $P \notin l$. Let t and m be lines passing through p such that $t \perp l$ and $m \perp t$ (see Figure 2.4). Then $m \parallel l$. If n is a line through P that is parallel to l, then t is a transversal to l and n. Let $Q = t \cap l$ and consider points $A \in n$ and $B \in l$ such that A and B are on opposite sides of t. Our assumption implies that $\angle BQP \simeq \angle APQ$ and hence $\angle APQ$ is a right angle. Therefore line n is perpendicular to t, implying that $m = n$, by the uniqueness of the perpendicular to t through P. $\square$

Proposition 2.1.5 *Euclidean parallel postulate $\Leftrightarrow$ "if line t is perpendicular to line l and $l \parallel m$ then $t \perp m$."*

Proof ($\Rightarrow$) First, notice that Proposition 2.1.3 implies that t intersects m. Further, it follows from Proposition 2.1.4 that the alternate interior angles, formed lines by l and m and cut by t, are congruent. Therefore $t \perp m$.

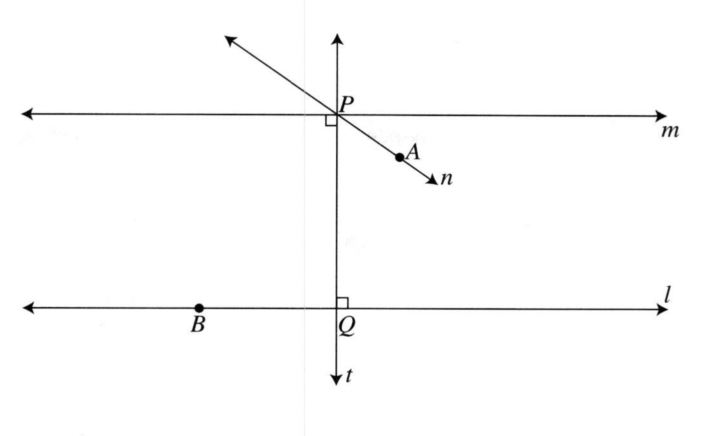

Figure 2.4

($\Leftarrow$) We consider line l, $P \notin l$, and lines t and m such that $t \perp l$ through P and $m \perp t$ also passing through P. If n is a line through P that is parallel to l, then our assumption implies that $t \perp n$ and hence $n = m$ by the uniqueness of the perpendicular. $\square$

Definition 2.1.6 *A* rectangle *is a quadrilateral in which all the angles are right angles.*

Proposition 2.1.7 *Euclidean parallel postulate* ($\Rightarrow$) *given any point P there exists a rectangle such that P is one of its vertices.*

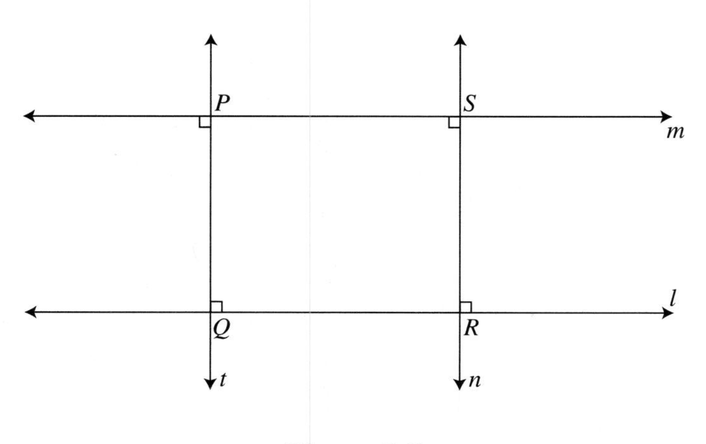

Figure 2.5

Proof Let l be any line not incident with P, and t and m lines such that $t \perp l$ through P and $m \perp t$ also passing through P (see Figure 2.5). Let Q denote the point where t intersects l and consider point $R \in l$, $R \neq Q$. Let n be the line through R that is perpendicular to m and S its foot. We then have a quadrilateral $\square PSRQ$ such that

$$m(\angle PQR) = m(\angle PSR) = m(\angle SPR) = 90^\circ.$$

Since line n is perpendicular to m and $m \parallel l$, from Proposition 2.1.5 we get that $n \perp l$ and then

$$m(\angle QRS) = 90^\circ.$$

$\square$

Theorem 2.1.8 *Euclidean parallel postulate* $(\Rightarrow)$ *the angle sum of every triangle is* 180°.

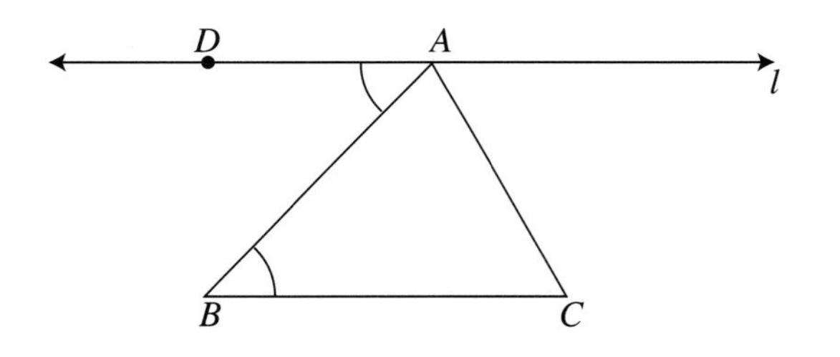

Figure 2.6

Proof Consider $\triangle ABC$. Let l be the line through point A that is parallel to line $\overleftrightarrow{BC}$. Let $D \in l$ such that D and B are on the same side of line $\overleftrightarrow{AC}$. Since we are assuming the parallel postulate, we conclude that $\angle DAC$ is congruent to the supplement of $\angle C$, since they form a pair of alternate interior angles for lines l and $\overleftrightarrow{BC}$ cut by line $\overleftrightarrow{AC}$. It follows that

$$m(\angle DAC) = 180^\circ - m(\angle C).$$

Moreover, $\angle DAB \simeq \angle B$, again because they are alternate interior angles of the same pair of lines cut by transversal $\overleftrightarrow{AB}$. Then

$$m(\angle DAC) = m(\angle DAB) + m(\angle A) = m(\angle B) + m(\angle A).$$

Combining the equations above, we get

$$m(\angle A) + m(\angle B) + m(\angle C) = 180^\circ.$$

□

Corollary 2.1.9 (Euclidean exterior angle theorem) *The measure of an exterior angle of a triangle is equal to the sum of the measures of the nonadjacent interior angles.*

It will be shown in Chapter 8 that the existence of a triangle of angle sum 180° implies the Euclidean parallel postulate, which in turn implies the existence of rectangles. We will also show that the existence of rectangles implies that the angle sum of a triangle is 180° and hence the Euclidean parallel postulate.

Exercises

Assume the *Euclidean parallel postulate* to do all problems of this chapter.

1. Show that opposite sides of a parallelogram are congruent.

2. Show that the diagonals of a parallelogram bisect each other.

3. Show that if a quadrilateral has two parallel and congruent opposite sides, then it is a parallelogram.

4. (a) Show that a rectangle is in particular a parallelogram.
   (b) Show that if a parallelogram has one right angle, then it is a rectangle.

5. If the endpoints of line segment $\overline{DE}$ are the midpoints of two sides $\overline{AB}$ and $\overline{AC}$ of $\triangle ABC$, then the line containing $\overline{DE}$ is parallel to line $\overleftrightarrow{BC}$ and $DE = (1/2)BC$.
   Hint: Consider a point $F \in \overrightarrow{ED}$ such that $DF = DE$. Observe that $\triangle ADE \simeq \triangle BDF$ and conclude that $\square FECB$ is a parallelogram.

6. Show that the median to the hypotenuse of a right triangle is one-half the length of the hypotenuse.
   Hint: Notice that there exists a rectangle whose diagonal is the hypotenuse and whose sides are the legs of the triangle.

7. In a right triangle, one of the angles measures $30°$ if and only if the side opposite this angle is one-half the length of the hypotenuse.

8. The sum of the measures of the interior angles of a convex n-gon is $(n-2)180°$.
   Hint: Prove by induction on n; use Theorem 2.1.8 and Proposition 1.3.7.

9. The sum of the measures of exterior angles (one at each vertex) of a convex n-gon is $360°$.

10. Show that a quadrilateral is a parallelogram if and only if opposite angles are congruent.

2.2 Euclidean Triangles

In the remaining sections of this chapter we will be assuming the Euclidean parallel postulate.

Proposition 2.2.1 *The perpendicular bisectors of the sides of any triangle all meet at a single point.*

Proof Let us consider $\triangle ABC$. Let lines l and m be the perpendicular bisectors of segments $\overline{AB}$ and $\overline{BC}$, respectively. We claim that l and m meet at some point O. In fact, suppose $l \parallel m$. Since $\overline{AB} \perp l$, Proposition 2.1.5 implies that $\overline{AB} \perp m$. Since $\overline{BC} \perp m$, the alternate interior angle theorem implies that $\overleftrightarrow{AB} \parallel \overleftrightarrow{BC}$, which is a contradiction, for they share point B.

We also claim that $OA = OB = OC$. In order to show this claim, let $L = \overline{AB} \cap l$ and $M = \overline{BC} \cap m$. Observe that $\triangle AOL \simeq \triangle BOL$, by SAS, and hence $OA = OB$. Further, $\triangle BOM \simeq \triangle COM$, by SAS, which implies $OB = OC$.

Let N be the midpoint of $\overline{AC}$. To complete this proof we need to show that $\overleftrightarrow{ON}$ is the perpendicular bisector of $\overline{AC}$, that is, $\overleftrightarrow{ON} \perp \overline{AC}$.

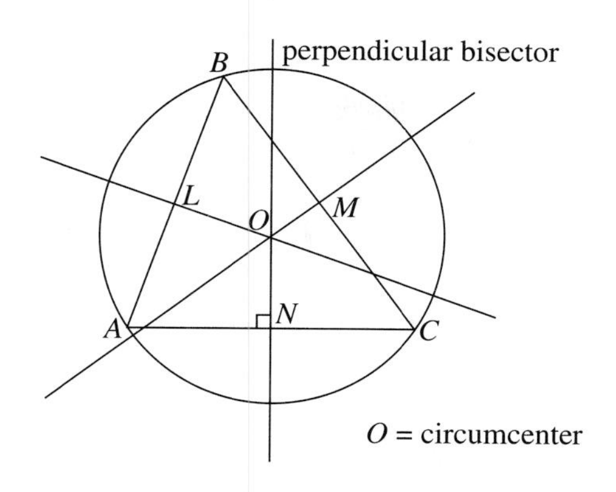

Figure 2.7

Since $\triangle AON \simeq \triangle CON$ by SSS, we obtain

$$m(\angle ONC) = m(\angle ONA) = 90°.$$

□

The point O above is called the *circumcenter* of the triangle, and we say that the circle centered at O and radius $\overline{OA}$ *circumscribes* the triangle, since it passes through all three vertices of the triangle. We observe that the last proposition of Chapter 1 implies that if two circles have three common points, then they are equal. It follows that the circumscribed circle is unique.

We recall that an *altitude* of a triangle is a line segment from a vertex of the triangle to the foot of the perpendicular dropped from the vertex to the line containing the side opposite that vertex.

Proposition 2.2.2 *The lines containing the three altitudes of a triangle all meet at a single point.*

Proof Given a triangle $\triangle ABC$, let us consider lines l, m, and n such that

$$\begin{array}{ll} l & \text{passes through } A \text{ and } l \parallel \overleftrightarrow{BC}, \\ m & \text{passes through } B \text{ and } m \parallel \overleftrightarrow{AC}, \\ n & \text{passes through } C \text{ and } n \parallel \overleftrightarrow{AB}. \end{array}$$

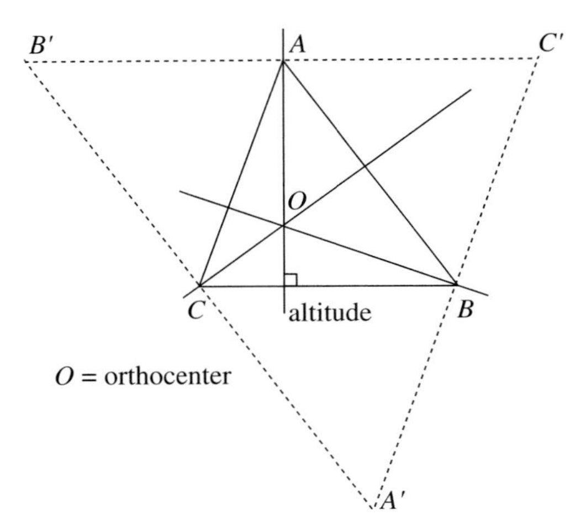

Figure 2.8

The reader can easily verify that that these lines intersect pairwise using Proposition 2.1.2. Let

$$A' = m \cap n, \quad B' = l \cap n, \quad \text{and} \quad C' = l \cap m.$$

It is easy to see that A', B', and C' are noncollinear. We will show that the lines containing the altitudes of $\triangle ABC$ are the perpendicular bisectors of $\triangle A'B'C'$ and then they all intersect at a point by Proposition 2.2.1. For that, notice that $\square AC'BC$ is a parallelogram and hence $AC' = BC$. Likewise $\square AB'CB$ is a parallelogram, implying that $AB' = BC$. Therefore A is the midpoint of $\overline{B'C'}$. In addition, the line containing the altitude through A is perpendicular to $\overleftrightarrow{BC} = \overleftrightarrow{B'C'} = l$, and then it is the perpendicular bisector of $\overline{B'C'}$. $\square$

The point of concurrence of the altitudes of a triangle is called the *orthocenter* of the triangle.

Definition 2.2.3 *A* median *of a triangle is a line segment whose endpoints are a vertex of the triangle and the midpoint of the side opposite that vertex.*

The next result states that medians are also concurrent, and this point of concurrence is called the *centroid* of the triangle. However, its proof depends on a result that will be proved in the next section. The idea of the proof is sketched in Exercise 7 of Section 2.3.

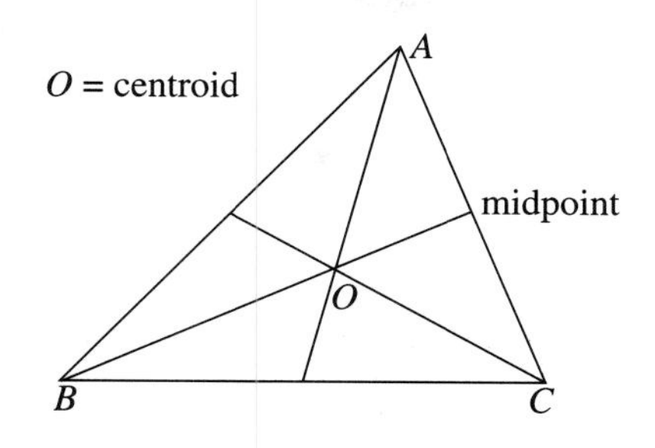

Figure 2.9

Proposition 2.2.4 *The three medians of a triangle are concurrent at a point that is 2/3 the length from each vertex to the midpoint of the opposite side.*

2.3 Similar Triangles

Definition 2.3.1 *Two triangles $\triangle ABC$ and $\triangle DEF$ are said to be* similar *and denoted by $\triangle ABC \sim \triangle DEF$, if there is a one-to-one correspondence between their vertices such that corresponding angles are congruent.*

Wallis' postulate *Given any triangle $\triangle ABC$ and given any line segment $\overline{DE}$, there exists a triangle $\triangle DEF$ such that $\overline{DE}$ is one of its sides and $\triangle ABC \sim \triangle DEF$.*

Theorem 2.3.2 *Euclidean parallel postulate $\Leftrightarrow$ Wallis' postulate.*

Proof ($\Rightarrow$) On one side of line through D and E we consider points G and H such that $\angle A \simeq \angle GDE$ and $\angle B = \angle HED$. We claim that ray $\overrightarrow{DG}$ intersects $\overrightarrow{EH}$ and let $F = \overrightarrow{DG} \cap \overrightarrow{EH}$. In fact, if $\overrightarrow{DG} \cap \overrightarrow{EH} = \phi$, then lines $\overleftrightarrow{DG}$ and $\overleftrightarrow{EH}$ are parallel, cut by the transversal line $\overleftrightarrow{DE}$. Since we are assuming the parallel postulate, we conclude that their alternate interior angles are congruent and therefore $\angle GDE$ is congruent

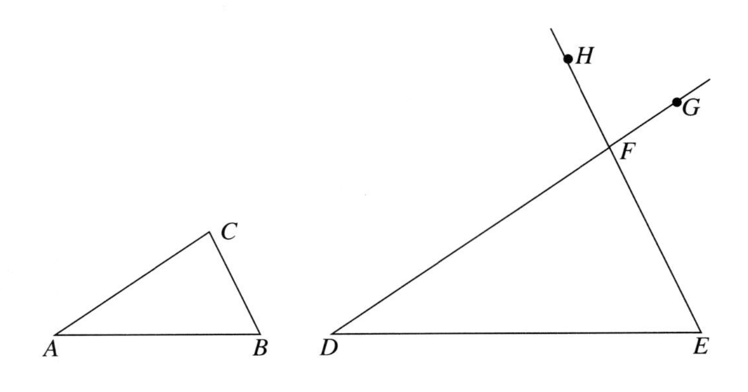

Figure 2.10

to the supplement of $\angle HED$. This implies that $m(\angle A)+m(\angle B) = 180$, and this is a contradiction. We have then $\triangle DEF \sim \triangle ABC$.
($\Leftarrow$) Let us consider a line l and $P \notin l$. Let t be a line through P such that $t \perp l$ and let $Q = l \cap t$ (see Figure 2.11). Let m be a line through P that is perpendicular to t and hence parallel to l, and n another line through P. We will show that n and l are concurrent. Let us consider $R \in n$ and $S \in t$ such that line $\overleftrightarrow{RS}$ is perpendicular to t. Wallis' postulate implies that there exists a point T such that $\triangle PSR \sim \triangle PQT$. Assume first that T is on the same side of t as

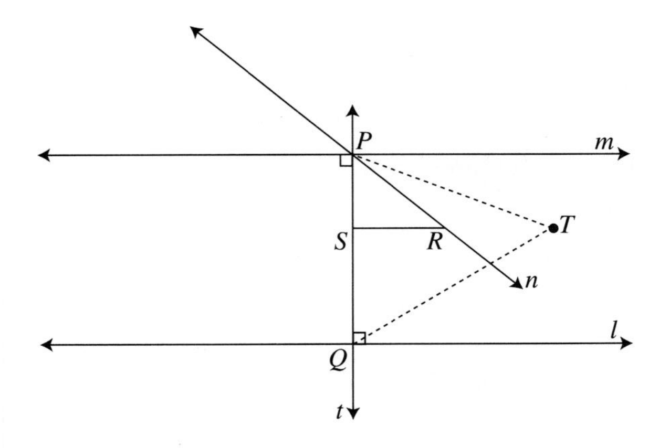

Figure 2.11

R. Then we have $\angle TPQ \simeq \angle RPS$, which in turn implies $\overrightarrow{PT} = \overrightarrow{PR}$ and hence $T \in n$. In addition, $\angle PQT \simeq \angle PSR$ implying that the line through Q and T is perpendicular to t. Then $T \in l$, and n and l are concurrent at T. If T and R are on opposite sides of t, we drop a perpendicular to t from T, intersecting t at point O. On the other side of t, we consider T' such that $OT = OT'$. Then $\triangle PQT \simeq \triangle PQT'$, and therefore $\triangle PQT' \sim \triangle PSR$. Now, since T' and R are on the same side, the previous argument implies that $T' \in n \cap l$. □

Theorem 2.3.3 (Parallel projection theorem) *Let l, m, and n be three parallel lines and t and t' be transversals to them. Let A, B, C and A', B', C' be the intersection points of these parallels with t and t', respectively* (see Figure 2.12). *Then*

$$\frac{AB}{BC} = \frac{A'B'}{B'C'}.$$

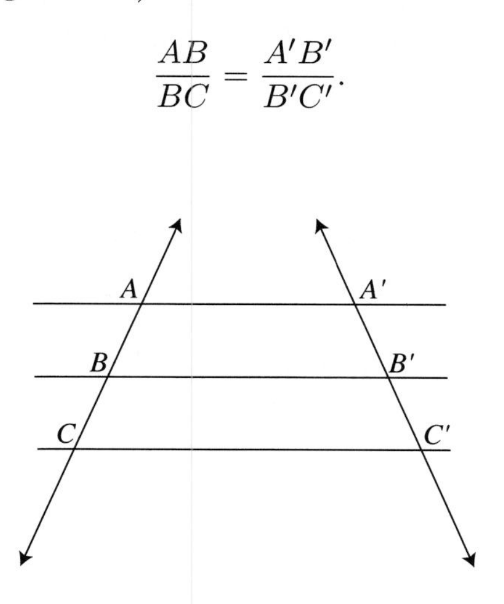

Figure 2.12

We first prove two preliminary lemmas that are weaker versions of Theorem 2.3.3.

Lemma 2.3.4 *Under the conditions of Theorem 2.3.3, if $AB = BC$ then $A'B' = B'C'$.*

Proof Let t'' be the line through B' that is parallel to t (see Figure 2.13). Let $A'' = l \cap t''$ and $C'' = n \cap t''$. Therefore, the quadrilaterals

$\square AA''B'B$ and $\square BB'C''C$ are parallelograms. Since we are assuming the parallel postulate, we obtain that $AB = A''B'$ and $BC = B'C''$ and hence $A''B' = B'C''$. Again from the parallel postulate we obtain that $\angle A'A''B' \simeq \angle C'C''B'$ (alternate interior angles). Then we have

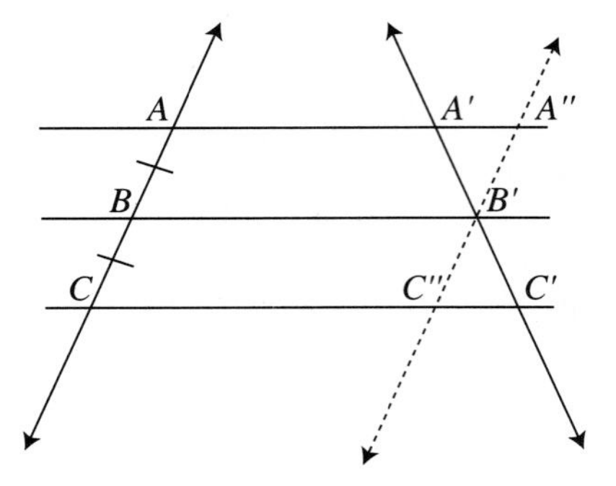

Figure 2.13

$\triangle A'B'A'' \simeq \triangle C'B'C''$ by ASA, implying $A'B' = B'C'$. $\square$

Lemma 2.3.5 *Under the conditions of Theorem 2.3.3, let us suppose that $AC = kAB$ where k is an integer. Then $A'C' = kA'B'$.*

Proof Our hypothesis implies that there exist points $A_0, \ldots, A_k$ on t such that $A_0 = A$, $A_1 = B$, $A_k = C$, and such that $A_0A_1 = A_1A_2 = \cdots = A_{k-1}A_k$. Let m_i be lines parallel to l and through A_i and $A_i' = t' \cap m_i$, for $i = 1, \ldots, k$. Observe that $m_1 = m$ and $m_k = n$. Then $A_1' = B'$ and $A_k' = C'$. Applying the previous lemma we obtain $A_0'A_1' = A_1'A_2' = \cdots = A_{k-1}'A_k'$ and then $A'C' = kA'B'$. $\square$

The proof of the parallel projection theorem in the general case requires the Archimedean property of the real numbers, stated in Proposition 1.7.2 of Chapter 1.

Proof of Theorem 2.3.3 It follows from Axiom $\mathbf{S_2}$ of Chapter 1 that, given a positive integer number N, there exists $P_1 \in t$ such that

$$\frac{AC}{N+1} < AP_1 < \frac{AC}{N}$$

(why?) which implies

$$NAP_1 < AC < (N+1)AP_1.$$

Supposing again that B is between A and C, we conclude that there exits a positive integer number $K \leq N$ such that

$$KAP_1 \leq AB \leq (K+1)AP_1.$$

Combining the inequalities above, we obtain

$$\frac{K}{N+1} < \frac{AB}{AC} < \frac{K+1}{N}.$$

Let $P_2, \ldots, P_r, \ldots$ be points on t such that $AP_i/AP_1 = i$ and consider lines $m_1, \ldots, m_r, \ldots$ through $P_1, \ldots, P_r, \ldots$, respectively, and parallel to l. Let $P_i' = t' \cap m_i$ for all i. Then from the previous lemma we obtain that $A'P_i'/A'P_1' = i$, which implies that $B' \in \overline{P_K' P_{K+1}'}$ and C' is between P_N' and P_{N+1}'. Similarly, we obtain

$$\frac{K}{N+1} < \frac{A'B'}{A'C'} < \frac{K+1}{N}.$$

Therefore,

$$\begin{aligned} \left|\frac{AB}{AC} - \frac{A'B'}{A'C'}\right| &< \frac{K+1}{N} - \frac{K}{N+1} \\ &= \frac{K+N+1}{N(N+1)} \leq \frac{2N+2}{N(N+1)} = \frac{2}{N}, \end{aligned}$$

where the last inequality above was implied by the fact that $K \leq N$. Since the inequality $|AB/AC - A'B'/A'C'| < 2/N$ holds for any positive integer N, we conclude that

$$\left|\frac{AB}{AC} - \frac{A'B'}{A'C'}\right| = 0,$$

by the Archimedean principle. Then we have

$$\frac{AB}{AC} = \frac{A'B'}{A'C'}.$$

Now, the same proof implies

$$\frac{CB}{CA} = \frac{C'B'}{C'A'}.$$

Solving for CA and substituting in the previous equality we obtain the result. $\square$

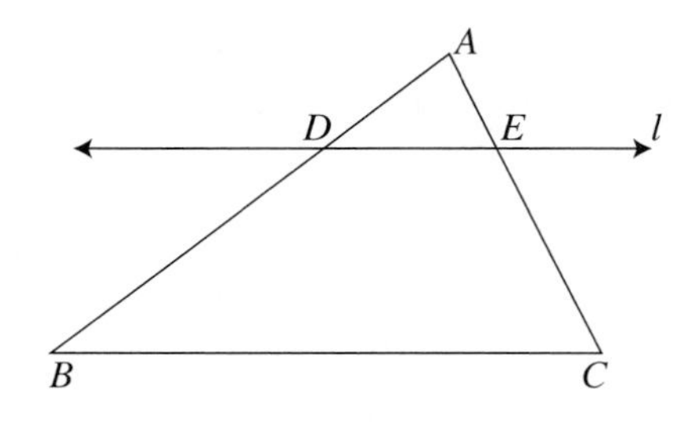

Figure 2.14

Corollary 2.3.6 *Suppose that line l intersects side $\overline{AB}$ of triangle $\triangle ABC$ and l is parallel to the line through B and C* (see Figure 2.14). *Let $D = l \cap \overline{AB}$. Then l intersects side $\overline{AC}$ and $AD/AB = AE/AC$, where $E = l \cap \overline{AC}$.*

Proof The first part follows from Pasch's theorem. For the second part, we consider a line m through A parallel to l and apply Theorem 2.3.3. □

Theorem 2.3.7 (Fundamental theorem of similar triangles) *Two triangles $\triangle ABC$ and $\triangle A'B'C'$ are similar if and only if the corresponding sides are proportional.*

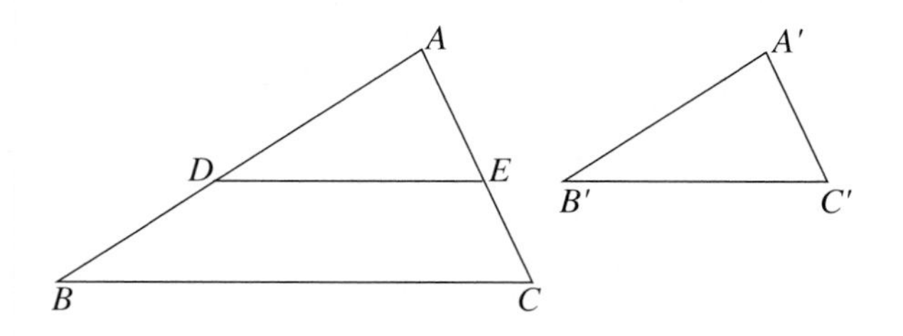

Figure 2.15

Proof (⇒) Without loss of generality, we suppose that $A'B' < AB$. Let $D \in \overline{AB}$ such that $\overline{AD} \simeq \overline{A'B'}$ and $E \in \overrightarrow{AC}$ such that $\overline{AE} \simeq \overline{A'C'}$. Since $\angle B \simeq \angle B'$ and $\triangle ADE \simeq \triangle A'B'C'$, we get $\angle B \simeq \angle D$, and then the line through D and E is parallel to side $\overline{BC}$ and E is between A and C. Now the corollary above implies that $AD/AB = AE/AC$, and since $AD = A'B'$ and $AE = A'C'$ we have $A'B'/AB = A'C'/AC$.

Now consider $F \in \overrightarrow{BC}$ such that $BF = B'C'$ and $D' \in \overline{AB}$ such that $BD' = B'A'$. Similarly, one obtains that $BF/BC = BD'/BA$ and hence $B'C'/BC = A'B'/AB$.
($\Leftarrow$) Let $D \in \overline{AB}$ such that $\overline{AD} \simeq \overline{A'B'}$. Let l be a line through D parallel to line $\overleftrightarrow{BC}$. By Pasch's theorem, l meets side $\overline{AC}$ at a point that we will denote by E. The corollary above implies that $AD/AB = AE/AC$. But $AD = A'B'$ and $A'B'/AB = A'C'/AC$, implying that $AE = A'C'$. Moreover, $\triangle ADE \sim \triangle A'B'C'$, and the first part of this proof implies that $AD/AB = DE/BC$. Using our assumption that $A'B'/AB = B'C'/BC$, we obtain that $DE = B'C'$. Therefore $\triangle ADE \simeq \triangle A'B'C'$, implying that $\triangle ABC \sim \triangle A'B'C'$. □

One of the most important consequences of the fundamental theorem of similar triangles is the well-known Pythagorean theorem.

Theorem 2.3.8 (The Pythagorean theorem) *Let $\triangle ABC$ be a right triangle with sides b and c and hypotenuse a. Then*

$$a^2 = b^2 + c^2.$$

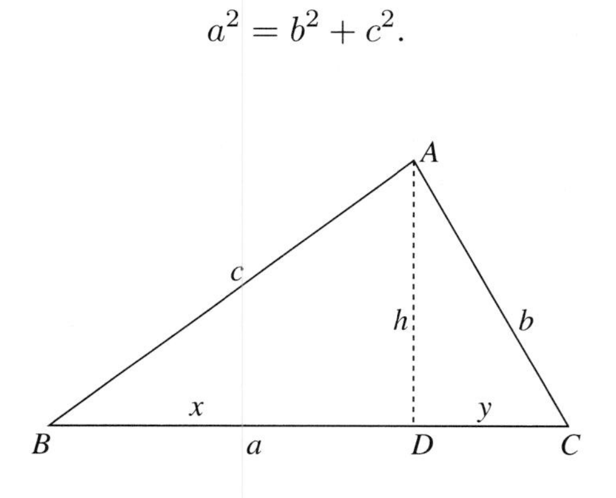

Figure 2.16

Proof Let A denote the right angle (see Figure 2.16). We have $a = BC$, $b = AC$, and $c = AB$. Let $\overline{AD}$ be the altitude to the hypotenuse. We claim that D is between B and C. In fact, suppose that C is between B and D. Then $\angle C$ is an exterior angle for $\triangle ADC$, and hence $m(\angle C) > m(\angle D)$. Since $\angle D$ is a right angle, we get that $\angle C$ is obtuse, contradicting that $\triangle ABC$ is a right triangle.

Now, observe that $\triangle ABC \sim \triangle DBA$ and $\triangle ABC \sim \triangle DAC$. Let $h = AD$, $x = CD$, and $y = DB$. From the similarity of the triangles we obtain that

$$\frac{x}{c} = \frac{c}{a} \quad \text{and} \quad \frac{y}{b} = \frac{b}{a}.$$

Therefore, $ax = c^2$ and $ay = b^2$. Then $ax + ay = b^2 + c^2$, and since $a = x + y$, we get $a^2 = b^2 + c^2$. □

Proposition 2.3.9 *Let $\triangle ABC$ be a triangle with sides of length a, b, and c. If $a^2 = b^2 + c^2$, then $\triangle ABC$ is a right triangle whose hypotenuse has length a.*

Proof $c = AB$, $b = AC$, and $a = BC$. On any line l, consider a line segment $\overline{DE}$ such that $\overline{DE} \simeq \overline{AB}$. Let $m \perp l$ and passing through point D. Let $F \in m$ such that $\overline{DF} \simeq \overline{AC}$. Now, $\triangle DEF$ is a right triangle and the Pythagorean relation implies

$$(FE)^2 = (DF)^2 + (DE)^2$$

and $\overline{FE} \simeq \overline{BC}$. Therefore, $\triangle ABC \simeq \triangle DEF$, and since $\angle D$ is a right angle, so is $\angle A$. □

Proposition 2.3.10 *Let $\triangle ABC$ be a right triangle whose right angle is A. Let $\overline{AD}$ be the altitude to the hypotenuse $\overline{BC}$. Let $AD = h$, $CD = x$, and $DB = y$. Then $h^2 = xy$.*

Exercises

1. Show that any two equilateral triangles are similar.

2. Find a necessary and sufficient condition that implies similarity between two isosceles triangles.

3. *Definition:* We say that two circles are *orthogonal* at a point P if their tangent lines at point P are perpendicular to each other. Let γ_1 and γ_2 be a pair of circles centered at A and B, respectively, and orthogonal at points P, Q. Find a pair of similar, but not congruent, triangles in this configuration and write the corresponding proportion among their sides.

4. Show that if $\angle A \simeq \angle A'$ and $AB/A'B' = AC/A'C'$, then $\triangle ABC$ is similar to $\triangle A'B'C'$.

5. (a) Let D be the midpoint of $\overline{AB}$ and E be the midpoint of $\overline{AC}$. Prove that $\triangle ABC \sim \triangle ADE$.
(b) Show that if $\square ABCD$ is any quadrilateral and E, F, G, H are midpoints of its sides, then $\square EFGH$ is a parallelogram.

6. *Definition:* A *trapezoid* is a quadrilateral that has *only one* pair of parallel opposite sides. The parallel sides are called *bases* of the trapezoid. A trapezoid is called *isosceles* if the nonparallel sides are congruent.
(a) Show that the diagonals of a trapezoid divide each other into proportional sides.
(b) Show that the diagonals of an isosceles trapezoid are congruent.
(c) Let $\square ABCD$ be a trapezoid such that $\overline{AD}$ and $\overline{BC}$ are the nonparallel sides. Show that the line segment $\overline{MN}$ connecting their midpoints is parallel to the bases $\overline{AB}$, $\overline{CD}$ and

$$MN = \frac{AB + CD}{2}.$$

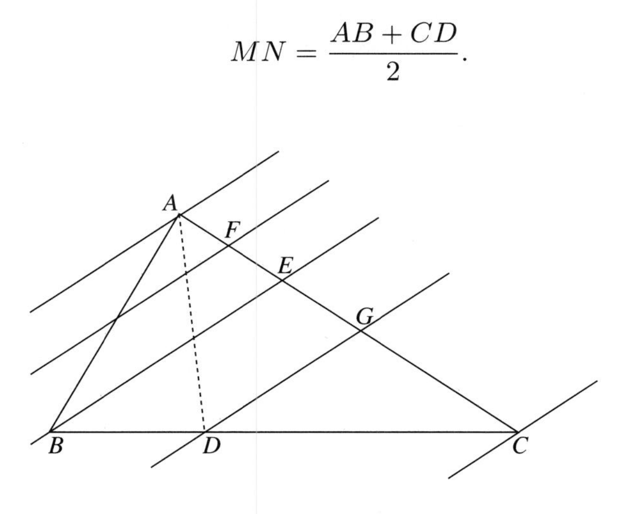

Figure 2.17

7. Prove Proposition 2.2.4.
Hint: Consider $\triangle ABC$ with medians $\overline{AD}$ and $\overline{BE}$ (see Figure 2.17). Let $F, G \in \overline{AC}$ that are the midpoints of $\overline{AE}$ and $\overline{EC}$, respectively. Let l be the line containing the median $\overline{BE}$. Let l_1, l_2, l_3, and l_4 be lines parallel to l and through points A, F, G,

and C, respectively. Let $X = l_2 \cap \overline{AD}$ and $Y = l \cap \overline{AD}$. Use the parallel projection theorem to conclude that l_3 intersects $\overline{BC}$ at point D and that $AX = XY = YD$. Then $AY = (2/3)AD$.

8. Show that if line $\overleftrightarrow{BD}$ bisects angle $\angle B$ of $\triangle ABC$ then

$$\frac{AD}{DC} = \frac{AB}{BC}.$$

9. State and show the converse of the result of Exercise 8.

10. Prove Proposition 2.3.10

2.4 Euclidean Circles

We start this section observing that in Euclidean geometry three non-collinear points A, B, and C determine a circle that circumscribes $\triangle ABC$. This fact is stronger than Proposition 1.7.13(d) of Chapter 1, since for proving it we assume the existence of two circles containing at least three common points. Notice that the proof of Proposition 2.2.1 uses the Euclidean parallel postulate to show that the circumscribed circle exists. In this section we will show some other results on circles that hold in Euclidean geometry only.

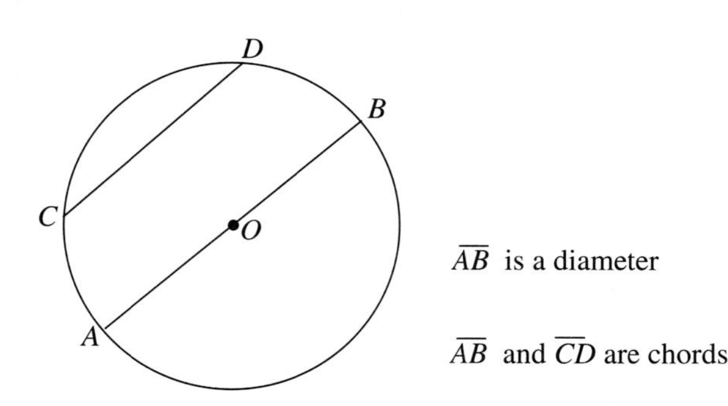

Figure 2.18

Recall that given two points A and B on circle γ, line $\overleftrightarrow{AB}$ divides the circle in two disjoint parts, and each part is called an open *arc* of γ

determined by A and B. In the case that A and B are endpoints of a diameter, such arcs are called *semicircles.*

If $\overline{AB}$ is not a diameter, then the arc that is on the same half-plane determined by $\overleftrightarrow{AB}$ as the center O is called the *major arc*, while the other is called the *minor arc.* The minor arc determined by A and B will be denoted by $\overset{\frown}{AB}$.

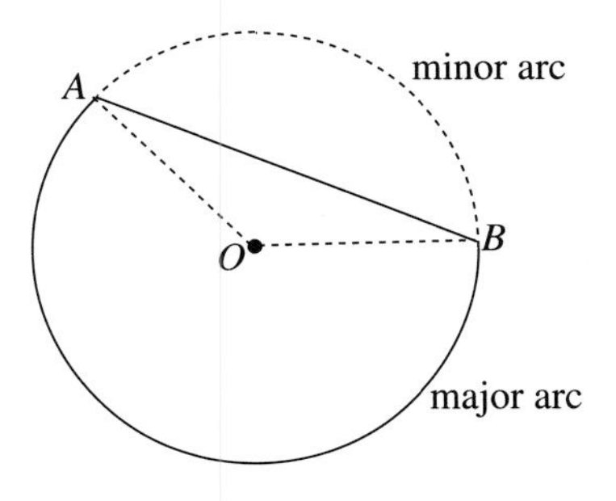

Figure 2.19

Definition 2.4.1 *Let $\overline{AB}$ be a chord of circle γ with center O. Then $\angle AOB$ is called a* central angle. *The* measure of the minor arc *is defined as $m(\angle AOB)$, i.e.,*

$$m(\overset{\frown}{AB}) = m(\angle AOB).$$

Observe that if $\overline{AB}$ is a diameter, then $\angle AOB$ is a straight angle. Observe also that if σ_1 and σ_2 are two disjoint semicircles of circle γ determined by diameter $\overline{AB}$, then $\gamma = \sigma_1 \cup \sigma_2 \cup \{A\} \cup \{B\}$. This motivates the following definition.

Definition 2.4.2 *(i) The measure of a semicircle is $180°$.*
(ii) The measure of a circle is $360°$.
(iii) The measure of the major arc *determined by a chord $\overline{AB}$ is*

$$360° - m(\angle AOB) = 360° - m(\overset{\frown}{AB}).$$

Definition 2.4.3 *An angle is said to be* inscribed *in a circle γ if its vertex A is on γ and its sides each intersect the circle in another point. Let B and C be the points on γ determined by the sides of the inscribed angle. The points B and C determine two arcs. The arc not containing* *the vertex A is called the* arc subtended *by angle $\angle BAC$. If the chord $\overline{BC}$ is a diameter, then $\angle BAC$ is said to be* inscribed in a semicircle.

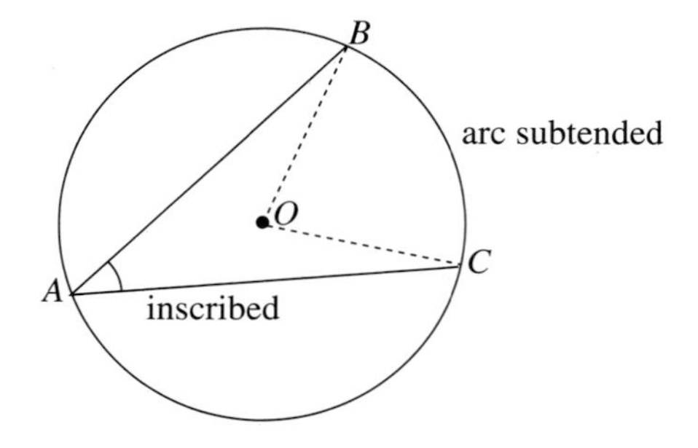

Figure 2.20

Proposition 2.4.4 *The measure of an inscribed angle in a circle is equal to half the measure of its subtended arc.*

Proof We consider first the case that one side of the angle is a diameter (see Figure 2.21). In this case the subtended arc is a minor arc. Let O be the center of the circle γ, A the vertex of the angle, and B and C the other two points of the angle on γ. Let us suppose that $\overline{AC}$ is a diameter. In this case, the measure of the subtended arc is $m(\angle BOC)$. Since $\triangle AOB$ is isosceles and $m(\angle BOC) = m(\angle OAB) + m(\angle OBA)$, we get $m(\angle BOC) = 2m(\angle CAB)$, which in turn implies

$$m(\angle CAB) = \frac{1}{2}m(\angle BOC) = \frac{1}{2}m(\overset{\frown}{CB}).$$

Now we suppose that neither of its sides is a diameter and consider point $D \in \gamma$ such that $\overline{AD}$ be a diameter. Then, the first part of this proof implies

$$m(\angle BOD) = 2m(\angle BAD) \quad \text{and} \quad m(\angle DOC) = 2m(\angle DAC).$$

We divide the rest of the proof into two cases:

Case 1: Ray $\overrightarrow{AD}$ is not between $\overrightarrow{AC}$ and $\overrightarrow{AB}$.

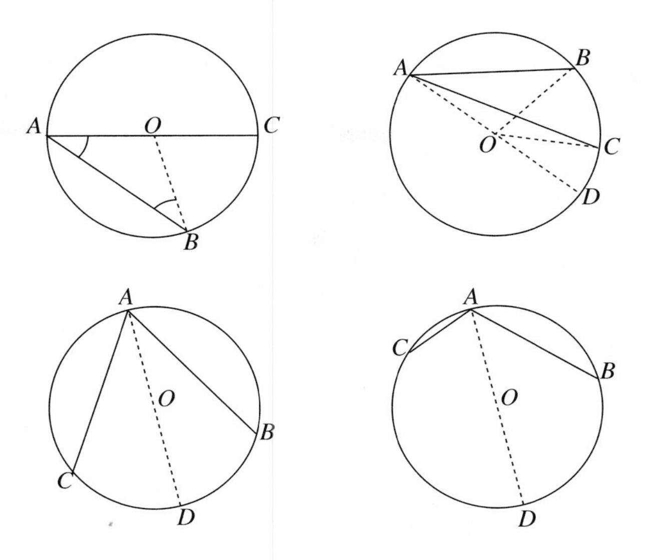

Figure 2.21

Here, we suppose without loss of generality that $\overrightarrow{AC}$ divides $\angle BAD$ and hence the subtended arc is a minor arc. Then

$$m(\angle BAC) = m(\angle BAD) - m(\angle CAD) = \frac{1}{2}[m(\angle BOD) - m(\angle COD)].$$

But $m(\overset{\frown}{CB}) = m(\angle COB)$ and $m(\angle COB) = m(\angle BOD) - m(\angle COD)$. Substituting above, we obtain

$$m(\angle BAC) = \frac{1}{2}m(\overset{\frown}{CB}).$$

Case 2: Ray $\overrightarrow{AD}$ is between $\overrightarrow{AC}$ and $\overrightarrow{AB}$.

Then $m(\angle BAC) = m(\angle BAD) + m(\angle DAC)$. Therefore

$$m(\angle BAC) = \frac{1}{2}[m(\angle BOD) + m(\angle DOC)]$$

If the subtended arc is a minor arc, then

$$m(\overset{\frown}{CB}) = m(\angle COB) = m(\angle BOD) + m(\angle DOC) = 2m(\angle BAC)$$

and hence

$$m(\angle BAC) = \frac{1}{2} m(\widehat{CB}).$$

Now we suppose that the subtended arc is a major arc. Observe that

$$\begin{aligned} m(\angle COB) &= m(\angle COA) + m(\angle AOB) \\ &= 180^\circ - m(\angle BOD) + 180^\circ - m(\angle DOC) \\ &= 360^\circ - 2m(\angle BAC) \end{aligned}$$

which implies

$$m(\angle BAC) = \frac{1}{2}[360^\circ - m(\angle COB)].$$

□

Corollary 2.4.5 *(a) If two angles inscribed in a circle subtend the same arc, then they are congruent.*
(b) An angle inscribed in a semicircle is a right angle.

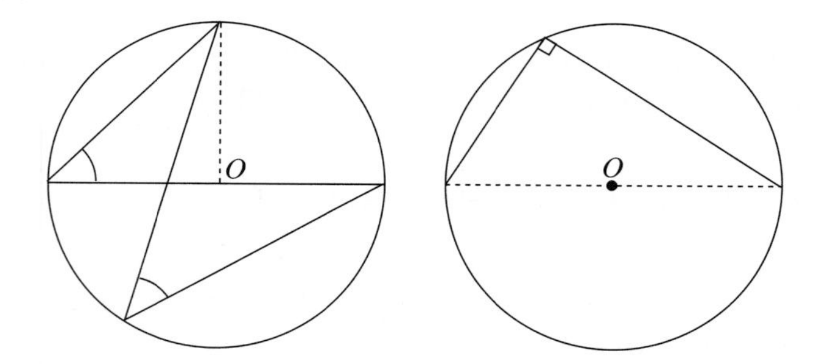

Figure 2.22

Proposition 2.4.6 *The measure of an angle formed by a line l tangent to a circle γ at point B and any chord $\overline{AB}$ is half of the measure of the arc intercepted by l and $\overline{AB}$.*

Proof Let B be the point of tangency. Consider $P \in l$ so that the arc subtended by $\angle PBA$ is the minor arc determined by $\overline{AB}$. The measure of such an arc is $m(\angle AOB)$. Therefore we want to show that

$$m(\angle PBA) = \frac{1}{2} m(\angle AOB).$$

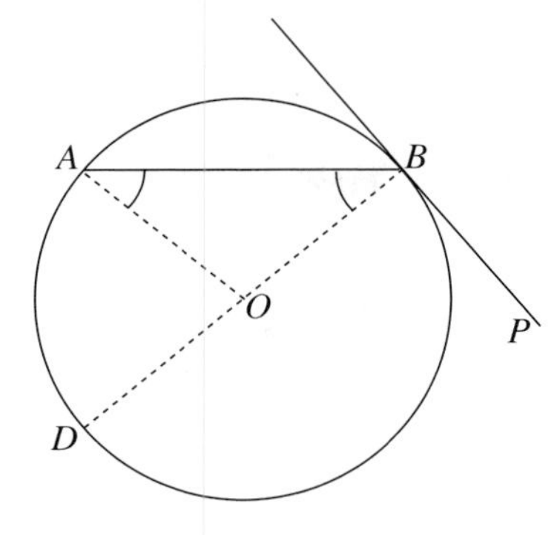

Figure 2.23

Since $\overline{BO} \perp \overline{PB}$, we have

$$m(\angle PBA) = 90^\circ - m(\angle ABO).$$

But $m(\angle ABO) = (1/2)m(\angle AOD)$, and since $m(\angle AOD) = 180^\circ - m(\angle AOB)$, substituting above, we obtain the result. Now if P' is on the other side of $\overleftrightarrow{AB}$ we have

$$m(\angle P'BA) = 180 - \frac{1}{2}m(\angle AOB) = \frac{1}{2}[360 - m(\angle AOB)],$$

which is the measure of the major arc determined by $\overline{AB}$. □

Lemma 2.4.7 *If a secant and a tangent line intersect outside a circle, then the measure of the angle formed is half the difference of the measures of the intercepted arcs.*

Proof Let P denote the point of concurrence of the tangent and the secant lines (see Figure 2.24). Let A denote the point of tangency and B, C the points where the secant intersects the circle. From triangle $\triangle APC$ we get

$$m(\angle P) = 180 - m(\angle A) - m(\angle C).$$

From Proposition 2.4.6 we get that $180 - m(\angle A) = (1/2)m(\widehat{AC})$, and from Proposition 2.4.4 we obtain $m(\angle C) = (1/2)m(\widehat{AB})$. Therefore

$$m(\angle P) = \frac{1}{2}[m(\widehat{AC}) - m(\widehat{AB})].$$

□

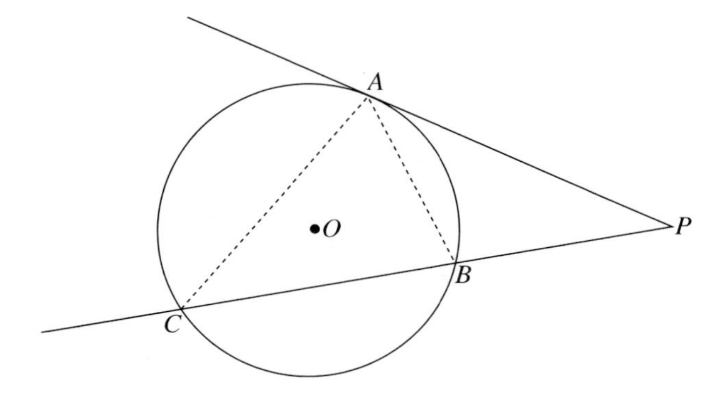

Figure 2.24

Proposition 2.4.8 *If tangent and secant lines are drawn to a circle from the same point in the exterior of the circle, then the length of the tangent segment is the mean proportional between the length of the external and interior secant segment.*

Proof With the notation used in the proof of Lemma 2.4.7, we will show that

$$\triangle PCA \sim \triangle PAB.$$

Since they have a common angle, it is enough to find another pair of congruent angles. We claim that $\angle PAB \simeq \angle C$. In fact, Proposition 2.4.6 implies that

$$m(\angle PAB) = \frac{1}{2}m(\widehat{AB}),$$

while Proposition 2.4.4 gives

$$m(\angle C) = \frac{1}{2}m(\widehat{AB}).$$

From the similarity of the triangles we obtain

$$\frac{PC}{PA} = \frac{PA}{PB}$$

and then

$$(PA)^2 = PB \cdot PC.$$

□

Exercises

1. Show that if $\angle ABC$ is a right angle, then there exists a unique circle γ such that $A, B, C \in \gamma$, and $\overline{AC}$ is a diameter.

2. *Definition:* A polygon is said to be *inscribed in a circle* if all its vertices lie on the circle.
   (a) Show that a parallelogram inscribed in a circle is a rectangle.
   (b) Show that a trapezoid inscribed in a circle is an isosceles trapezoid.

3. Show that if two chords $\overline{AB}$ and $\overline{CD}$ of the same circle intersect at point P then
$$PA \cdot PB = PC \cdot PD.$$

4. Show that if two secants intersect outside a circle, then the measure of the angle of intersection is half the difference of the measures of the intercepted arcs.

5. Show that if two secants intersect outside a circle, then the product of the lengths of two of the segments formed on one secant is equal to the product of the lengths of the corresponding segments on the other secant.

2.5 Trigonometric Functions

Let α be an angle whose sides are rays $\overrightarrow{AB}$ and $\overrightarrow{AC}$. Consider the circle γ centered at A and radius $\overline{AB}$ (see Figure 2.25). Let l be the line passing through B that is perpendicular to line $\overleftrightarrow{AC}$. Let D denote the point where l intersects $\overleftrightarrow{AC}$.

Definition 2.5.1 *The* sine of angle α, *denoted by* $\sin\alpha$, *is defined by*

$$\sin\alpha = \frac{BD}{AB}.$$

The cosine of angle α, *denoted by* $\cos\alpha$, *is defined by*

$$\cos\alpha = \frac{AD}{AB} \quad \text{if } \alpha \text{ is acute} \qquad \cos\alpha = \frac{-AD}{AB} \quad \text{if } \alpha \text{ is obtuse.}$$

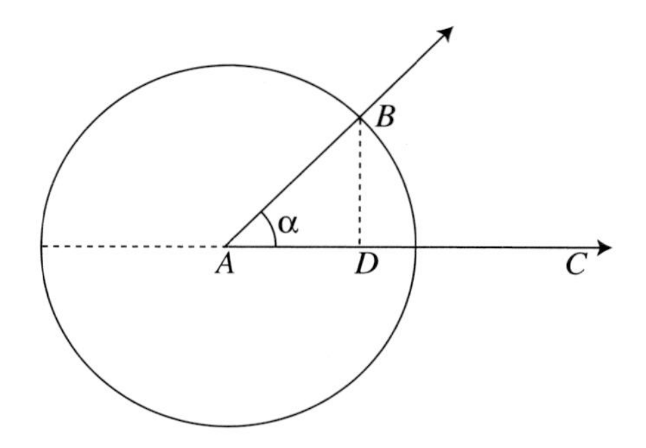

Figure 2.25

It follows from this definition that

$$\sin 0^\circ = 0, \quad \sin 90^\circ = 1, \quad \text{and} \quad \sin 180^\circ = 0$$

$$\cos 0^\circ = 1, \quad \cos 90^\circ = 0, \quad \text{and} \quad \cos 180^\circ = -1.$$

We point out that the definitions of $\sin\alpha$ and $\cos\alpha$ do not depend on the choice of point $B \in \overrightarrow{AB}$, that is, on the radius of the circle γ. This is shown using the fundamental theorem of similar triangles and it is left to the reader. It is also a consequence of the same theorem that if $\alpha \simeq \beta$, then $\sin\alpha = \sin\beta$ and $\cos\alpha = \cos\beta$. The study of the functions *sine* and *cosine* is called *trigonometry.* Its fundamental identities are listed in the result below.

Theorem 2.5.2 (i) *For any angle* α *we have:*
(a) $\sin^2\alpha + \cos^2\alpha = 1$
(b) $\sin(180^\circ - \alpha) = \sin\alpha$
(b) $\cos(180^\circ - \alpha) = -\cos\alpha.$
(ii) *For any acute angle* α *we have:*
(a) $\sin(90^\circ - \alpha) = \cos\alpha$
(b) $\cos(90^\circ - \alpha) = \sin\alpha.$
(iii) *For all angles* α *and* β *we have:*
(a) $\sin(\alpha \pm \beta) = \sin\alpha\; cos\beta\; \pm\; \sin\beta\; \cos\alpha$
(b) $\cos(\alpha \pm \beta) = \cos\alpha\; cos\beta\; \mp\; \sin\alpha\; \cos\beta.$

Notice that $\triangle ABD$ is a right triangle. Therefore, given a right triangle $\triangle ABC$ such that $\overline{AB}$ is its hypotenuse, one can determine

$\sin \angle A$ and $\cos \angle A$ by considering the circle centered at A with radius $\overline{AB}$.

Proposition 2.5.3 (The law of cosines) *Let $\triangle ABC$ be a triangle with sides of length a, b, and c and let C denote the vertex opposite side c. Then*

$$c^2 = a^2 + b^2 - 2ab\, \cos(\angle C).$$

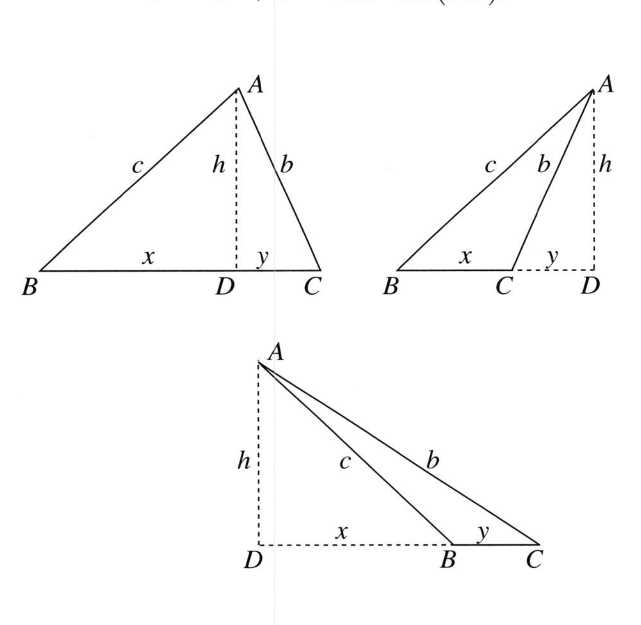

Figure 2.26

Proof Let $\overline{AD}$ be the altitude to the side $\overline{BC}$. If $D = B$, then $\triangle ABC$ is a right triangle and $\cos(\angle A) = c/b$. Then, the relation above is the Pythagorean relation $a^2 = b^2 - c^2$. Similarly, if $D = C$, we obtain the Pythagorean relation again. Then we suppose that B, C, and D are distinct points. Let $AD = h$, $BD = x$, and $DC = y$. Since triangles $\triangle ADB$ and $\triangle ADC$ are right triangles, we have

$$c^2 = x^2 + h^2 \quad \text{and} \quad b^2 = y^2 + h^2.$$

Subtracting these equations, we get

$$c^2 - x^2 = b^2 - y^2.$$

We consider now the three possible cases:
(a) D is between B and C.

Here, $a = x + y$ and then $a^2 = x^2 + 2xy + y^2$. Substituting above, we get

$$c^2 + 2xy + y^2 - a^2 = b^2 - y^2$$

$$c^2 = a^2 + b^2 - 2y(x + y) = a^2 + b^2 - 2ay.$$

Notice that $\cos(\angle C) = y/b$ and then

$$c^2 = b^2 + a^2 - 2ab\ \cos(\angle C).$$

(b) C is between B and D.

In this case $x = y + a$ and then $x^2 = y^2 + 2ya + a^2$. Substituting above and canceling y^2, we obtain

$$c^2 = b^2 + a^2 + 2ya.$$

But $y = b\ \cos(\angle ACD)$ and $\cos(\angle C) = -\cos(\angle ACD)$, which in turn gives

$$c^2 = b^2 + a^2 - 2ab\ \cos(\angle C).$$

(c) B is between C and D.

This case is proved in a similar manner. □

Proposition 2.5.4 (The law of sines) *Let $\triangle ABC$ be a triangle with sides of length $a, b,$ and c and opposite vertices $A, B,$ and C respectively. Then*

$$\frac{a}{\sin(\angle A)} = \frac{b}{\sin(\angle B)} = \frac{c}{\sin(\angle C)} = d,$$

where d is the diameter of the circle that circumscribes the triangle.

Proof Let γ denote the circle that circumscribes $\triangle ABC$ (see Figure 2.27). Let $D \in \gamma$ be such that $\overline{BD}$ is a diameter of γ. If points A and D are on the same side of line $\overleftrightarrow{BC}$, then $\angle D = \angle BDC \simeq \angle BAC = \angle A$, since they subtend the same arc. If A and D are on opposite sides of line $\overleftrightarrow{BC}$, we claim that $m(\angle BAC) + m(\angle BDC) = 180°$. In fact, $m(\angle BAC) = (1/2)(360 - m(\angle COB))$ by Proposition 2.4.4. The same result implies that $m(BDC) = (1/2)m(\angle COB)$ and the claim is proved.

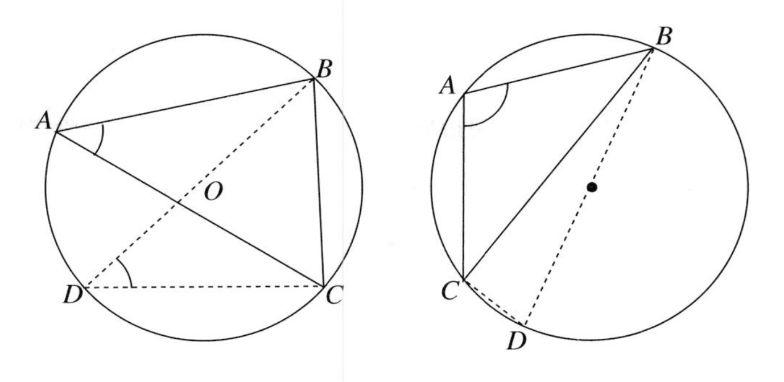

Figure 2.27

In both cases, we have that $\sin(\angle A) = \sin(\angle D)$. Since $\triangle BCD$ is a right triangle, we have that

$$BC = a = BD \sin(\angle D) = d \sin(\angle A).$$

A similar argument shows that

$$b = d \sin(\angle B) \quad \text{and} \quad c = d \sin(\angle C).$$

□

2.6 Euclidean Constructions

As in Chapter 1, we will be using only a compass and straightedge.

Construction 1 *To construct a parallel to a line.*
Given: a line l and a point $P \notin l$.

We want to find a point $Q \notin l$ such that line $\overleftrightarrow{PQ} \parallel l$.
1. Draw any line m through P intersecting line l at a point A; center the compass at point A and draw an arc of radius AP; this arc intersects line l at B.
2. Lift the compass and, keeping the same opening, draw one arc centered at point P and another with B. These arcs intersect at point Q. Draw line $\overleftrightarrow{PQ}$.
Justification: Quadrilateral $\square PABQ$ is a rhombus (see Exercise 2 of Section 1.5) and hence opposite sides $\overline{AB}$ and $\overline{PQ}$ are parallel.

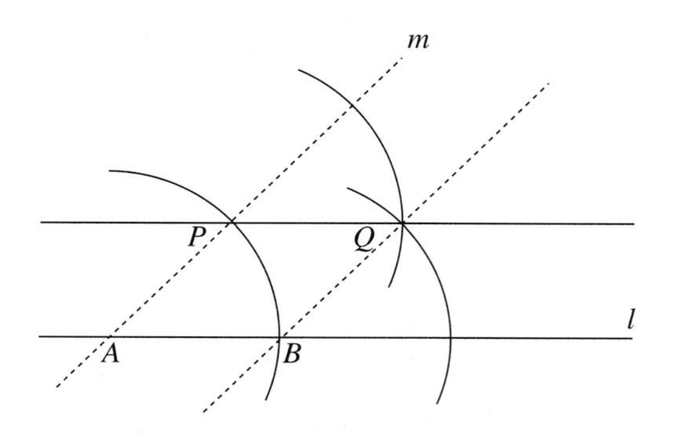

Construction 1

Remark: The construction above could have been included among the basic constructions of Chapter 1, since it is justified by theorems of neutral geometry. However, since it produces only one parallel line, doing it here is more appropriate.

Construction 2 *To partition a segment into any number of congruent segments.*
Given: a line segment $\overline{AB}$ and a positive integer number n.

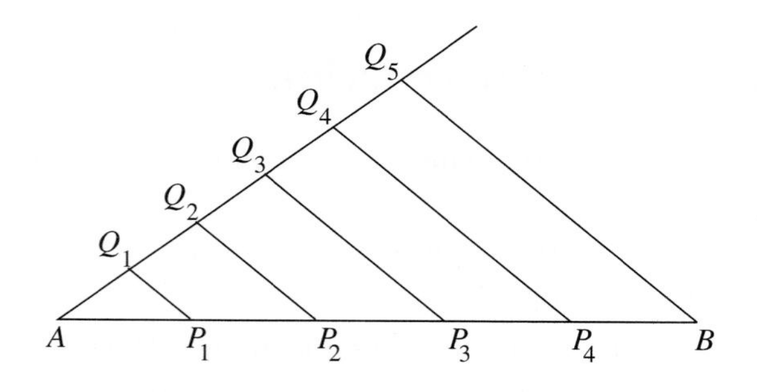

Construction 2

We want to find a points $P_0, P_1, \ldots, P_n \in \overline{AB}$ such that $P_0 = A$, $P_n = B$, and $P_{i-1}P_i = P_iP_{i+1}$ for $i = 1, \ldots, n-1$.

1. Draw any ray r, other than $\overrightarrow{AB}$, emanating from point A; using a fixed opening for the compass, plot n points $Q_1, \ldots, Q_n$.
2. Draw a line through B and Q_n. Now draw lines $m_1, \ldots, m_{n-1}$ parallel

to line $\overleftrightarrow{BQ_n}$, using Construction 1. These lines intersect $\overline{AB}$ at points $P_1, \ldots, P_{n-1}$.
Justification: Since r and $\overline{AB}$ are cut by the parallel lines $m_1, \ldots, m_{n-1}$, $\overleftrightarrow{BQ_n}$, using the parallel projection theorem we obtain

$$1 = \frac{P_{i-1}P_i}{P_iP_{i+1}} = \frac{Q_{i-1}Q_i}{Q_iQ_{i+1}}.$$

Construction 3 *To construct a line tangent to a circle from a point outside the circle.*
Given: a circle γ and a point $P \notin \gamma$.

We want to find a point $Q \in \gamma$ such that line $\overleftrightarrow{PQ}$ is tangent to γ.
1. Draw line segment $\overline{OP}$, where O is the center of the circle; find the midpoint M of $\overline{OP}$, using Construction 6 of Chapter 1.
2. Draw a circle δ centered at M and radius OM; circle δ intersects γ at points Q_1 and Q_2. Draw lines $\overleftrightarrow{PQ_1}$ and $\overleftrightarrow{PQ_2}$.

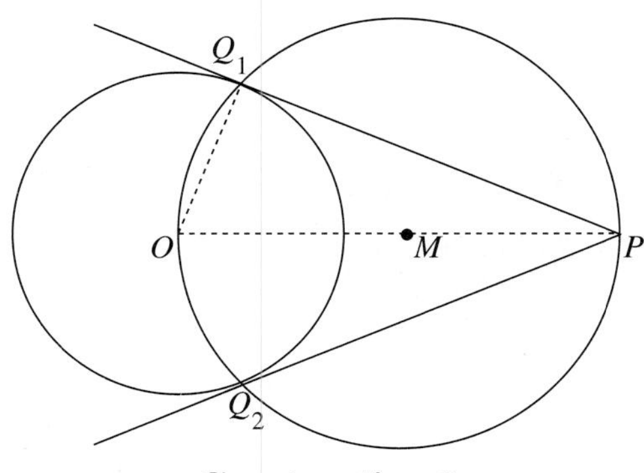

Construction 3

Justification: Notice that circle δ has point P outside γ and a point O inside γ. Then circle δ intersects γ in two points, by the circular continuity principle. Now, $\angle Q_1OP$ and $\angle Q_2OP$ are inscribed in semicircles and therefore are right angles. Then both lines are tangent to γ, for $\overleftrightarrow{PQ_1} \perp \overline{Q_1O}$ and $\overleftrightarrow{PQ_2} \perp \overline{Q_2O}$.
Remark: The construction above is strictly Euclidean, since it is justified using Corollary 2.4.5. Compare to this construction to one suggested in Exercise 14 of Section 1.8.

Construction 4 *To construct a line that is a common internal tangent to two circles.*
Given: two external circles γ_1 and γ_2 with centers O and O' and radii r and r', respectively.

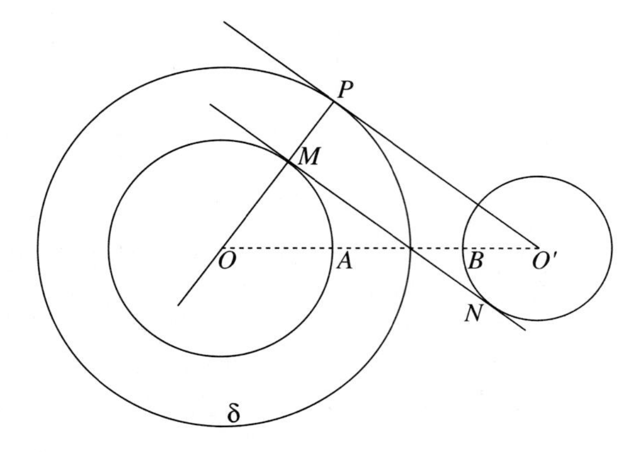

Construction 4

We want to find a line $\overleftrightarrow{MN}$ that is tangent to both circles and intersects line $\overleftrightarrow{OO'}$ at a point between O and O'.

1. Draw line $\overleftrightarrow{OO'}$; it intersects circle γ at point A and circle γ' at point B; draw circle δ of center O and radius $\rho = OA + O'B$ (see the figure above). Point O' is outside δ.
Justification: Since γ_1 and γ_2 are external to each other, $OO' > OA + O'B = \rho$.

2. Find point $P \in \delta$ such that line $\overleftrightarrow{PO'}$ is tangent to δ, using Construction 3. Draw line segment $\overline{OP}$ and let $M = \overline{OP} \cap \gamma$.

3. Draw line m perpendicular to line $\overleftrightarrow{PO'}$ and through O'. Let $N = m \cap \gamma'$ such that N is on the same side of $\overleftrightarrow{PO'}$ as O. Draw line $\overleftrightarrow{MN}$.
Justification: Observe that $\overleftrightarrow{PM} \parallel \overleftrightarrow{O'N}$, since they are both perpendicular to $\overleftrightarrow{O'P}$; moreover, $\rho - r = PM = r' = O'N$. Then $\square PO'NM$ is a parallelogram by Exercise 3 of Section 2.1, which imples that $\overleftrightarrow{PO'} \parallel \overleftrightarrow{MN}$. Therefore $\overleftrightarrow{MN} \perp \overline{O'N}$ and $\overleftrightarrow{MN} \perp \overline{OM}$.

Construction 5 *To construct a line that is a common external tangent to two circles.*
Given: two external circles γ_1 and γ_2 with centers O and O' and radii r and r', respectively.

We want to find a line $\overleftrightarrow{MN}$ that is tangent to both circles and does not intersect line $\overleftrightarrow{OO'}$ at a point between O and O'.
Case 1: $r > r'$.
1. Draw line $\overleftrightarrow{OO'}$; it intersects circle γ at point A and circle γ' at point B; draw circle δ of center O and radius $\rho = OA - O'B$. Point O' is outside δ.
Justification: Observe that O' is outside γ_1 and δ is internal to γ_1.
2. Find point $P \in \delta$ such that line $\overleftrightarrow{PO'}$ is tangent to δ, using Construction 3. Draw ray $\overrightarrow{OP}$ and let $M = \overrightarrow{OP} \cap \gamma$.
3. Draw line m perpendicular to line $\overleftrightarrow{PO'}$ and through O'. Let $N = m \cap \gamma'$ such that N and O are on opposite sides of line $\overleftrightarrow{PO'}$. Draw line $\overleftrightarrow{MN}$.

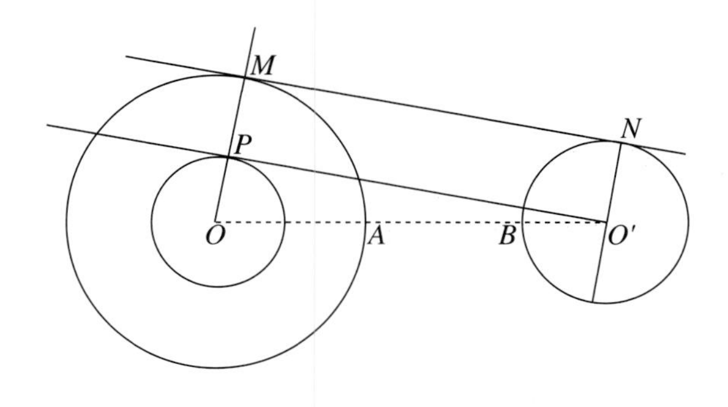

Construction 5

Justification: Observe again that $\overleftrightarrow{PM} \parallel \overleftrightarrow{O'N}$, since they are both perpendicular to $\overleftrightarrow{O'P}$; moreover, $OP - \rho = PM = O'N$. Then $\square PO'NM$ is a parallelogram by Exercise 3 of Section 2.1, which implies that $\overleftrightarrow{PO'} \parallel \overleftrightarrow{MN}$. Therefore $\overleftrightarrow{MN} \perp \overline{O'N}$ and $\overleftrightarrow{MN} \perp \overline{OM}$.

Case 2: $r = r'$.

This part is left as an exercise.

Construction 6 *To construct a segment that represents a positive rational number.*

Given: a segment $\overline{OU}$ that represents number 1 and two positive integer numbers a, b.

We want to draw a segment $\overline{OX}$ such that

$$OX = \frac{a}{b}\,OU.$$

Procedure 1: Divide segment $\overline{OU}$ into b congruent segments, using Construction 2. On any ray, starting at the vertex, copy one of these congruent segments a times. The resulting segment has length a/b.

Procedure 2: Consider ray $\overrightarrow{OU}$ and another ray r emanating from O; on $\overrightarrow{OU}$ copy the unit $\overline{OU}$ b times, obtaining point B. Similarly, obtain

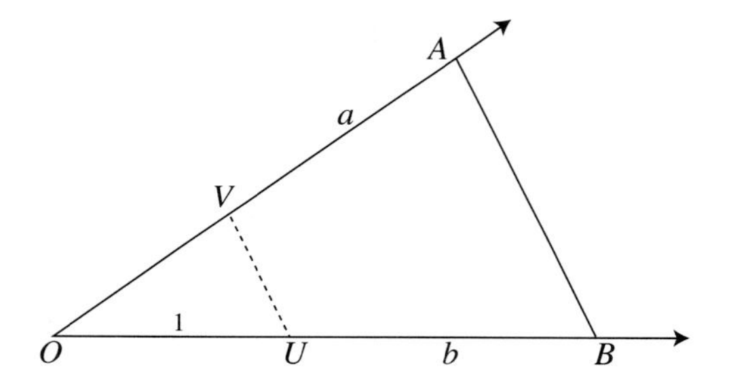

Construction 6

point A on ray r, copying the unit a times. Through point U, draw line $l \parallel \overleftrightarrow{AB}$ and let $V = l \cap r$. Then $\triangle OVU \sim \triangle OAB$, and then

$$\frac{OV}{a} = \frac{1}{b}$$

and then $OV = a/b$.

Using only a compass and straightedge, we can construct segments whose lengths represent some of the irrational numbers (see Exercise 2) of this section. However, a general technique for an arbitrary irrational number does not exist. Nevertheless, a procedure exists for $\sqrt{x}$, where x is any positive real number, provided that we are given not only a segment of unit length but a segment of length x.

Construction 7 *To construct a segment that represents $\sqrt{x}$.*
Given: a unit segment $\overline{OU}$ and segment $\overline{AB}$ such that $AB = x$.
1. Copy line segment $\overline{OU}$ on ray $\overrightarrow{AB}$, starting at point B and obtaining segment $\overline{AC}$. Draw line $l \perp \overline{AC}$ through B.
2. Find the midpoint M of $\overline{AC}$ and draw a circle γ centered at M and radius MC. Circle γ intersects line l at point D. Draw segment $\overline{BD}$.
Justification: Line l has point B inside γ and hence intersects it by the

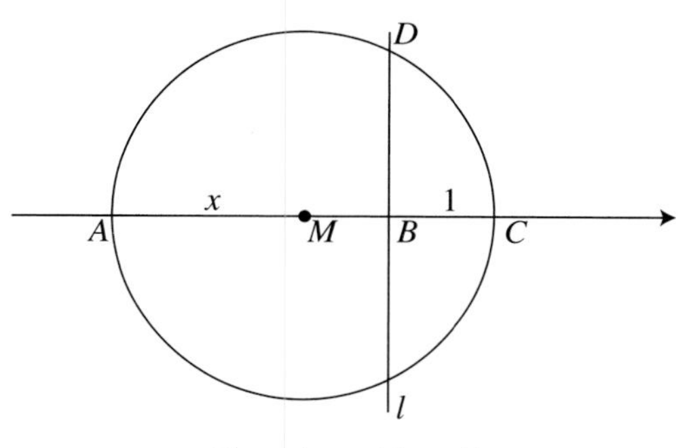

Construction 7

elementary continuity principle. In addition, $\angle ADC$ is inscribed in a semicircle, and then it is a right angle. It follows that $\triangle ADC$ is a right triangle of hypotenuse $\overline{AC}$ and altitude $\overline{BD}$. From Proposition 2.2.4, we get that

$$(BD)^2 = AB \cdot BC = x \cdot 1$$

and then $BD = \sqrt{x}$.

Exercises

1. Describe a procedure for Construction 5, case 2.

2. Given a unit segment $\overline{OU}$, describe a method that uses only compass and straightedge for:
(a) constructing line segments that represent the numbers $\sqrt{2}$, $\sqrt{5}$, and $\sqrt{13}$.

(b) constructing line segments that represent numbers of the type $a + b\sqrt{2}$, where a and b are positive rational numbers.

3. *Definition:* A rectangle of sides a and b is called a *golden rectangle* if

$$\frac{a}{b} = \frac{1+\sqrt{5}}{2},$$

and the number $(1+\sqrt{5})/2$ is called a *golden ratio.*

Show that a golden rectangle can be constructed with a compass and straightedge.
Hint: Draw a square $\square ABCD$; find the midpoint M of $\overline{AB}$ and draw a circle γ centered at M and radius MC; circle γ intersects ray $\overrightarrow{AB}$ at point E. Let F be the foot of the perpendicular dropped from E to $\overleftrightarrow{DC}$. Show that $\square AEFD$ and $\square BEFC$ are both golden rectangles.

Chapter 3

GEOMETRIC TRANSFORMATIONS

3.1 Rigid Motions

Throughout this chapter, unless otherwise stated, $\mathcal{P}$ will denote a Euclidean plane, that is, one single plane where we have assumed the incidence, betweenness and congruence axioms together with the Euclidean parallel postulate.

We will study functions that carry points of $\mathcal{P}$ to points of $\mathcal{P}$ and have some geometric significance. Although the reader must have studied the concept of *function* before, we have included at the end of this chapter an appendix that contains its definition and related concepts.

Definition 3.1.1 *Let $\mathcal{S}$ be a subset of $\mathcal{P}$. A* transformation *is a map $T : \mathcal{S} \to \mathcal{P}$.*

In other words, a *transformation* assigns to every point $P \in \mathcal{S}$ another point $P' \in \mathcal{P}$. The point $P' = T(P)$ is called the *image* of P.

Definition 3.1.2 *A transformation $T : \mathcal{P} \to \mathcal{P}$ is called a* rigid motion *or a* Euclidean isometry *if $T(A)T(B) = AB$, for all $A, B \in \mathcal{P}$.*

The next proposition shows the main properties of a rigid motion. It also justifies the use of the word *rigid*, because a rigid motion preserves the size and shape of every geometric figure.

Proposition 3.1.3 *Let $T : \mathcal{P} \to \mathcal{P}$ be a rigid motion. Then:*
(a) T is injective.
(b) $T(\overline{AB}) = \{T(P) \mid P \in \overline{AB}\} = \overline{A'\,B'}, \ \ \forall\, A, B \in \mathcal{P}$.
(c) T preserves angle measure.
(d) T is onto.

Proof *(a)* Suppose that there exist points A and B with the same image. Let A' and B' denote the images of A and B, respectively. Then we have $A' = B'$, which implies $A'B' = 0$ and hence $AB = A'B' = 0$. From Axiom $\mathbf{S_1}$ we conclude that $A = B$, and this shows that T is one-to-one.
(b) Consider a point $C' \in T(\overline{AB})$ such that C', A', and B' are three distinct points. Let C be the only point on $\overline{AB}$ such that $T(C) = C'$. Since T is one-to-one, we have that $C \neq A$ and $C \neq B$. Therefore C is between A and B, and hence $AB = AC\ +\ CB$, by Axiom $\mathbf{S_3}$. Then from the hypothesis we obtain

$$A'B' = A'C'\ +\ C'B'.$$

Therefore C', A', and B' are collinear, because if not, the triangle inequality would give

$$A'\,B' < A'\,C'\ +\ C'\,B',$$

contradicting the equality above. Now the fact that C' is between A' and B' follows from Proposition 1.4.3. This shows each $C' \in T(\overline{AB})$ is in $\overline{A'\,B'}$, and that T preserves betweenness.

Now we will show that $\overline{A'\,B'} \subset \{T(P) \mid P \in \overline{AB}\} = T(\overline{AB})$, and then we conclude $\overline{A'\,B'} = T(\overline{AB})$. Let P' be between A' and B'. Let $x = P'A'$. Since $x < A'B' = AB$, there exists a point P between A and B such that $AP = x$. From the first part of this proof we conclude that $T(P)$ is between A' and B' and $A'T(P) = x$. Therefore $T(P) = P'$ by Axiom $\mathbf{S_2}$.
(c) Consider an angle $\angle ACB$ formed by the rays $\overrightarrow{AC}$ and $\overrightarrow{CB}$. If $m(\angle ACB) = 0$ or $180°$, the result follows from (b). Otherwise, connecting points A, B, and C, we obtain the triangle $\triangle ACB$. It follows from part (b) that its image is the triangle $\triangle A'\,C'\,B'$, and since T preserves length we conclude that

$$\triangle A'\,C'\,B' \simeq \triangle ACB$$

by SSS. Therefore $\angle ACB \simeq \angle A'C'B'$.
(d) Let P' be any point of $\mathcal{P}$, A', B' two distinct points in the image of T (notice that since T is injective, the image of T is not a single point), and $x = m(\angle P'A'B')$. Let A and B be the preimages of A' and B', respectively. Axiom $\mathbf{A_2}$ implies that there is only one ray $\overrightarrow{AP}$ on one side of line $\overleftrightarrow{AB}$ such that $m(\angle PAB) = x$. Moreover, point P can be chosen so that $AP = A'P'$, by Axiom $\mathbf{S_2}$. Similarly, on the other side of $\overleftrightarrow{AB}$ we find point Q such that $m(\angle QAB) = x$ and $AQ = A'P'$. We claim that either $T(P) = P'$ or $T(Q) = P'$. If not, since T preserves length and angle measure, there would be a point $P'' \neq P'$ on the same side of $\overleftrightarrow{A'B'}$ as P' such that $m(\angle P''A'B') = x$ and $A'P'' = A'P'$, contradicting Axioms $\mathbf{S_2}$ and $\mathbf{A_2}$. □

Definition 3.1.4 *Let $T : \mathcal{P} \to \mathcal{P}$ be a transformation. A point $P \in \mathcal{P}$ is called a* fixed point *of T if $T(P) = P$. A subset $\mathcal{X}$ of $\mathcal{P}$ is called* invariant *by T if*

$$T(\mathcal{X}) = \{T(P) \mid P \in \mathcal{X}\} = \mathcal{X}.$$

Proposition 3.1.5 *Let T and S be rigid motions of a plane $\mathcal{P}$. Then*
(a) If T has three noncollinear fixed points, then T fixes all points of $\mathcal{P}$; that is, T is the identity transformation.
(b) If $T(A_1) = S(A_1)$, $T(A_2) = S(A_2)$, $T(A_3) = S(A_3)$, and A_1, A_2, A_3 are noncollinear points, then $T = S$; that is; if T and S agree on three noncollinear points, then they agree everywhere.

Proof *(a)* Let A_1, A_2, and A_3 be the three noncollinear fixed points. Let P be an arbitrary point of $\mathcal{P}$. Then we have

$$T(P)T(A_i) \;=\; T(P)A_i \;=\; PA_i, \quad \forall\, i = 1, 2, 3.$$

If $T(P) \neq P$, then P and $T(P)$ belong to the circle σ_i centered at A_i and radius A_iP for $i = 1, 2, 3$ (see Figure 3.1). But this implies that A_1, A_2, and A_3 lie on the perpendicular bisector of $\overline{PT(P)}$, contradicting that they are noncollinear. Then $T(P) = P$ for all $P \in \mathcal{P}$.
(b) It follows from Exercise 2 that S^{-1} is a rigid motion, and then so is $S^{-1} \circ T$. Moreover,

$$S^{-1} \circ T(A_i) \;=\; S^{-1}(T(A_i))$$

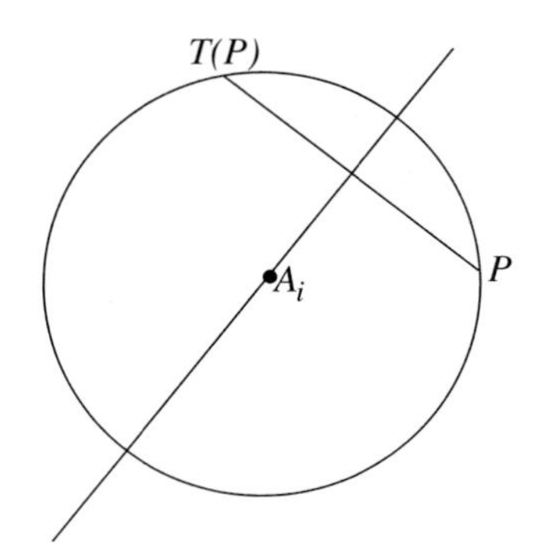

Figure 3.1

$$
\begin{aligned}
&= S^{-1}(S(A_i)) \\
&= (S^{-1} \circ S)(A_i) = A_i, \quad \forall\, i = 1, 2, 3.
\end{aligned}
$$

Therefore $S^{-1} \circ T$ fixes the three noncollinear points A_1, A_2, A_3, and part (a) of this proposition implies that $S^{-1} \circ T = I$, where I denotes the identity map. Composing $S^{-1} \circ T$ and I with S, we have

$$S \circ S^{-1} \circ T = S \circ I,$$

which implies that $T = S$. □

In the rest of this section we shall describe the fundamental types of rigid motions of Euclidean plane.

Reflection through a line l

Let $P \in \mathcal{P}$. If $P \in l$, the reflection of P is itself. If $P \notin l$, consider m

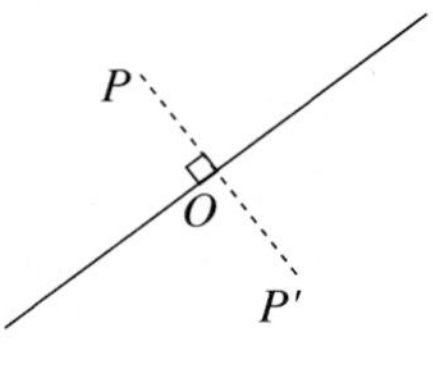

Figure 3.2

to be a line perpendicular to l and through point P and let $O = l \cap m$.

The *reflection of* P in l is the point $P' \in m$ such that P and P' are on opposite sides of l and $OP = OP'$ (see Figure 3.2). Observe that every point lying on l is a fixed point of the reflection through l and every fixed point of the reflection lies on l.

Proposition 3.1.6 *Let $\mathcal{P}$ be a plane and $S : \mathcal{P} \to \mathcal{P}$ the reflection of $\mathcal{P}$ through line l. Then*
(a) $(S \circ S)(P) = S^2(P) = P$ for all $P \in \mathcal{P}$.
(b) S is a rigid motion.
(c) S is invertible and $S^{-1} = S$.

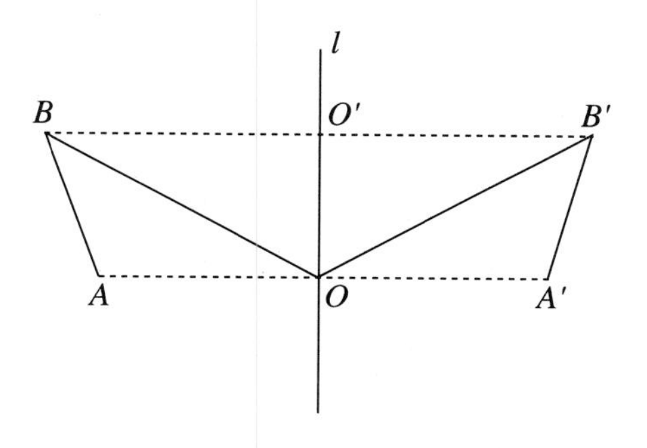

Figure 3.3

Proof *(a)* Let l be the line that S reflects through and $P' = S(P)$. By definition of reflection, P' lies on line m that is perpendicular to l. Let $S(P') = P''$. Then P'' also lies on m, since m is the only line that is perpendicular to l and passes through P'. Let $O = l \cap m$. We have

$$OP = OP' = OP''.$$

This implies that $P'' = P$; that is, $S(S(P)) = P$ for all $P \in \mathcal{P}$.
(b) Let A, B be two points of $\mathcal{P}$ with respective images A', B' (see Figure 3.3). Let O, O' denote the midpoints of line segments $\overline{AA'}$ and $\overline{BB'}$, respectively. Then $\triangle AOO' \simeq \triangle A'OO'$ by SAS. It follows that $\angle AO'O \simeq \angle A'O'O$, which in turn gives $\angle BO'A \simeq \angle B'O'A'$. Therefore $\triangle ABO' \simeq \triangle A'B'O'$ by SAS. This congruence implies that $\overline{AB} \simeq \overline{A'B'}$, which implies $AB = A'B'$.
(c) Proposition 3.1.3 implies that S is invertible. Since $S \circ S = I$, we have

that $S = S^{-1}$ by the uniqueness of the inverse function (Proposition 3.5.10). This fact can also be seen directly, since

$$S^{-1} = S^{-1} \circ I = S^{-1} \circ S \circ S = I \circ S = S.$$

$\square$

Definition 3.1.7 *A geometric figure $\mathcal{F}$ is said to have a* line of symmetry *l if it is invariant under the reflection through l. A rigid motion that leaves a geometric figure invariant is called a* symmetry *of $\mathcal{F}$.*

Example 1
(a) $\mathcal{F}$ is a circle and l a line through its center.
(b) $\mathcal{F}$ is a rectangle and l a line through the midpoints of opposite sides.
(c) $\mathcal{F}$ is an isosceles trapezoid and l a line through the midpoints of the two parallel bases.

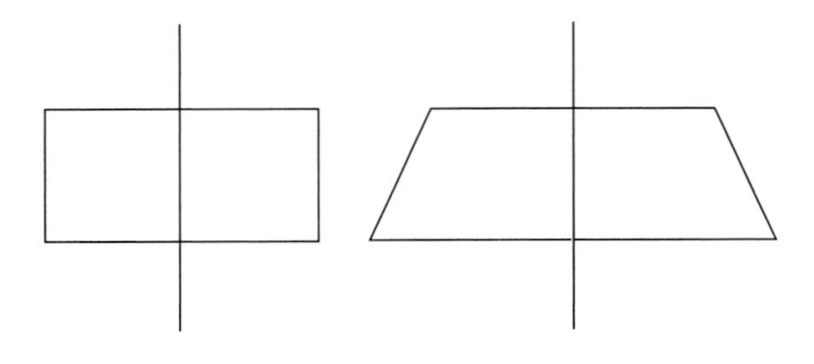

Figure 3.4

The reader is asked to verify that in (a), (b), and (c) above, l is a line of symmetry of $\mathcal{F}$.

Lemma 3.1.8 *Suppose that $\triangle ABC$ is an isosceles triangle such that $\overline{AB} \simeq \overline{AC}$. Then the line which bisects $\angle A$ is line of symmetry of $\triangle ABC$.*

Proof Let S denote the reflection through l. Since triangle $\triangle ABC$ is isosceles, l is the perpendicular bisector of line segment $\overline{BC}$, and then S maps B to C. Notice that A is on line l, and then $S(A) = A$. Therefore $S(\overline{AB}) = \overline{AC}$ and $S(\overline{BC}) = \overline{BC}$ by Proposition 3.1.3(b). $\square$

Definition 3.1.9 *A polygon is called a* regular polygon *if all its sides are congruent and all its angles are congruent.*

Theorem 3.1.10 *Let P_n be a regular n-gon. Then P_n has a line of symmetry.*

Proof Let $A_1, \ldots, A_n$ be the vertices of P_n. We reorder them so that A_2, A_1, A_3 are three consecutive vertices. Let l be a line that bisects angle A_1 (see Figure 3.5). Since triangle $\triangle A_1A_2A_3$ is isosceles, l is the perpendicular bisector of line segment $\overline{A_2A_3}$. Let S denote the reflection through line l. We will show that P_n is invariant by S and hence l is a line of symmetry. It suffices to show that S maps a vertex of P_n to a vertex of P_n (see Exercise 3 of this section). We have that S maps A_1 to itself and A_i to A_j, for $\{i, j\} = \{2, 3\}$.

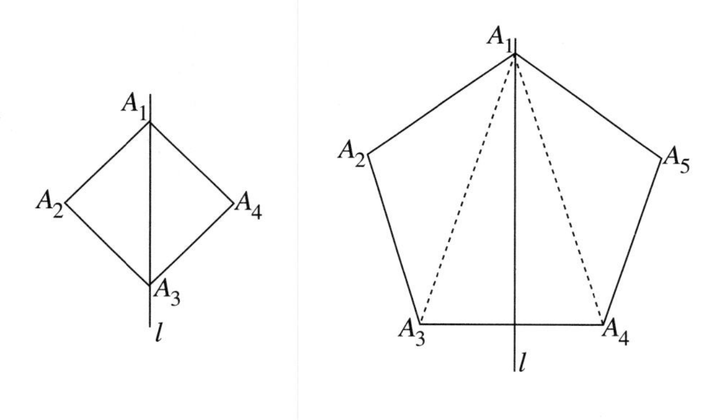

Figure 3.5

If $n = 4$, we consider l' bisecting angle A_4. Since triangle $\triangle A_4A_2A_3$ is isosceles, l' is the perpendicular bisector of line segment $\overline{A_2A_3}$ and hence $l = l'$, by the uniqueness of the perpendicular bisector. It follows that $S(A_4) = A_4$ and P_4 is invariant by S. If $n = 5$, consider triangle $\triangle A_1A_4A_5$. We claim it is isosceles. In fact, observe that because P is a regular n-gon, $\triangle A_1A_2A_4 \simeq \triangle A_1A_3A_5$ by SAS. Then $\overline{A_1A_4} \simeq \overline{A_1A_5}$. Moreover, line l bisects $\angle A_4A_1A_5$, since $\angle A_4A_1A_2 \simeq \angle A_5A_1A_3$. Therefore l is the perpendicular bisector of line segment $\overline{A_4A_5}$ and $S(A_i) = A_j$, for $\{i, j\} = \{4, 5\}$.

The cases $n = 4, 5$ show how to prove the general case by induction.

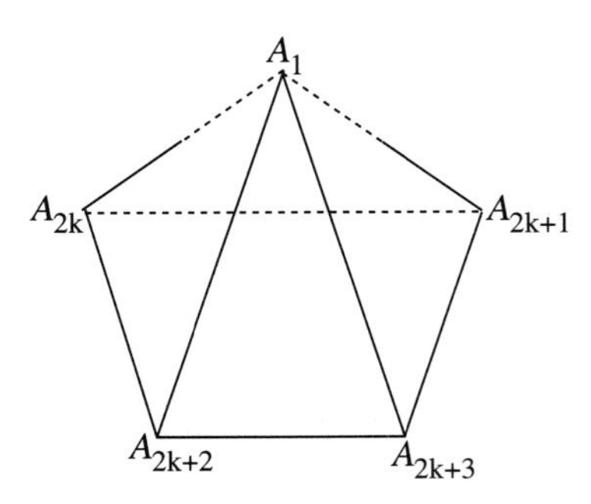

Figure 3.6

Suppose by induction that $S(A_{2k}) = A_{2k+1}$ and $S(A_{2k+1}) = A_{2k}$; that is, l is the perpendicular bisector of line segment $\overline{A_{2k}A_{2k+1}}$. If $n = 2k+2$, let l' be the line that bisects angle A_{2k+2} (see Figure 3.6). Again we obtain that l' is the perpendicular bisector of line segment $\overline{A_{2k}A_{2k+1}}$ and hence $l = l'$. It follows that P_n is invariant by S. If $n = 2k+3$, we consider triangle $\triangle A_1A_{2k+2}A_{2k+3}$. Since S preserves length and angle measure, the hypothesis of induction implies:

$$\begin{aligned} \overline{A_1A_{2k}} &\simeq \overline{A_1A_{2k+1}}, \\ \angle A_1A_{2k}A_{2k+1} &\simeq \angle A_1A_{2k+1}A_{2k}, \\ \angle A_1A_{2k}A_{2k-2} &\simeq \angle A_1A_{2k+1}A_{2k-1}. \end{aligned}$$

Therefore $\angle A_1A_{2k}A_{2k+2} \simeq \angle A_1A_{2k+1}A_{2k+3}$, and since

$$\overline{A_{2k}A_{2k+2}} \simeq \overline{A_{2k+1}A_{2k+3}},$$

we conclude that $\triangle A_1A_{2k}A_{2k+2} \simeq \triangle A_1A_{2k+1}A_{2k+3}$ by SAS. It follows that triangle $\triangle A_1A_{2k+2}A_{2k+3}$ is isosceles and that line l is perpendicular bisector of the line segment $\overline{A_{2k+2}A_{2k+3}}$. Then $S(A_{2k+2}) = A_{2k+3}$, and this finishes the proof. □

Corollary 3.1.11 *Every regular polygon P_n has n lines of symmetry.*

Proof This proof is Exercise 4. □

Rotation about C through an angle θ

Fix a point C of a plane $\mathcal{P}$. Revolve plane $\mathcal{P}$ about C through an angle of measure θ. Such a motion is called *rotation of* $\mathcal{P}$ *about* C *through the angle* θ. The point C is called *the center of the rotation* and the number θ *the angle of the rotation.* Notice that if $P \in \mathcal{P}$ and P' is the image of P by such a transformation, then $CP = CP'$ and $m(\angle PCP') = \theta$. But the plane can be revolved clockwise or counterclockwise. Therefore, in order to completely detemine a rotation, a direction should be given for the turn. For that we make the following definition.

Definition 3.1.12 *A* directed angle *is an angle where one of the rays is identified as the initial ray and the other ray is the terminal ray. An angle can be directed either* clockwise *or* counterclockwise. *An angle* $\angle BAC$ *is said to be* directed as $\angle EDF$ *if they are both directed clockwise or both counterclockwise.*

The orientation of a directed angle $\angle BAC$ is usually indicated by saying *the angle from* $\overrightarrow{AB}$ *to* $\overrightarrow{AC}$. A directed angle will be denoted by $\angle^{\mathcal{D}}BAC$ and will indicate from $\overrightarrow{AB}$ to $\overrightarrow{AC}$.

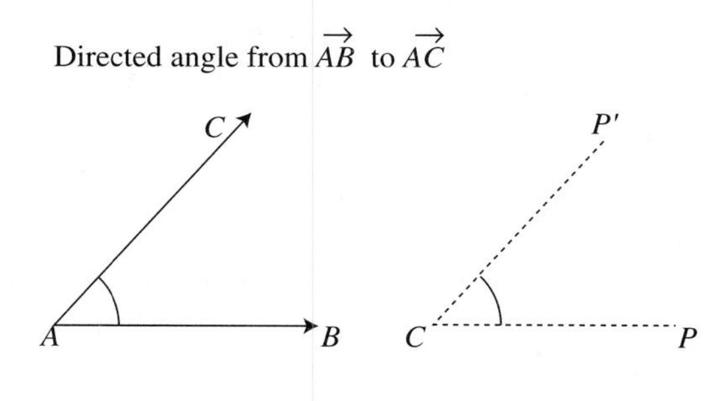

Figure 3.7

Let us now consider the composition of two rotations R_1 and R_2 through angles θ_1 and θ_2, respectively. Suppose that they are both counterclockwise and $\theta_1+\theta_2 > 180$. Recall that angle measure has been defined as a number in the interval $[0, 180°]$. But observe that the image of a point P by $R_2 \circ R_1$ coincides with the image of P by another rotation, namely, a clockwise rotation R through an angle $360 - (\theta_1 + \theta_2)$, and hence $R_2 \circ R_1 = R$. Therefore $R_2 \circ R_1$ is a rotation about C in the counterclockwise direction that has angle $\theta' > 180$ given by $\theta_1+\theta_2$. With this in mind, we say that the angle of rotation θ satisfies $0 \leq \theta \leq 360°$.

It is clear that if $\theta = 0$ or $\theta = 360$, then the rotation is the identity transformation and then it fixes all points of the plane. If $\theta \neq 0, 360$, then C is the only point fixed by such a transformation. Observe also that if R is a rotation about C through the directed angle $\angle^{\mathcal{D}} BAC$, then R is an invertible map and R^{-1} is the rotation about C through the directed angle $\angle^{\mathcal{D}} CAB$ (see Figure 3.7). It is also easy to see that if R_1 and R_2 have the same center, angles θ and φ, and R_1 is a counterclockwise rotation while R_2 is clockwise, then both $R_1 \circ R_2$ and $R_2 \circ R_1$ are rotations through the angle $|\theta - \varphi|$.

Proposition 3.1.13 *Let R be a rotation of a plane $\mathcal{P}$ centered at C through the angle θ. Then R is a rigid motion of $\mathcal{P}$.*

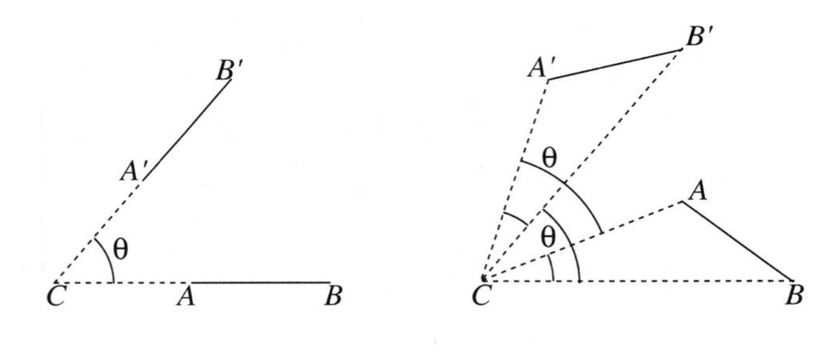

Figure 3.8

Proof Let A, B be two points of $\mathcal{P}$ with respective images A', B' (see Figure 3.8). Let us suppose first that A, B, and C are collinear points. Since $AC = A'C$ and $BC = B'C$ and the line through A and B is taken to the line A' and B', it follows from Proposition 1.4.3 that

$$AB = CB - CA = CB' - CA' = A'B'.$$

If A, B, and C are noncollinear, we have that $\angle ACB \simeq \angle A'CB'$, since $m(\angle ACA') = \theta = m(\angle BCB')$. Therefore $\triangle ACB \simeq \triangle A'CB'$ by SAS, which implies $AB = A'B'$. □

Proposition 3.1.14 *Let l_1 and l_2 be two intersecting lines at a point C. Let S_1 and S_2 be reflections through l_1 and l_2, respectively. Consider* *points $P \in l_1$ and $Q \in l_2$ such that $m(\angle PCQ) = \phi \leq 90°$. Then the transformation $R = S_2 \circ S_1$ is a rotation about C through the angle directed as $\angle^{\mathcal{D}} PCQ$ and of measure 2ϕ.*

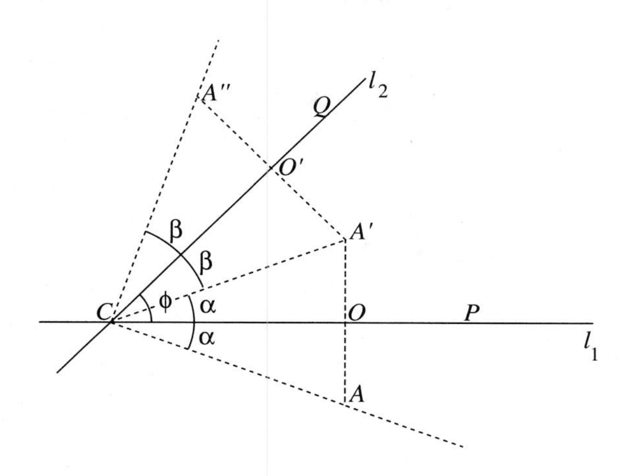

Figure 3.9

Proof Let A be a point of the plane such that its image $A' = S_1(A)$ is an interior point of angle $\angle PCQ$ (see Figure 3.9). Let $A'' = S_2(A')$, $O = \overline{AA'} \cap l_1$, and $O' = \overline{A'A''} \cap l_2$. Let

$$\alpha = m(\angle ACO) \quad \text{and} \quad \beta = m(\angle A'CO').$$

Then we have

$$m(\angle A'CO) = \alpha \quad \text{and} \quad m(\angle A''CO') = \beta.$$

Since $\overrightarrow{CA'}$ is between l_1 and l_2, from Axiom $\mathbf{A_3}$ we obtain that

$$\alpha + \beta = \phi \quad \text{and} \quad m(\angle ACA'') = 2\alpha + 2\beta = 2\phi.$$

Now, let B be another point of the plane such that A, B, C are non-collinear and the ray $\overrightarrow{CB'}$, for $B' = S_1(B)$, also lies between between l_1 and l_2. Similarly we obtain that $m(\angle BCB'') = 2\phi$, where $B'' = S_2(B')$. Let R be a rotation about C through the angle 2ϕ. Then $R(A) = A''$, $R(B) = B''$, and $R(C) = C$. Applying Proposition 3.1.5(b), we conclude that $R = S_2 \circ S_1$. □

Proposition 3.1.15 *(a) Let R and S be a rotation and a reflection of a plane $\mathcal{P}$, respectively, with a common fixed point C. Then $R \circ S$ and $S \circ R$ are reflections of $\mathcal{P}$.*
(b) Every rotation can be decomposed into the composition of two reflections.

Proof *(a)* If R is the identity transformation, then $R \circ S = I \circ S = S$ and $S \circ R = S \circ I = S$, and then both are reflections. If $R \neq I$, then C is the only fixed point of R and then it is the center of the rotation. The fixed points of reflection S are the points of a line that we denote by l. Let $m = R(l)$; that is, the line that is the image of l by rotation R. Then $l \cap m = C$. Let us consider $P \in l$, $P' = R(P) \in m$, and l' the line that bisects angle $\angle P'CP$. Let S' be the reflection through l'. By Proposition 3.1.14, $R = S' \circ S$. Therefore

$$R \circ S = S' \circ S \circ S = S',$$

and then $R \circ S$ is a reflection.

Now from Exercise 2(b) we get that $(S \circ R)^{-1} = R^{-1} \circ S^{-1} = R^{-1} \circ S$. Since R^{-1} is also a rotation centered at C, it follows from the first part of this proof that $R^{-1} \circ S$ is a reflection. Therefore $(S \circ R)^{-1}$ is a reflection and so is $S \circ R$, since the inverse of a reflection is itself, that is, $(S \circ R)^{-1} = S \circ R$.
(b) Let R be a rotation centered at C and l a line through the point C. Let S be a reflection through line l. Then from the proof of part (a) we conclude that there exists a reflection S' such that $S' = R \circ S$, which in turn implies that $R = S' \circ S$. □

Definition 3.1.16 *A geometric figure is said to have θ-rotational symmetry if it is invariant by a rotation through an angle θ.*

Proposition 3.1.17 *Let P_n be a regular n-gon. Then P_n has $(360°/n)$-rotational symmetry.*

Proof Let l_1, l_2 be lines of symmetry through vertices A_1, A_2, respectively (see Figure 3.10). From Theorem 3.1.10 we know that line l_i bisects $\angle A_i$ for $i = 1, 2$. We claim that l_1 and l_2 are intersecting lines. In fact, suppose that $l_1 \parallel l_2$. Let $P_1 \in l_1$ and $P_2 \in l_2$ such that P_1

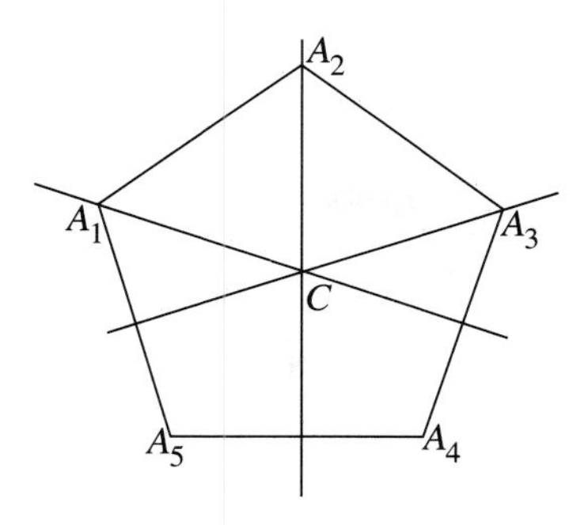

Figure 3.10

and P_2 are on opposite sides of $\overleftrightarrow{A_1A_2}$. Then from the converse of the alternate interior angle theorem we get that

$$\angle P_1A_1A_2 \simeq \angle P_2A_2A_1.$$

Since $\angle A_1 \simeq \angle A_2$ and l_i bisects $\angle A_i$, we also have that the supplement of $\angle P_1A_1A_2$ is congruent to $\angle P_2A_2A_1$. Therefore we would obtain that $\angle P_1A_1A_2$ is a right angle and hence $\angle A_1$ is a straight angle. This is clearly a contradiction, since A_1 is a vertex of a polygon.

Now, let $C = l_1 \cap l_2$, θ the measure of $\angle A_1CA_2$ and S_i a reflection through line l_i, $i = 1, 2$. Then $R = S_2 \circ S_1$ is rotation through the angle 2θ. Further, the polygon P_n is invariant by S_1 and S_2, and hence

$$R(P_n) = S_2(S_1(P_n)) = S_2(P_n) = P_n,$$

which implies that P_n is invariant by R. Since C is the center of R, we have that

$$A_iC = A_jC, \quad \forall\, i, j, = 1, \ldots, n.$$

Consider now three consecutives vertices, A_1, A_2, A_3. We have that $\triangle A_1CA_2 \simeq \triangle A_2CA_3$ by SSS and hence $\angle A_1CA_2 \simeq \angle A_2CA_3$. It follows that all angles whose vertex is C and whose sides pass through two consecutive vertices are congruent. Since there are n such angles, the measure of each is $360°/n$. Then a rotation R' centered at C through the angle $360°/n$ leaves P_n invariant. □

Corollary 3.1.18 *Every regular polygon is inscribed in a circle. The center of this circle is called the* center of the polygon.

Translations

A *translation* of a plane $\mathcal{P}$ is a motion that moves all points of $\mathcal{P}$ the same length in the same direction.

In order to provide a precise characterization of such a transformation, we recall that given a point P and a line l through P, the line separation property (see Chapter 1) implies that P divides l in two opposite rays. We shall call each of these rays a *direction*. Then we call a *directed line segment* or an *arrow*, denoted by $\overset{\mapsto}{PQ}$, a line segment $\overline{PQ}$ that is in the direction of $\overrightarrow{PQ}$.

Definition 3.1.19 *We say that two rays $\overrightarrow{A_1B_1}$ and $\overrightarrow{A_2B_2}$ have the* same direction *if* (see Figure 3.11):
(i) $\overleftrightarrow{A_1B_1}$ is parallel to $\overleftrightarrow{A_2B_2}$.
(ii) For some $i = 1, 2$ and for every point $E \in \overrightarrow{A_iB_i}$, the point E', given by $\overleftrightarrow{A_jB_j} \cap m$, lies on $\overrightarrow{A_jB_j}$, where m is the line perpendicular to $\overleftrightarrow{A_jB_j}$ dropped from E and $j \neq i$.

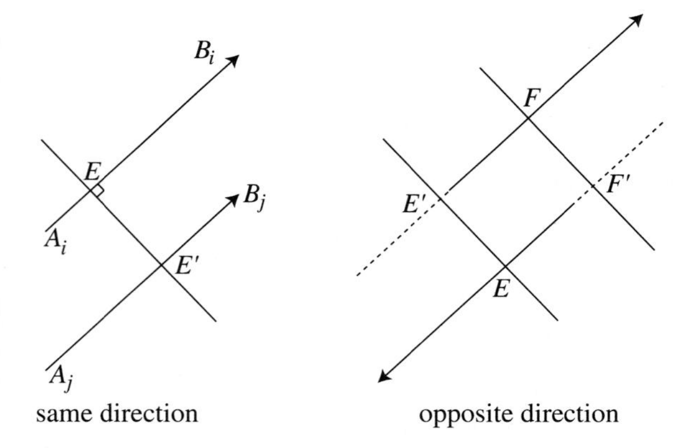

Figure 3.11

Let $P \in \mathcal{P}$ and $\overset{\mapsto}{AB}$ be a directed line segment. Consider a line m parallel to $\overleftrightarrow{AB}$ and through P. Then, the *translation* of P along $\overset{\mapsto}{AB}$, denoted by $L_{\overset{\mapsto}{AB}}$, is the point P' on m such that $PP' = AB$ and $\overrightarrow{AB}$

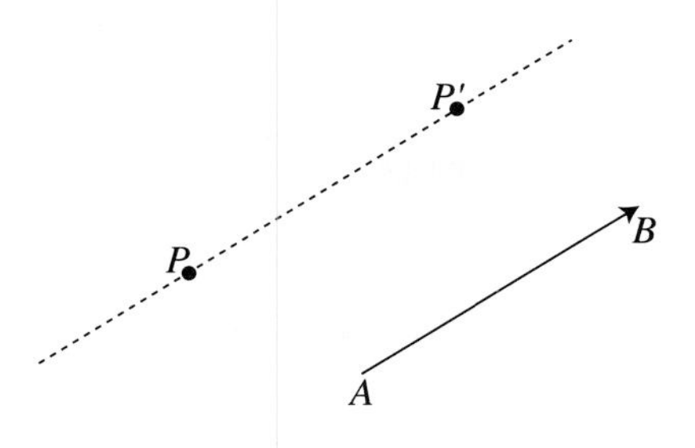

Figure 3.12

and $\overrightarrow{PP'}$ have the same direction (see Figure 3.12).

Observe that a directed line segment defines a translation. Conversely, given a translation L, the directed line segment $\overset{\mapsto}{PP'}$, where P' is the image of P, completely describes the translation L. Also, if two directed line segments $\overset{\mapsto}{AB}$ and $\overset{\mapsto}{CD}$ have same length and same direction, then

$$L_{\overset{\mapsto}{AB}} = L_{\overset{\mapsto}{CD}}.$$

Therefore there is *not* a one-to-one correspondence between translations and directed line segments. Such a correspondence exists between translations and *vectors*, a concept that will be introduced in the next chapter. It is obvious that a translation is a rigid motion of the plane with no fixed points. A translation $L_{\overset{\mapsto}{AB}}$ is an invertible map whose inverse is the translation $L_{\overset{\mapsto}{BA}}$.

In the following a directed line segment in the same direction of directed segment $\overset{\mapsto}{AB}$ and with length rAB will be denoted by $r\,\overset{\mapsto}{AB}$.

Proposition 3.1.20 *Let l_1 and l_2 be two parallel lines and m a line perpendicular to both, l_1 and l_2. Let $A_1 = l_1 \cap m$, $A_2 = l_2 \cap m$, and S_1, S_2 be reflections through l_1 and l_2, respectively. Then $S_2 \circ S_1$ is translation by the directed line segment $2\,\overset{\mapsto}{A_1A_2}$.*

Proof Consider a point P on line m such that $PA_1 < A_1A_2$ (see Figure 3.13). Then we have

$$\begin{aligned} P' &= S_1(P) \in m \quad \text{and} \quad PP' = 2PA_1, \\ P'' &= S_2(P') \in m \quad \text{and} \quad P''P' = 2P'A_2. \end{aligned}$$

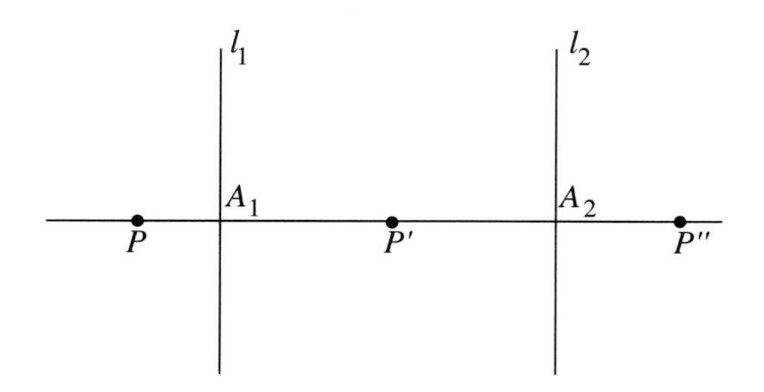

Figure 3.13

But $A_1P' + P'A_2 = A_1A_2$, and then

$$PP'' = PA_1 + A_1A_2 + A_2P'' = 2A_1A_2.$$

Let m_1 and m_2 be lines parallel to m and consider the points

$$\begin{aligned} B_1 &= m_1 \cap l_1 \quad \text{and} \quad B_2 = m_1 \cap l_2, \\ C_1 &= m_2 \cap l_1 \quad \text{and} \quad C_2 = m_2 \cap l_2. \end{aligned}$$

We conclude that $\overrightarrow{B_1B_2}$, $\overrightarrow{C_1C_2}$, and $\overrightarrow{A_1A_2}$ have the same direction. Choosing points P_1 and P_2 on m_1 and m_2, respectively, such that $P_1B_1 < B_1B_2$, $P_2B_2 < C_1C_2$, and P, P_1, P_2 are noncollinear, we obtain again that

$$P_1P_1'' = 2A_1A_2 \quad \text{and} \quad P_2P_2'' = 2A_1A_2.$$

Now let L be the translation that moves all points the length $2A_1A_2$ in the direction of $\overrightarrow{A_1A_2}$. Then $L(P) = P'', L(P_1) = P_1''$, and $L(P_2) = P_2''$. Therefore $L = S_2 \circ S_1$ by Proposition 3.1.5(b). □

Definition 3.1.21 *A rotation of* 180° *is called a* half-turn. *A half-turn centered at point C will be denoted by H_C.*

Proposition 3.1.22 *Let A and B be two points of $\mathcal{P}$. Then $H_B \circ H_A$ is translation by the directed line segment $2\overset{\mapsto}{AB}$* (see Figure 3.14).

Proof Let l denote the line through A and B. It is clear that $A'' = H_B(H_A(A)) = H_B(A)$ lies in l and $A''A = 2AB$. It is also easy to see

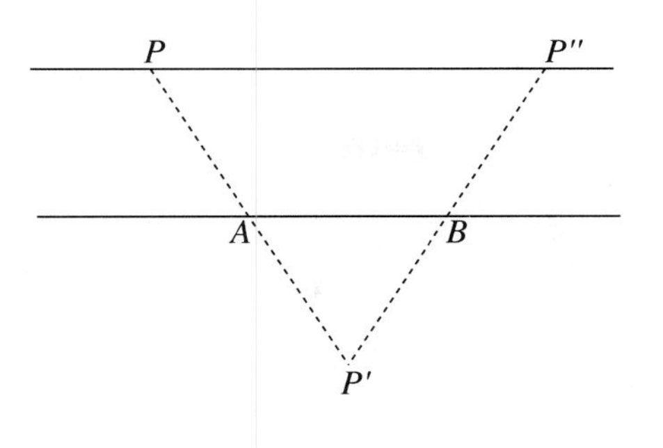

Figure 3.14

that $B'' = H_B(H_A(B))$ is on line l and $B''B = 2AB$. Now we consider a point P not on l. Let $P' = H_A(P)$ and $P'' = H_B(P')$. Notice that we have

$$\frac{PA}{PP'} = \frac{P''B}{P''P'} = 2$$

and $\triangle P'PP''$ and $\triangle P'AB$ have a common angle $\angle P'$. Then from Exercise 4 of Section 2.3 we obtain that $\triangle P'PP''$ is similar to $\triangle P'AB$. This implies that $\angle P''PA \simeq \angle BAP'$ and hence line $\overleftrightarrow{PP''}$ is parallel to line l. Moreover, $PP'' = 2AB$. Now let L be a translation in the direction of $\overrightarrow{AB}$ of a length $2AB$. Since L and $H_B \circ H_A$ coincide on three non-collinear points, we conclude that $L = H_B \circ H_A$, by Proposition 3.1.5.□

Glide reflections

Definition 3.1.23 *Let $\overset{\mapsto}{AB}$ be a directed line segment and l a line such that $l \parallel \overleftrightarrow{AB}$. Let S_l denote the reflection through l. The transformation given by*

$$G_l = L_{\overset{\mapsto}{AB}} \circ S_l$$

is called a glide reflection along line l (see Figure 3.15).

Proposition 3.1.24 *(a) A glide reflection G_l is a rigid motion.*
(b) $G_l = L_{\overset{\mapsto}{AB}} \circ S_l = S_l \circ L_{\overset{\mapsto}{AB}}$.
(c) The inverse of G_l is also a glide reflection along l.

Proof (a) G_l is the composition of two rigid motions and hence a rigid motion, by Exercise 2(a) of this section.

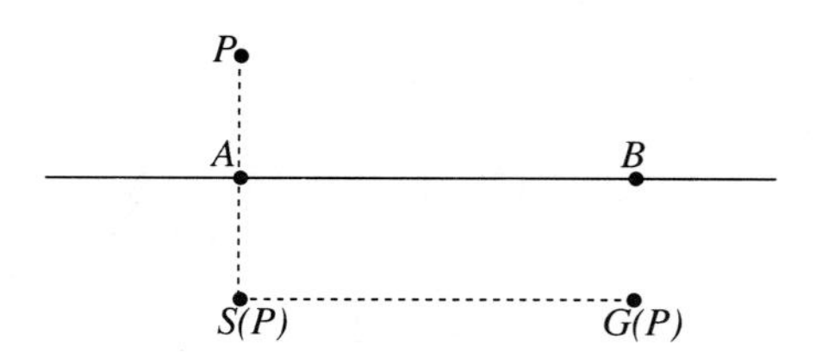

Figure 3.15

(b) Let P be an arbitrary point of $\mathcal{P}$ not lying on l (see Figure 3.16). Let

$$P' = S_l(P) \quad \text{and} \quad P'' = L(P).$$

The point $S_l(P'')$ lies on line $m \perp l$, and $P''S_l(P'') = PP'$. On the other hand, point $L(P')$ lies on $l' \parallel l$, and hence $l' \perp m$, and $P'L(P') = PP''$. Now let $Q = l' \cap m$. Then the quadrilateral $\square PP''QP'$ is a rectangle, and hence

$$PP' = P''Q \quad \text{and} \quad PP'' = P'Q.$$

Therefore

$$Q = S_l(P'') = S_l(L(P) \quad \text{and} \quad Q = L(P') = L(S_l(P).$$

Since this argument can be repeated for three noncollinear points, we conclude that

$$G_l = S_l \circ L = L \circ S_l.$$

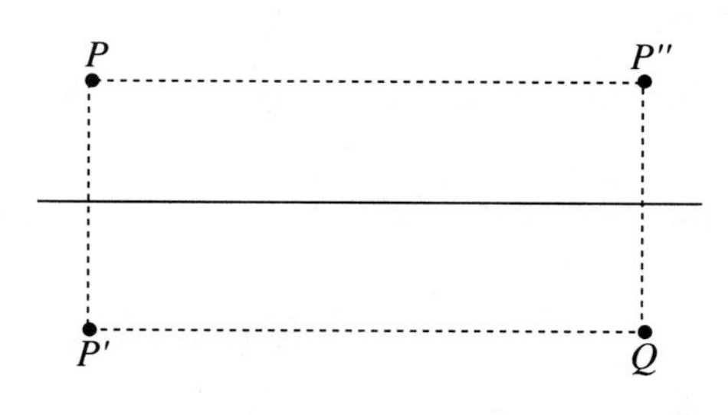

Figure 3.16

(c) Now we have

$$G_l^{-1} = L_{\overrightarrow{AB}}^{-1} \circ S_l^{-1} = L_{\overrightarrow{BA}} \circ S_l = S_l \circ L_{\overrightarrow{BA}}$$

and hence a glide reflection along l. □

Proposition 3.1.25 *Let l, m, and n be nonconcurrent lines such that two of them meet at some point. Then $T = S_l \circ S_m \circ S_n$ is a glide reflection.*

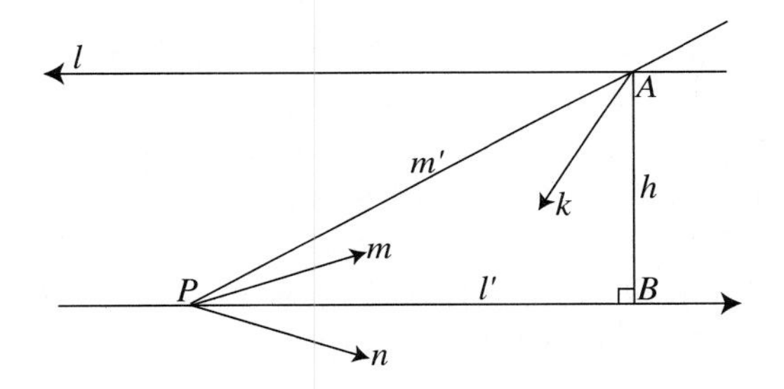

Figure 3.17

Proof First we suppose that m and n are not parallel. Let $P = m \cap n$ and let m' a line through P and A, where A is any arbitrary point of l. Then by Exercise 8 there exists a line l' through P such that

$$S_{l'} = S_{m'} \circ S_m \circ S_n.$$

Let h be the line through A that is perpendicular to l'; let h intersect l' at point B. Then, again by Exercise 8 we obtain that there exists line k through A such that

$$S_k = S_l \circ S_{m'} \circ S_h.$$

We claim that $B \notin k$; in fact, suppose that $B \in k$; since $B \in h$ and A lies on both lines k and h, we would conclude that $k = h$. Substituting above, we get

$$S_h = S_l \circ S_{m'} \circ S_h,$$

that is, $S_l \circ S_{m'}$ is the identity, which in turn implies that $S_l^{-1} = S_{m'}$. But $S_l^{-1} = S_l$, which gives $S_l = S_{m'}$, implying $l = m'$. This is a contradiction, since we are assuming that l does not pass through P. Now, since $B \notin h$, let t be the perpendicular to k through B. From Exercise 9 we get that $S_k \circ H_B$ is a glide reflection along t. We will show now that $T = S_l \circ S_m \circ S_n$ is the map $S_k \circ H_B$, completing this part of the proof. Observe first that $S_k \circ H_B = S_k \circ S_h \circ S_l'$ by Proposition 3.1.14, for the angle of a half-turn is $180°$ and $h \perp l'$. Now we obtain by substitution

$$\begin{aligned} S_k \circ H_B &= S_k \circ S_h \circ S_l' \\ &= S_l \circ S_{m'} \circ S_h \circ S_h \circ S_{m'} \circ S_m \circ S_n \\ &= S_l \circ S_{m'} \circ I \circ S_{m'} \circ S_m \circ S_n \\ &= S_l \circ S_m \circ S_n = T, \end{aligned}$$

where we used that $S_h \circ S_h = I$ and $S_{m'} \circ I \circ S_{m'} = I$, with I denoting the identity transformation.

To finish this proof we need to assume that m and n are parallel; then the hypothesis implies that l and m are not parallel. From the paragraph above we get that $S_n \circ S_m \circ S_l$ is a glide reflection. Since

$$T^{-1} = (S_l \circ S_m \circ S_n)^{-1} = S_n \circ S_m \circ S_l$$

we conclude that T^{-1} is a glide reflection and then so is T. □

Exercises

1. Let T be rigid motion of a plane $\mathcal{P}$. Show that T^{-1} is also a rigid motion of $\mathcal{P}$.
   Hint: Let A and B be two points of $\mathcal{P}$. Since T is onto, there exist points C and D such that $A = T(C)$ and $B = T(D)$. Further, $AB = CD$. Now show that

   $$T^{-1}(A)T^{-1}(B) = AB.$$

2. Let T and S be rigid motions of a plane $\mathcal{P}$.
   (a) Show that $T \circ S$ is a rigid motion of $\mathcal{P}$.
   (b) Show that $(T \circ S)^{-1} = S^{-1} \circ T^{-1}$.
   Hint: Show that $S^{-1} \circ T^{-1} \circ T \circ S = I$ and $T \circ S \circ S^{-1} \circ T^{-1} = I$. Then use Proposition 3.5.10 of the appendix.

3. Let T be rigid motion of a plane $\mathcal{P}$ and P a regular polygon in $\mathcal{P}$. Show that if T maps vertices of P to vertices of P, then P is invariant by T.
   Hint: Use Proposition 3.1.3(b) to conclude that sides of P are mapped to sides of P.

4. Prove Corollary 3.1.11.

5. (a) Show that if a parallelogram has a line of symmetry, then it is either a rectangle or a rhombus.
   (b) Show that if a rhombus has more than two lines of symmetry, then it is a square.

6. Let T be a rigid motion of a plane $\mathcal{P}$ which is not the identity.
   (a) Show that if T has only one fixed point, then T is a rotation centered at the fixed point.
   (b) Show that if T fixes two points A, B, then it is the reflection through line $\overleftrightarrow{AB}$.
   (c) Let S_l and S_m be reflections of the plane through lines l and m, respectively. Show that if $S_l = S_m$, then $l = m$.

7. Show that every rigid motion of a plane is the composition of at most three reflections.
   Hint: Consider two cases. For the case that the rigid motion T has a fixed point use parts (a) and (b) of Exercise 6. If T has no fixed points, consider an arbitrary point P and its image P'. Let m be the perpendicular bisector of $\overline{PP'}$ and S the reflection through m. Now $S \circ T$ fixes point P.

8. Let l, m, and n be three distinct lines passing through point P and let S_l, S_m, and S_n denote reflections through l, m, and n, respectively. Show that there exists line t through P such that $S_t = S_l \circ S_m \circ S_n$ is the reflection through line t.

9. Let l and m be two perpendicular lines and $A = l \cap m$. Show the following:
   (a) If S is the reflection through line l and H_A the half-turn centered at A, then $S \circ H_A$ is the reflection through line m.
   (b) If S is the reflection through line l, $B \in m$ and $B \neq A$, then $S \circ H_B$ is a glide reflection along m.

10. Show that a rigid motion of $\mathcal{P}$ is one of the four types: reflection, rotation, translation, or glide reflection.
Hint: Use Exercises 7 and 9.

11. Let A, B, and C be three distinct points of the plane. Show that the composition $H_C \circ H_B \circ H_A$ is a half-turn.

12. For each pair below, say whether the rigid motions must commute or not. Justify your answer with a proof or a counterexample.
(a) R_1 and R_2, where R_1 and R_2 are both rotations centered at the same point C.
(b) S_l and S_m, where l and m are two concurrent lines.
(c) S_l and S_m, where l and m are two parallel lines.
(d) S_l and R_C, where R_C is a rotation centered at point $C \in l$.
(e) L_l and L_m, where l and m are two parallel lines and L denotes a translation.
(f) L_l and L_m, where l and m are two concurrent lines.
(g) H_A and S_l, where $A \in l$.
(h) H_A and H_B.
(i) H_A and a translation L_l.
(j) A translation L_l and a rotation R_θ, where $\theta < 180°$.

13. Show that if $\triangle ABC \simeq \triangle DEF$, then there exists a rigid motion T such that $T(\triangle ABC) = \triangle DEF$.

The next set of exercises constitutes just a brief example of the beautiful interaction between geometry and algebra, showing how geometric transformations appear naturally in *group theory*.

14. Let G be a nonempty set. Recall that an associative operation $\circ : G \times G \to G$ defines a *group structure* on G if:
(i) $g_1 \circ g_2 \in G$, for all $g_1, g_2 \in G$.
(ii) There exists an element $e \in G$ such that
$$e \circ g = g \circ e = g, \quad \forall g \in G.$$
(iii) For each $g \in G$ there exists $g^{-1} \in G$ such that
$$g \circ g^{-1} = g^{-1} \circ g = e.$$

Show that the set of rigid motions of a plane $\mathcal{P}$ is a group whose operation is the composition of transformations. We denote this group by Iso $(\mathcal{P})$.

15. Let G be a group whose structure is given by the operation $\circ$ and $H \subset G$. If $\circ$ induces a group structure on H, then we say that H is a *subgroup* of G. Let

$$\mathcal{O}(\mathcal{P}, C) = \{T \in \text{Iso}\,(\mathcal{P}) \mid T(C) = C\}.$$

Show that $\mathcal{O}(\mathcal{P}, C)$ is a subgroup of Iso $(\mathcal{P})$. Conclude that the set of rotations of a plane $\mathcal{P}$ about a fixed point C form a subgroup of Iso $(\mathcal{P})$.

16. Let G be a group and Σ a subset of G. Then the *subgroup generated* by Σ is the smallest subgroup H of G that contains Σ. The generated subgroup H is denoted by $H = \langle \Sigma \rangle$.
(a) Let R be a rotation about a point C through an angle $\theta = 360°/n$, where $n \geq 2$ is a natural number. Let R^k denote the transformation $R \circ \overset{k \text{ times}}{\cdots} \circ R$. Define

$$\mathcal{C}_n = \{R^k \mid k \in \mathbf{N}\}.$$

Observe that $\mathcal{C}_n$ is a subgroup of $\mathcal{O}(\mathcal{P}, C)$ generated by R, that is, $\mathcal{C}_n = \langle R \rangle$. Show that $R^n = I$, where I is the identity transformation, and that $\mathcal{C}_n$ has n elements.
(b) A group G is said to be *cyclic* if there is some element $g \in G$ such that $G = \langle g \rangle$. In particular, if G is a cyclic group with n elements, then $g^n = e$, where e denotes the identity of G. Conclude that $\mathcal{C}_n$ defined in (a) is a finite cyclic group.

17. Let $R \in \mathcal{O}(\mathcal{P}, C)$ be a rotation through an angle $360°/n$, where $n \geq 2$ is a natural number. Let $S \in \mathcal{O}(\mathcal{P}, C)$ be a reflection. Show that the set

$$\langle R, S \rangle = \{I, R, ..., R^{n-1}, S, S \circ R, ..., S \circ R^{n-1}\}$$

is a group whose operation satisfies the following

$$R \circ S = S \circ R^{-1} = S \circ R^{n-1}.$$

Hint: Use and justify the following facts:
(1) $R \circ S$ is a reflection and then $(R \circ S)^{-1} = R \circ S$.
(2) $R^n = I$ and then $R^{n-1} = R^{-1}$.
(3) $(R \circ S)^{-1} = S^{-1} \circ R^{-1} = S \circ R^{-1}$, which implies

$$R \circ S = S \circ R^{-1} = S \circ R^{n-1}.$$

(4) The relations above imply that

$$\langle R, S \rangle = \{I, R, ..., R^{n-1}, S, S \circ R, ..., S \circ R^{n-1}\}.$$

The group $\langle R, S \rangle$ is called a *dihedral group* and it is denoted by D_{2n}.

18. Let l_1 and l_2 be two intersecting lines at a point P. Let S_1 and S_2 be reflections in l_1 and l_2, respectively. Show that the group generated by S_1 and S_2 is the dihedral group D_{2n}, where $n \geq 3$ and the angle formed by l_1 and l_2 is $180°/n$.
Hint: Let $R = S_2 \circ S_1$. Notice that $R^n = I$.

19. Let G be a finite subgroup of $\mathcal{O}(\mathcal{P}, C)$. Prove that G is either a cyclic group $\mathcal{C}$ or a dihedral group D_{2n}.
Here is the proof. Justify the steps.
(1) If $T \in G$, then T is either a rotation or a reflection.
(2) G contains a rotation (recall that a composition of reflections with a common fixed point is a rotation).
(3) The set of all rotations in G constitutes a subgroup $\mathcal{C}$ of G.
(4) For each $T \in \mathcal{C}$, let $\theta(T)$ denote the angle of the rotation T. Choose a rotation $R \neq I$ in $\mathcal{C}$ for which $\theta(R)$ is minimal.
(5) Choose an integer n such that

$$n\theta(R) \leq \theta(T) < (n+1)\theta(R), \quad \forall T \in G.$$

(6) For each $T \in \mathcal{C}$, consider now the rotation $(R^n)^{-1} \circ T$ (this transformation can be thought of as a counterclockwise rotation through an angle $\theta(T)$ followed by n clockwise rotations, each through an angle $\theta(R)$). Then

$$\theta((R^n)^{-1} \circ T) = \theta(T) - n\theta(R).$$

(7) It follows that $\theta\Big((R^n)^{-1} \circ T\Big) < \theta(R)$.
(8) Therefore $\theta\Big((R^n)^{-1} \circ T\Big) = 0$ and then $(R^n)^{-1} \circ T = I$.
(9) Hence $T = R^n$, and $\mathcal{C}$ is cyclic group generated by R.
(10) If $G = \mathcal{C}$, we have proved that G is a cyclic group. Suppose that $\mathcal{C} \neq G$. Then G contains a reflection S. Let us consider the set

$$S\mathcal{C} = \{S, S \circ R, ..., S \circ R^{n-1}\}.$$

Then $S\mathcal{C}$ contains only reflections.
(11) If $T \in G$ is a reflection, then $S \circ T$ is a rotation, implying that $S \circ T = R^k$.
(12) Therefore $T = S^{-1} \circ R^k = S \circ R^k$, implying that $T \in S\mathcal{C}$.
(13) We then conclude that

$$G = \{I, R, ..., R^{n-1}, S, S \circ R, ..., S \circ R^{n-1}\} = D_{2n}.$$

20. Show that the set of all rigid motions that leave a geometric figure $\mathcal{F}$ invariant is a group. This group is called the *symmetry group* of $\mathcal{F}$.

21. *Definition:* A group G of rigid motions of the plane is called a *reflection group* if it is generated by reflections, that is, if there exists a subset $\Sigma \subset G$ of reflections such that each element of G is a composition of elements of Σ.

    Give examples of finite groups of rigid motions of the plane that are not reflection groups.

22. Give an example of a finite group of rigid motions of the plane that does not have a fixed point, that is, there does not exist a point P that is fixed by all transformations of the group.

23. Let G be a finite reflection group of the plane. Show that G is the symmetry group $\mathcal{S}$ of a regular polygon.
    Here is the proof. Justify all steps from (2) through (6).
    (1) There exists a point C such that $T(C) = C$, for all $T \in G$, that is, G has a fixed point (a proof of this step is in Exercise 17 of Section 5.4).
    (2) From Exercise 19 we conclude that G is a dihedral group, since G contains reflections.

(3) If G has m elements, let $n = m/2$; m is an integer.
(4) Then G is generated by a rotation R of the plane through an angle $360°/n$ and a reflection S in a line l through C.
(5) Let P_n be a regular n-gon whose center is point C and such that l bisects one of its vertices. Then $G \subset \mathcal{S}$.
(6) But we also have that $\mathcal{S} \subset G$ (use Theorems 3.1.10 and 3.1.17 and Exercise 18). It follows that $G = \mathcal{S}$.

3.2 Similarities

Definition 3.2.1 *Let O be a point of plane $\mathcal{P}$ and r a nonzero real number. A* homothety $\mathcal{H}_O^r : \mathcal{P} \to \mathcal{P}$ *is a transformation that maps O to itself and*

(i) if $r > 0$, $P \neq O$ is mapped to the only point $P' \in \overrightarrow{OP}$ such that $OP' = r\,OP$.

(ii) if $r < 0$, $P \neq O$ is mapped to the only point P' on the ray opposite to $\overrightarrow{OP}$ such that $OP' = |r|OP$.

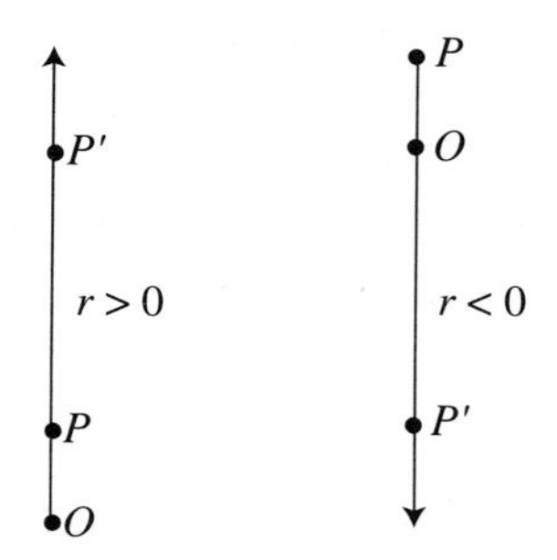

Figure 3.18

The point O is called *center of the homothety* and the number r the *ratio of the homothety.* Notice that if $r = 1$, then $\mathcal{H}_O^1$ is the identity transformation, while if $r = -1$, then $\mathcal{H}_O^1$ is half-turn centered at O. A homothety for which $r > 1$ is also called a *dilation*, and one for which $0 < r < 1$, a *contraction.*

Proposition 3.2.2 *Let $T : \mathcal{P} \to \mathcal{P}$ be a map with only one fixed point O and r a nonzero real number. Suppose that $\overset{\mapsto}{OP'} = |r| \overset{\mapsto}{OP}$, $\forall P \in \mathcal{P}$, where $P' = T(P)$. Then:*
(a) T is injective.
(b) $T(\overline{AB}) = \overline{A'B'}$, $\forall\, A, B \in \mathcal{P}$.
(c) $\triangle OAB$ is similar to $\triangle OA'B'$.
(d) T preserves angle measure.
(e) T is onto.
(f) T is a homothety centered at O and ratio r.

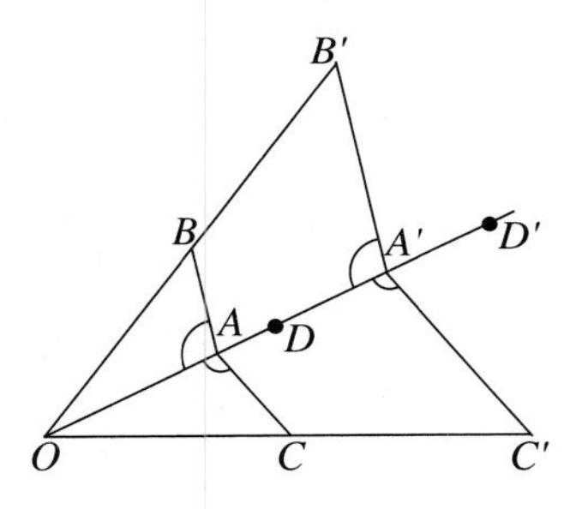

Figure 3.19

Proof Parts *(a)* and *(b)* of this proposition are proved similarly to *(a)* and *(b)* of Proposition 3.1.3.
(c) Observe that $\triangle OAB$ and $\triangle OA'B'$ have a common angle, namely, $\angle O$, and

$$\frac{OA'}{OA} = \frac{OB'}{OB} = |r|$$

and then the result follows from Exercise 4 of Section 2.3.
(d) Consider $\angle BAC$. If $m(\angle BAC) = 0$ or $180°$, the result follows from (b). We then suppose that $m(\angle BAC) \neq 0, 180°$ (see Figure 3.19). Since $\triangle OAB \sim \triangle OA'B'$, we have

$$\angle OAB \simeq \angle OA'B'.$$

We also have $\triangle OAC \sim \triangle OA'C'$ and hence

$$\angle OAC \simeq \angle OA'C'.$$

Consider $D \in \overrightarrow{OA}$ such that D is an interior point of $\angle BAC$. Then $D' \in \overrightarrow{OA}$ is an interior point of $\angle B'A'C'$. We have that $\angle BAD$ and $\angle B'A'D'$ are the supplements of $\angle OAB$ and $\angle OA'B'$, respectively, and $\angle CAD$ and $\angle C'A'D'$ are the supplements of OAC and $OA'C'$, respectively. Since

$$m(\angle BAC) = m(\angle BAD) + m(\angle DAC)$$

and

$$m(\angle B'A'C') = m(\angle B'A'D') + m(\angle D'A'C'),$$

we conclude the result.

The proofs of *(e)* and *(f)* are left as exercises. □

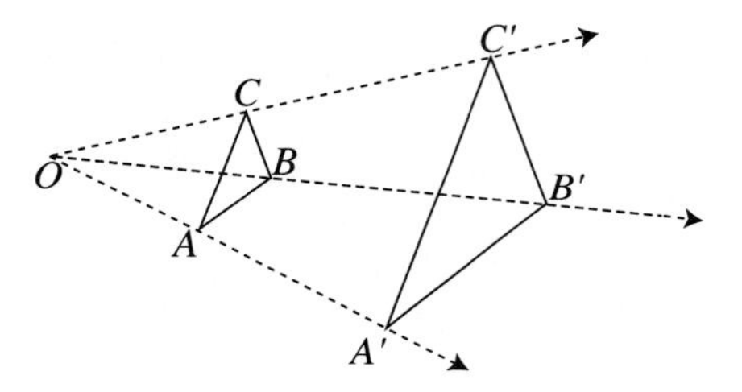

Figure 3.20

Corollary 3.2.3 *Let $\mathcal{H}_O^r : \mathcal{P} \to \mathcal{P}$ be the homothety of center O and ratio r. Then $\triangle ABC$ is similar to $\triangle A'B'C'$* (see Figure 3.20).

Definition 3.2.4 *A* similarity $\mathcal{S}_r$ *is a transformation obtained by composing a rigid motion with a homothety of ratio r.*

Proposition 3.2.5 *Let $\triangle ABC$ and $\triangle DEF$ be similar and noncongruent triangles. Then there exists a similarity $T : \mathcal{P} \to \mathcal{P}$ such that $T(\triangle ABC) = \triangle DEF$.*

Proof Without loss of generality we suppose that $DE < AB$. Let $B' \in \overline{AB}$ such that $AB' = DE$ and $C' \in \overline{AC}$ such that $AC' = DF$. Then $\triangle AB'C' \simeq \triangle DEF$ by SAS. From Exercise 13 we get that there exists a rigid motion T' such that $T'(\triangle AB'C') = \triangle DEF$. Since $\triangle ABC \sim \triangle DEF$, their sides are proportional. Let r denote the ratio

$$\frac{DE}{AB} = \frac{DF}{AC} = r.$$

The homothety $\mathcal{H}_A^r$ maps $\triangle ABC$ onto $\triangle AB'C'$. Then the composition

$$T = T' \circ \mathcal{H}_A^r$$

maps $\triangle ABC$ onto $\triangle DEF$. $\square$

Exercises

1. Prove (a), (b), (e), and (f) of Proposition 3.2.2.

2. (a) It follows from Theorem 3.2.2 that $\mathcal{H}_O^r$ is an invertible map. What is its inverse?
(b) Is a similarity an invertible map?

3. Show that a homothety maps lines onto lines.

4. Let $\mathcal{H}_O^r$ be a homothety of center O and ratio r. Show that if $\triangle A'B'C'$ is the image of $\triangle ABC$ by $\mathcal{H}_O^r$, then:
(a) the incenter of $\triangle ABC$ is mapped to the incenter of $\triangle A'B'C'$.
(b) the circumcenter of $\triangle ABC$ is mapped to the circumcenter of $\triangle A'B'C'$.
(c) the orthocenter of $\triangle ABC$ is mapped to the orthocenter of $\triangle A'B'C'$.
Hint: Consider first the case that the center of the homothety is one vertex of the triangle and use Execise 3. For the general case, consider a rigid motion that maps the center of the homothety to a vertex of the triangle.

5. Let γ be a circle centered at $C \neq O$ and radius s. Show that $\mathcal{H}_O^r$ maps γ to the circle γ' centered at $C' = \mathcal{H}_O^r(C)$ and radius $|r|s$. Moreover, a line tangent to γ at point Q is parallel to the line tangent to γ' at Q'.

6. Given a triangle $\triangle ABC$, let P, G, and O denote its *centroid*, *circumcenter*, and *orthocenter*, respectively (see Chapter 2 for their definitions). Show that P, G, O lie on the same line, called the *Euler line* of $\triangle ABC$.
Hint: Let $\triangle A'B'C'$ be the image of $\triangle ABC$ by $\mathcal{H}_P^{-1/2}$. Use Proposition 2.2.4 of Chapter 2 to conclude that A', B', and C' are the midpoints of $\overline{CB}$, $\overline{AC}$, and $\overline{AB}$, respectively. Then apply Exercise 4(c) to show that $\mathcal{H}_P^{-1/2}(O) = G$.

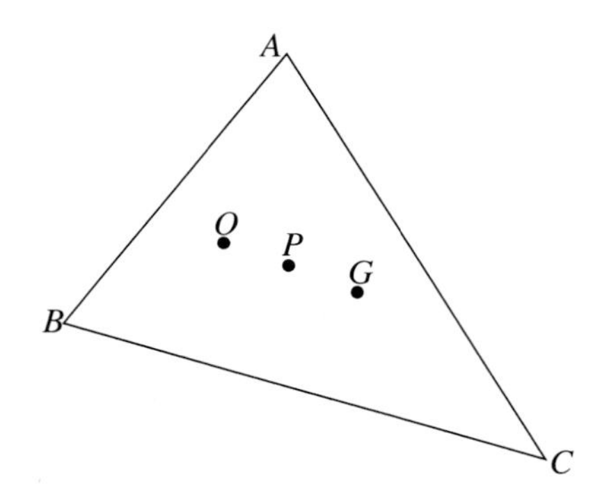

Figure 3.21

7. Given a triangle $\triangle ABC$, let O and G denote its orthocenter and circumcenter, respectively. Let D, E, F be the feet of the altitudes, R, P, Q the midpoints of the sides, and X, Y, Z the midpoints of the segments $\overline{OA}$, $\overline{OB}$, and $\overline{OC}$ (see Figure 3.22). Show that there exists a circle γ centered at the midpoint H of $\overline{OG}$ that contains all nine points $D, E, F, R, P, Q, X, Y, Z$. This circle is called the *nine-point circle of* $\triangle ABC$.
Hint: Let δ be the circle that circumscribes $\triangle ABC$. Consider the homothety $\mathcal{H}_O^2$. Then show the following:
(1) $\mathcal{H}_O^2$ maps H to G.
(2) $\mathcal{H}_O^2$ maps X to A, Y to B, and Z to C.
(3) $\mathcal{H}_O^2$ maps points D, E, and F onto circle δ.
(4) $\mathcal{H}_O^2$ maps points R, P, and Q onto circle δ; to prove this fact for point R, consider first point A' on δ diametrically opposite to A and show that $\square OCA'B$ is a parallelogram whose the diagonals intersect each other at point R. Use a similar procedure for points P and Q.
Now consider the homothety $\mathcal{H}_O^{1/2}$ and show the following:
(1) $\mathcal{H}_O^{1/2}$ maps G to H.
(2) Apply Exercise 5 to show that the circle $\mathcal{H}_O^{1/2}(\delta)$ is the nine-point circle.

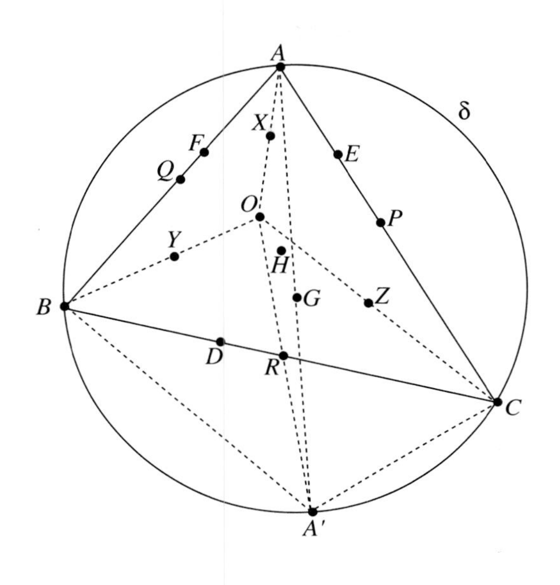

Figure 3.22

3.3 Inversions

In Section 3.1 we studied geometric transformations that preserve lengths of line segments. In Euclidean geometry, the distance between two points A and B is defined as the length of the segment connecting these two points. This is why rigid motions are called *Euclidean isometries.* In this section we shall study a significantly different type of geometric transformation of a plane, called *inversion.* These transformations have interesting properties such as transforming certain lines into circles, and vice versa. In addition, they have interactions with the transformations that preserve length in some models of *hyperbolic geometry.*

Let $\mathcal{P}$ be a Euclidean plane and $O \in \mathcal{P}$. The set $\mathcal{P} - \{O\}$ will be called a *punctured plane.*

Definition 3.3.1 *Let C be a circle with a center at O and radius r. The inversion of $\mathcal{P} - \{O\}$ in circle C is a transformation*

$$f : \mathcal{P} - \{O\} \longrightarrow \mathcal{P} - \{O\}$$

that maps point P to $P' = f(P)$ such that P' lies on the ray $\overrightarrow{OP}$ and

$OP \cdot OP' = r^2$.

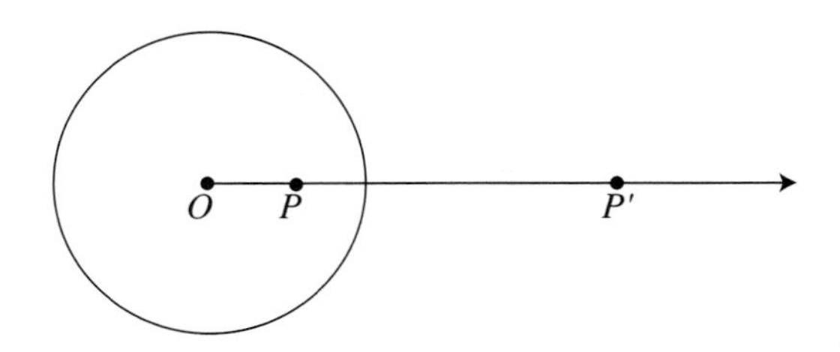

Figure 3.23

The point O is called the *center of inversion* and the number r the *radius of inversion*. It follows that the points lying on C are fixed points of the inversion.

Proposition 3.3.2 *Let $f : \mathcal{P} - \{O\} \to \mathcal{P} - \{O\}$ be an inversion in circle C centered at O. Then*
(a) $f(P) = P$ for all points $P \in C$.
(b) f is invertible and $f^{-1} = f$.

Proof *(a)* Let $P \in C$ and $P' = f(P)$. Since $OP\ OP' = r^2$ and $OP = r$, we get that $OP' = r$. But, Axiom $\mathbf{S}_2$ implies that P is the only point on ray $\overrightarrow{OP}$ such that $OP = r$ and hence $P = P'$.
(b) If $P' = f(P) = f(Q) = Q'$, then by the definition of inversion we have that $\overrightarrow{OP} = \overrightarrow{OQ}$. Moreover,

$$OP = \frac{r^2}{OP'} = \frac{r^2}{OQ'} = OQ$$

and then $P = Q$, by Axiom $\mathbf{S}_2$. This shows that f is one-to-one. Now let Q be any point on $\mathcal{P} - \{O\}$. There exists $P \in \overrightarrow{OQ}$, such that $OP \cdot OQ = r^2$. Therefore, $Q = f(P)$ and f is onto. Now to finish this proof, notice that $f \circ f = I$ and then $f^{-1} = f$. □

Proposition 3.3.3 *Let $f : \mathcal{P} - \{O\} \to \mathcal{P} - \{O\}$ be the inversion in circle C centered at O. Let P, Q be points in $\mathcal{P} - \{O\}$ such that O, P, and Q are noncollinear. Then $\triangle OPQ$ is similar to $\triangle OQ'P'$, where P' and Q' denote the images of P and Q, respectively.*

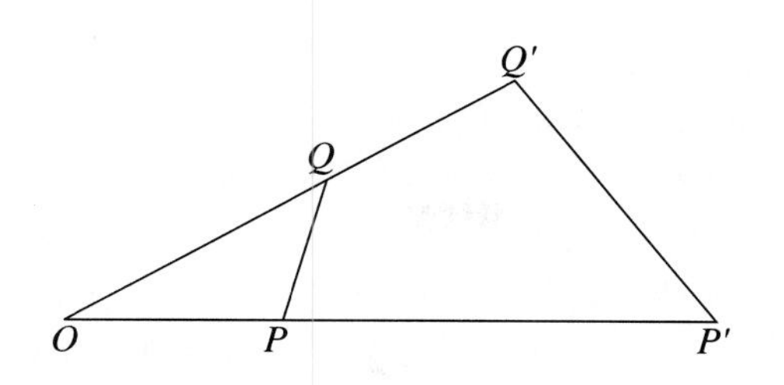

Figure 3.24

Proof Since $OP \cdot OP' = r^2$ and $OQ \cdot OQ' = r^2$, we obtain the relation

$$\frac{OP}{OQ'} = \frac{OQ}{OP'}.$$

Notice that $\triangle OPQ$ and $\triangle OQ'P'$ share angle $\angle O$. Then the result follows from Exercise 4 of Section 2.3. $\square$

Let $\mathcal{P} - \{O\}$ be a punctured plane. If l is a line through point O, $l - \{O\}$ will be called a *punctured line*. Similarly, if γ is a circle through O, $\gamma - \{O\}$ will be called a *punctured circle*.

Proposition 3.3.4 *Let $f : \mathcal{P} - \{O\} \to \mathcal{P} - \{O\}$ be the inversion in circle C centered at O. Then f takes punctured circles to lines and lines to punctured circles.*

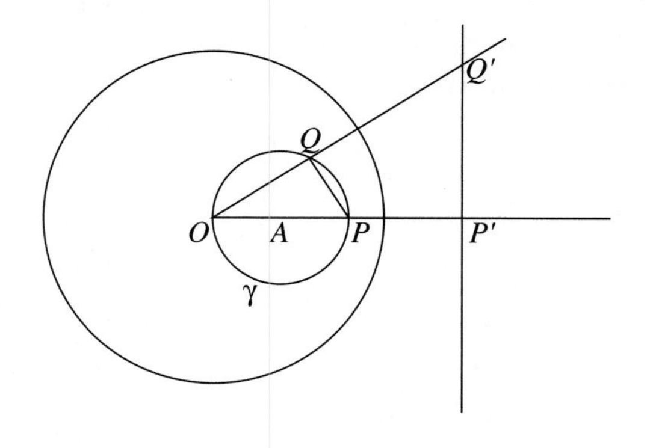

Figure 3.25

Proof Let γ be a circle centered at point A passing through O (see Figure 3.25). Let $P \in \gamma$ such that $\overline{OP}$ is a diameter of γ. By the definition of inversion P' lies on ray $\overrightarrow{OP}$. Let Q be any other point on γ. Notice that $\angle OQP$ is inscribed in a semicircle, and then it is a right angle. It follows from the proof of Proposition 3.3.3 that $\angle OP'Q'$ is also a right angle, and then Q' lies on line m that is perpendicular to line $\overleftrightarrow{OP}$ and passes though Q'. Since this fact is true for any arbitrary point of γ, then $Q' \in m$ for all $Q \in \gamma$, for there is only one line through P' that is perpendicular to $\overleftrightarrow{OP}$. Now to complete the proof we must show that any point X on line m is the image of a point on γ. Let $X' = f(X)$. Again from Proposition 3.3.3 we get that $\angle OX'P$ is a right angle. This implies that X' lies on the unique circle whose diameter is $\overline{OP}$, and then $X' \in \gamma$.

The proof that f carries lines to punctured circles is similar and is left as an exercise. □

Proposition 3.3.5 *Let $f : \mathcal{P} - \{O\} \to \mathcal{P} - \{O\}$ be the inversion in circle C centered at O. Then f carries circles to circles.*

Proof Let γ be a circle centered at point A. Then A' lies on ray $\overrightarrow{OA}$. Let P and Q be the points on γ obtained by the intersection of line $\overleftrightarrow{OA}$ with γ. Then $\overline{PQ}$ is a diameter of γ and P' and Q' also lie on $\overleftrightarrow{OA}$. Now let X be a point on γ not lying on $\overleftrightarrow{OA}$. By Proposition 3.3.3, $\triangle OPX \sim \triangle OX'P'$ and $\triangle OQX \sim \triangle OX'Q'$. From Exercise 2 of this section we obtain that $\angle PXQ \simeq \angle P'X'Q'$ and hence $\angle P'X'Q'$ is a right angle. Therefore X' lies on the circle whose diameter is $\overline{P'Q'}$. Since X is an arbitrary point of γ, we conclude that $f(\gamma)$ is the circle of diameter $\overline{P'Q'}$. □

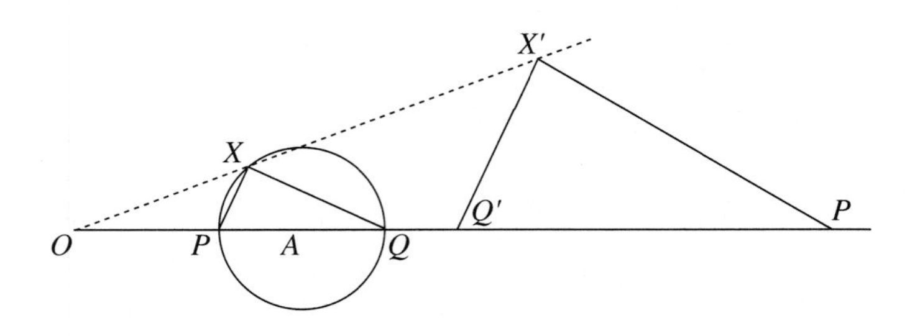

Figure 3.26

Definition 3.3.6 *We say that circle γ_1 is* orthogonal *to circle γ_2 if their tangent lines at the point of intersection are perpendicular to each other.*

Theorem 3.3.7 *Let $f : \mathcal{P} - \{O\} \to \mathcal{P} - \{O\}$ be the inversion in circle C centered at O and γ be any other circle. Then γ is invariant — that is,*

$$f(\gamma) = \{f(P) \mid P \in \gamma)\} = \gamma$$

if and only if γ is orthogonal to C.

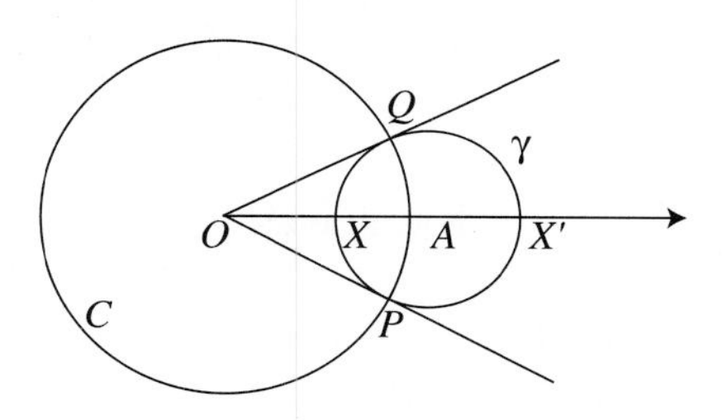

Figure 3.27

Proof (⇐) Let P and Q denote the points of intersection of C and γ. Then $f(P) = P$ and $f(Q) = Q$. Let X be another point on γ. We want to show that $X' = f(X) \in \gamma$ (see Figure 3.27) and then, since three noncollinear points uniquely determine a circle, we conclude that $f(\gamma) = \gamma$. For that, consider ray $\overrightarrow{OX}$ and let X'' be the intersection point of $\overrightarrow{OX}$ with circle γ. Since γ is orthogonal to C, we have that $\overleftrightarrow{OQ}$ is tangent to γ. Applying Proposition 2.4.8, we obtain

$$OX \cdot OX'' = OQ^2 = r^2.$$

Therefore $X'' = X'$.
(⇒) Let γ denote a circle, other than C, that is invariant by f. Since points inside of C are mapped to points outside of C, if γ is invariant by f, then γ has points inside and points outside C. Then the circular continuity principle implies that γ intersects C at two points, which we denote by P and Q.

We claim that the center of the inversion O is exterior to γ. If not, we consider a point X on $\overleftrightarrow{OA}$, where A is the center of γ, such that

O is between X and A. Since X' lies on ray $\overrightarrow{OX}$, X' is not on γ, contradicting that γ is invariant by f.

Now, since O is exterior to γ, there exists a tangent line l from O to γ. Let us suppose that l intersects γ at point R and let us consider $Y \in \gamma$. Since $Y' \in \gamma$, and $Y' \in \overrightarrow{OY}$, we have again by Proposition 2.4.8 that $OY\ OY' = OR^2$. But from the definition of inversion we get that $OY\ OY' = r^2 = OP^2 = OQ^2$ and therefore either $R = P$ or $R = Q$. In either case the radii are orthogonal. □

Corollary 3.3.8 *Let $f : \mathcal{P} - \{O\} \to \mathcal{P} - \{O\}$ be an inversion in circle C centered at O. Let P' be the image of point P. Then, any circle containing both P and P' is orthogonal to C.*

Exercises

1. Show that the composition of inversions in concentric circles with center O and of radii r_1 and r_2 is the homothety $\mathcal{H}_O^c$, where $c = (r_2/r_1)^2$.

2. Let f be an inversion of center O. Consider $\angle BAC$ such that B is between O and C. Show that $\angle BAC \simeq \angle B'A'C'$, where A', B', and C' are the images of A, B, and C by the inversion f.
   Hint: Use Proposition 3.3.3.

3. Let γ be a circle centered at the point O and radius r. Let C be a circle meeting γ orthogonally. Let f be the inversion in C and denote $P' = f(P)$. Show that if $OP < r$, then $OP' < r$.
   Hint: Use Theorem 3.3.7.

4. Prove Corollary 3.3.8.

5. Let C be a circle with center O and radius r. Let P and Q be two points inside the circle such that O, P, and Q are noncollinear. Show that there exists only one circle γ containing P and Q that meets C orthogonally.
   Hint: Consider P' the image of P under the inversion in γ; show that there is a circle C containing P, P', Q that is invariant by the inversion. Then use Corollary 3.3.8.

3.4 Coordinate Systems

The study of rigid motions can often be simplified by introducing a rectangular coordinate system. In studying more properties of inversions, it is convenient to use both rectangular and polar coordinates.

We start this section describing how to construct a rectangular coordinate system. Select a line l in the plane. Choose a point $O \in l$ and call it the *origin*. Let m be the unique line through O that is perpendicular to l. These lines will be called the *x-axis* and the *y-axis*.

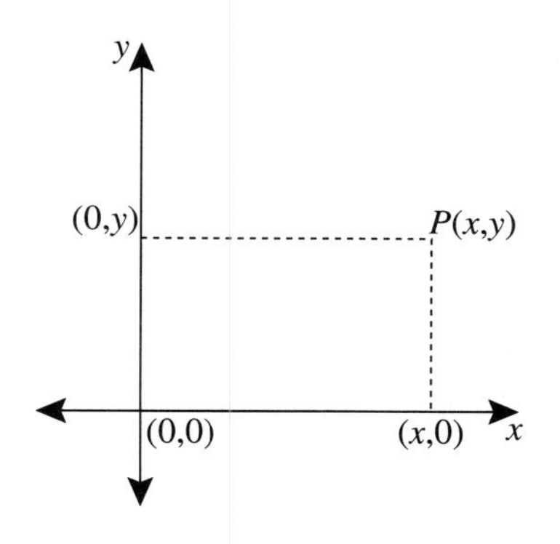

Figure 3.28

Recall that to each point of a line corresponds a real number, called the *coordinate* of the point by Axiom $\mathbf{S_2}$. We then choose the same unit of length for both lines, and from remark (iii) of Axiom $\mathbf{S_2}$ in Chapter 1 we conclude that the correspondence with the real numbers can be made so that both lines have the number 0 at point O. Now, it is clear that every point on lines x and y can be represented by an ordered pair of real numbers, namely, $(x, 0)$ for points on the x-axis and $(0, y)$ for points on the y-axis. In particular, the point O is represented by the pair $(0, 0)$. Now, let P be any arbitrary point of $\mathcal{P}$. There is a unique line l' through P that is perpendicular to x-axis, meeting the axis at a point $(x, 0)$. The alternate interior angle theorem implies that l' is parallel to y-axis. Similarly, there exists a unique line through P that is perpendicular to the y-axis, and hence parallel to the x-axis, and meeting the y-axis at a point $(0, y)$. These perpendicular lines create a rectangle that assigns in a unique way a pair of real numbers (x, y) to the point P. The numbers x and y are called *rectangular coordinates* of the point P, and the coordinatization of the plane by rectangular coordinates is called a *rectangular coordinate system.*

A rectangular coordinate system establishes a *one-to-one* correspondence between a plane and the set

$$\mathbf{R}^2 = \{(x, y) \mid x, y \in \mathbf{R}\}.$$

Using rectangular coordinates, Axiom $\mathbf{S_2}$, and the parallel postulate we can now determine a formula for the *Euclidean* length of a line segment in the plane.

Euclidean length

We first consider the case that both endpoints P and Q of the line segment lie on lines parallel to the x-axis or y-axis. Let us suppose that $P = (a, b)$ and $Q = (a, c)$. Then points $P' = (0, b), Q' = (0, c), (a, c)$, and (a, b) form a rectangle. It follows from the parallel postulate that $\overline{P'Q'} \simeq \overline{PQ}$, and then Axiom $\mathbf{S_2}$ implies that $PQ = |b - c|$. Similarly, if we have $P = (a, b)$ and $Q = (c, b)$, $PQ = |a - c|$. Now, if the line

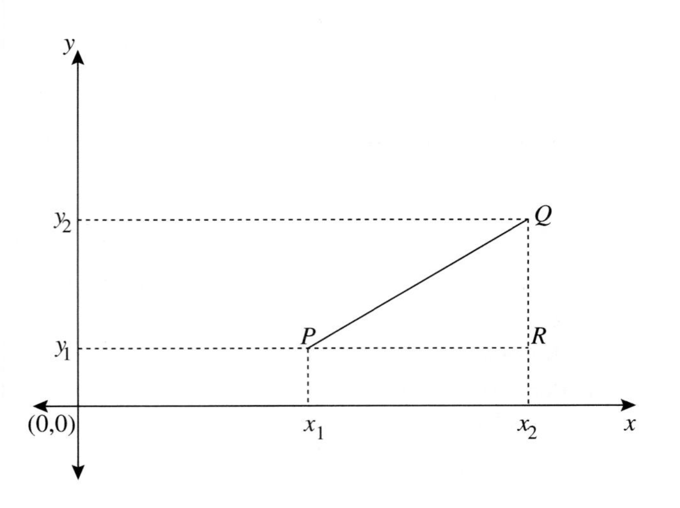

Figure 3.29

through P and Q is not parallel to either of the axes, let

$$P = (x_1, y_1), \quad Q = (x_2, y_2), \quad \text{and} \quad R = (x_2, y_1).$$

Then $\triangle RPQ$ is a right triangle and hence the Pythagorean theorem implies

$$PQ = \sqrt{(x_1 - x_2)^2 + (y_1 - y_2)^2}$$

Observe that the formula above is the general formula, for it also applies to the particular case when the endpoints lie on the coordinate axes.

A rectangular coordinate system also gives an analytical approach to study lines, circles, and rigid motions of a plane.

Analytic equations of lines and circles

It follows from the construction of the rectangular system that the equation $x = a$ describes a line parallel to y-axis and intersecting the x-axis at point $(a, 0)$. Similarly, the equation of a line parallel to x-axis and intersecting the y-axis at point $(0, b)$ is $y = b$.

Definition 3.4.1 *Let α be an angle. The* tangent of angle α, *denoted by* $\tan\alpha$, *is defined by*

$$\tan\alpha = \frac{\sin\alpha}{\cos\alpha}.$$

Let us consider now a line l, which is not parallel to the x-axis. Let α be the smallest angle measured counterclockwise from the x-axis to l. Notice that if $P = (x_1, y_1)$ and $Q = (x_2, y_2)$ are two arbitrary points on l (see Figure 3.30),then

$$\frac{y_2 - y_1}{x_2 - x_1} = \tan\alpha.$$

Let $m = \tan\alpha$. Therefore if $A = (x, y)$ is any other point on l, we have

$$\frac{y - y_1}{x - x_1} = \frac{y_2 - y_1}{x_2 - x_1} = m$$

which implies

$$y - y_1 = m(x - x_1) \qquad \text{or} \qquad y = mx + b$$

where $b = y_1 - mx_1$.

Now, let us consider a circle γ centered at $C = (a, b)$. Let $A = (x_1, y_1) \in \gamma$ and r denote the length OA. It follows from the above that

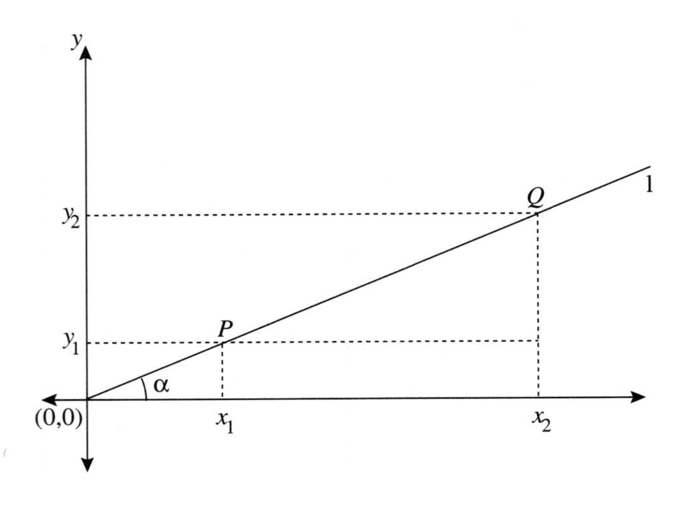

Figure 3.30

$r^2 = (x_1 - a)^2 + (y_1 - b)^2$. Given any point $P = (x, y)$ on γ, we have that $OP = OA$, and therefore the equation

$$r^2 = (x - a)^2 + (y - b)^2.$$

describes all points $P = (x, y)$ on γ.

Rigid motions in rectangular coordinates

We now show how to analytically describe the rigid motions defined in the beginning of this chapter.

Example 1

Let $T : \mathbf{R}^2 \to \mathbf{R}^2$ be the reflection through x-axis (see Figure 3.31). It is easy to verify that if $P = (x, y)$ then

$$T(P) = T((x, y)) = (x, -y).$$

Example 2

Let $T : \mathbf{R}^2 \to \mathbf{R}^2$ be a counterclockwise rotation about the origin $O = (0, 0)$ through an angle θ (see Figure 3.32). If $P = (x, y)$ and r

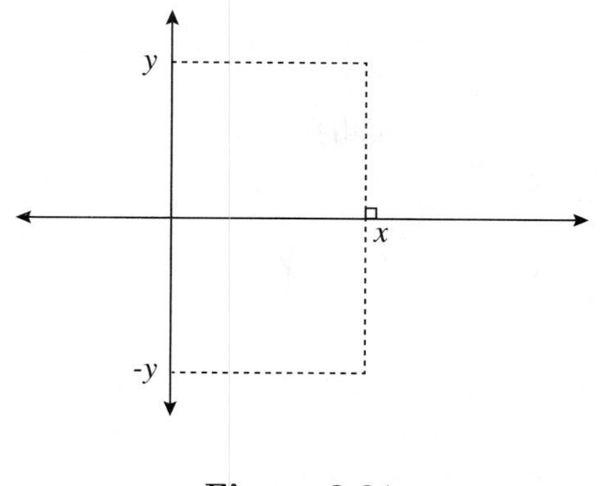

Figure 3.31

denotes the length of OP, then

$$x = r\cos\phi, \qquad y = r\sin\phi$$

where ϕ is the angle between $\overrightarrow{OP}$ and the ray given by the positive side of the x-axis. Since $\overline{OT(P)}$ has the same length as $\overline{OP}$, denoting $T(P) = (x', y')$, we have

$$x' = r\cos(\phi + \theta) \qquad y' = r\sin(\phi + \theta).$$

Using matrix notation we write

$$T(P) = \begin{pmatrix} x' \\ y' \end{pmatrix} = \begin{pmatrix} r\cos(\phi + \theta) \\ r\sin(\phi + \theta) \end{pmatrix} = \begin{pmatrix} r\cos\theta\,\cos\phi - r\sin\theta\,\sin\phi \\ r\sin\theta\,\cos\phi + r\cos\theta\,\sin\phi \end{pmatrix}$$

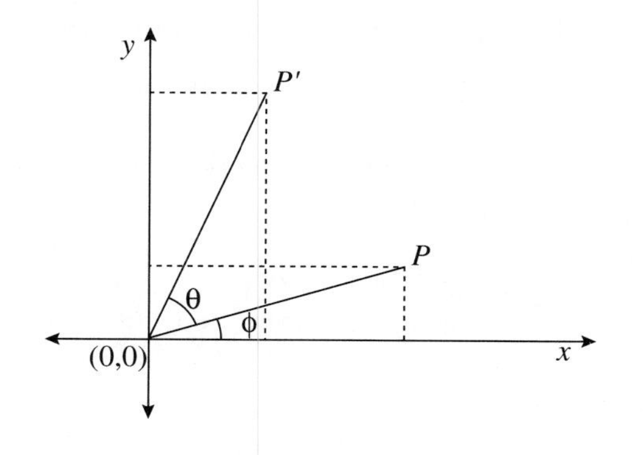

Figure 3.32

$$= \begin{pmatrix} x\cos\theta & - & y\sin\theta \\ x\sin\theta & + & y\cos\theta \end{pmatrix} = \begin{pmatrix} \cos\theta & -\sin\theta \\ \sin\theta & \cos\theta \end{pmatrix} \begin{pmatrix} x \\ y \end{pmatrix}.$$

The matrix $\begin{pmatrix} \cos\theta & -\sin\theta \\ \sin\theta & \cos\theta \end{pmatrix}$ is called the *rotation matrix of* $\mathbf{R}^2$ *through the angle* θ.

Example 3

Let $T : \mathbf{R}^2 \to \mathbf{R}^2$ be a clockwise rotation about the origin $O = (0, 0)$ through an angle θ. If R denotes a counterclockwise rotation through the same angle, then $T \circ R = R \circ T = I$, where I denotes the identity transformation. It follows that the matrix of a clockwise rotation is the inverse of the matrix of R and hence

$$\begin{pmatrix} \cos\theta & \sin\theta \\ -\sin\theta & \cos\theta \end{pmatrix}.$$

Therefore

$$T((x, y)) = (x\cos\theta \;+\; y\sin\theta,\; -x\sin\theta \;+\; y\cos\theta).$$

Example 4

In this example we want to find a formula for the transformation $T : \mathbf{R}^2 \to \mathbf{R}^2$ that reflects the plane through the line $y = mx,\ m > 0$.

For that, let $S : \mathbf{R}^2 \to \mathbf{R}^2$ be a reflection through the y-axis and $R : \mathbf{R}^2 \to \mathbf{R}^2$ a counterclockwise rotation about the origin $O = (0, 0)$ through an angle θ, where θ is the acute angle formed by the line and the y-axis. Observe that T is the following composition of rigid motions:

$$T = R^{-1} \circ S \circ R.$$

Now, let ϕ denote the acute angle formed by the line and the x-axis. Then the matrix of R is given by

$$\begin{pmatrix} \cos(90^\circ - \phi) & -\sin(90^\circ - \phi) \\ \sin(90^\circ - \phi) & \cos(90^\circ - \phi) \end{pmatrix} = \begin{pmatrix} \sin\phi & -\cos\phi \\ \cos\phi & \sin\phi \end{pmatrix}.$$

Then, applying R to (x, y), we obtain

$$\begin{pmatrix} \sin\phi & -\cos\phi \\ \cos\phi & \sin\phi \end{pmatrix} \begin{pmatrix} x \\ y \end{pmatrix} = \begin{pmatrix} x\sin\phi - y\cos\phi \\ x\cos\phi + y\sin\phi \end{pmatrix}.$$

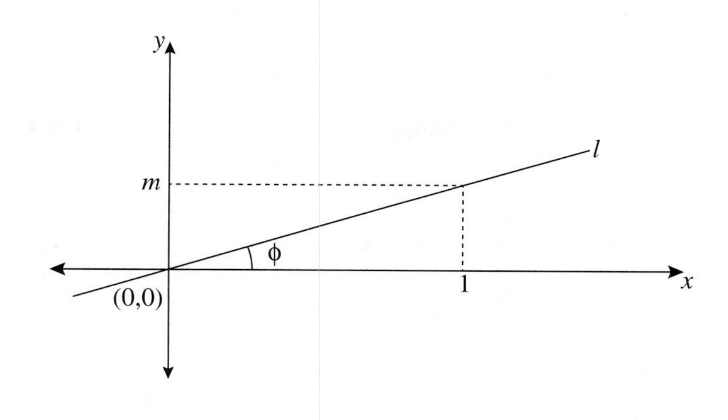

Figure 3.33

Now, reflecting through the y-axis, we get

$$(-x\sin\phi + y\cos\phi,\ x\cos\phi + y\sin\phi).$$

Since the matrix of R^{-1} is given by

$$\begin{pmatrix} \sin\phi & \cos\phi \\ -\cos\phi & \sin\phi \end{pmatrix},$$

we have now

$$\begin{pmatrix} \sin\phi & \cos\phi \\ -\cos\phi & \sin\phi \end{pmatrix} \begin{pmatrix} -x\sin\phi + y\cos\phi \\ x\cos\phi + y\sin\phi \end{pmatrix} =$$

$$= \begin{pmatrix} -x\sin^2\phi + x\cos^2\phi + 2y\sin\phi\cos\phi \\ y\sin^2\phi - y\cos^2\phi + 2x\sin\phi\cos\phi \end{pmatrix}.$$

But $\sin\phi$ and $\cos\phi$ are given by

$$\sin\phi = \frac{m}{\sqrt{1+m^2}} \quad \text{and} \quad \cos\phi = \frac{1}{\sqrt{1+m^2}}.$$

Substituting above, we obtain

$$T((x,y)) = \frac{1}{1+m^2}((1-m^2)x + 2my,\ (m^2-1)y + 2mx).$$

Example 5

Let $A = (a, b)$. Find a formula for the transformation $T : \mathbf{R}^2 \to \mathbf{R}^2$ that translates the plane along $\overset{\longmapsto}{OA}$.

Let $P = (x, y)$ be an arbitrary point of $\mathbf{R}^2$ and $P' = T(P) = (x', y')$. Since $\overleftrightarrow{PP'}$ and $\overleftrightarrow{OA}$ are parallel lines, we have

$$\frac{y' - y}{x' - x} = \frac{b}{a}.$$

Writing the expression for $(PP')^2$, we get

$$\begin{aligned}(PP')^2 &= (y' - y)^2 + (x' - x)^2 \\ &= \frac{b^2}{a^2}(x' - x)^2 + (x' - x)^2 \\ &= \frac{(x' - x)^2}{a^2}(b^2 + a^2).\end{aligned}$$

Since $PP' = OA$ we have

$$(PP')^2 = OA^2 = a^2 + b^2$$

which implies

$$\frac{(x' - x)^2}{a^2} = 1,$$

and hence $x' - x = a$. It follows that $y' - y = b$. Therefore

$$T((x, y)) = (x + a,\ y + b).$$

Example 6

Let $C = (a, b)$. Find a formula for the transformation $T : \mathbf{R}^2 \to \mathbf{R}^2$ that counterclockwise rotates the plane about C through an angle θ.

Let $L : \mathbf{R}^2 \to \mathbf{R}^2$ be the translation that maps C to the origin and R a rotation about the origin through the angle θ. Then $T = L^{-1} \circ R \circ L$. Combining Examples 2 and 5, we obtain

$$T(x, y) = \begin{pmatrix} (x - a)\cos\theta & - & (y - b)\sin\theta \\ (x - a)\sin\theta & + & (y - b)\cos\theta \end{pmatrix} + \begin{pmatrix} a \\ b \end{pmatrix}$$

Inversions in polar coordinates

Let $P \in \mathbf{R}^2$ with rectangular coordinates $P = (x, y)$ and let ϕ denote the measure of the angle between $\overrightarrow{OP}$ and the ray given by the positive side of the x-axis. In the following, by *angle* θ we shall mean:
(i) $\theta = \phi$ if $y \geq 0$.

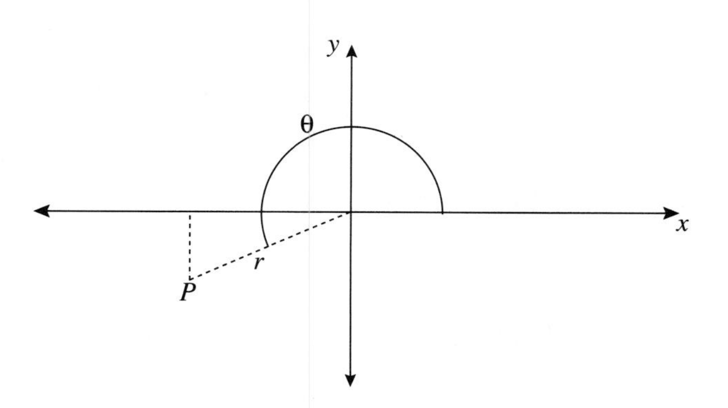

Figure 3.34

(ii) $\theta = 360° \;-\; \phi$ if $y < 0$.
Then $0 \leq \theta < 360°$. Further, if $\theta > 180°$, we define

$$\sin\theta = -\sin\phi \qquad \text{and} \qquad \cos\theta = \cos\phi$$

so that the identities of Theorem 2.5.2 of Chapter 2 still hold.

Definition 3.4.2 *Let $P \in \mathbf{R}^2$. We say that P has polar coordinates (r, θ) if*

$$x = r\cos\theta$$

$$y = r\sin\theta$$

where x and y are its rectangular coordinates and $0 \leq \theta < 360°$.

Notice that there is *not* a one-to-one correspondence between a plane and the set $\{(r, \theta) \mid r \geq 0, \;\; 0 \leq \theta < 360°\}$, for any pair of the type $(0, \theta)$ describes the polar coordinates of the origin.

Polar equations for lines and circles

We now collect a few simple formulae for lines and circles in polar coordinates.
(a) Line passing through the origin: $\theta = \theta_0$.
(b) Vertical line: $r\cos\theta = x$
(c) Horizontal line: $r\sin\theta = y$
(d) Circle centered at the origin and radius a: $r = a$.
(e) Circle centered at $C = (a, b)$ passing through the origin (see Figure 3.35). Let $P = (r, \theta)$ and (r_0, θ_0), be the polar coordinates of the center

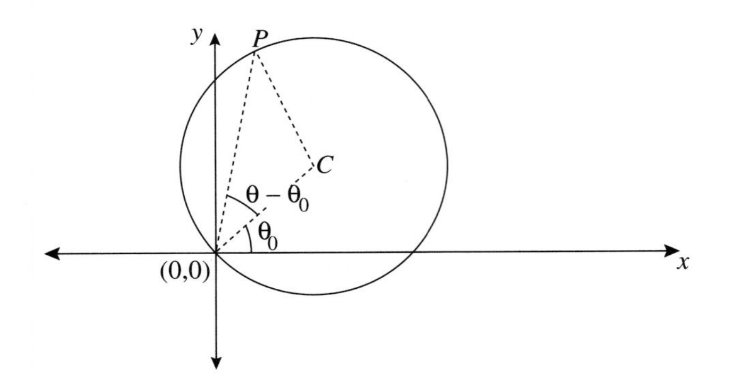

Figure 3.35

C. Since the circle passes through $(0, 0)$ and P, then by the law of cosines we have $r_0^2 = r^2 + r_0^2 - 2rr_0 \cos(\theta - \theta_0)$, which simplifies to

$$r = 2r_0 \cos(\theta - \theta_0).$$

In particular, if C lies on the positive side of the x-axis, we have that $\theta_0 = 0$, yielding

$$r = 2r_0 \cos\theta.$$

Therefore the equation above describes the circle centered at $(r_0, 0)$ and radius r_0.

The advantage of polar coordinates is that they allow us to describe the inversion in the simple form: Let f denote an inversion in circle C, centered at point O and radius ρ. We place a polar coordinate system such that the origin $(0, 0)$ is the center of the inversion O. Then we have

$$f : \mathbf{R}^2 - \{(0,0)\} \longrightarrow \mathbf{R}^2 - \{(0,0)\},$$
$$(r, \theta) \mapsto (r', \theta)$$

where $rr' = \rho^2$.

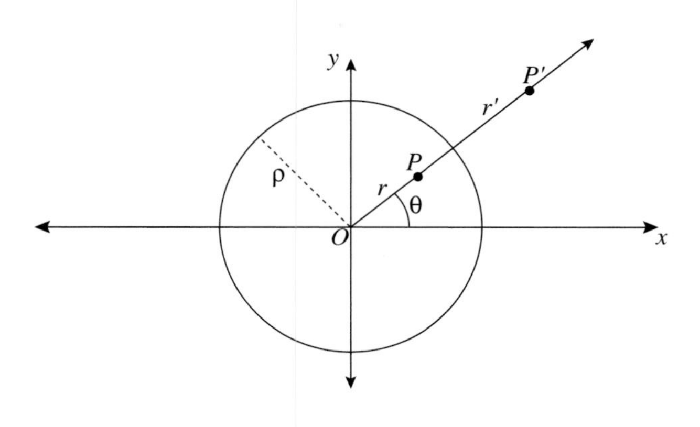

Figure 3.36

Using polar coordinates, we may give an alternative proof of Proposition 3.3.4:
(a) If l is a line in $\mathbf{R}^2 - \{(0,0)\}$, then $f(l)$ is a punctured circle.
Proof Let us place a rectangular coordinate system such that the x-axis is perpendicular to line l. Let P be a point on l whose Cartesian equation is $x = k$. Let $P = (r, \theta)$ be its polar coordinates. Then $r \cos\theta = k$. If P' has polar coordinates (r', θ), then $rr' = \rho^2$. Solving for r and substituting above, we get $(\rho^2/r') \cos\,\theta = k$, that is,

$$r' = \frac{\rho^2}{k} \cos\theta$$

which is a circle centered at a point of rectangular coordinates $(x, 0)$, where x is the radius of the punctured circle and is given by

$$x = \frac{\rho^2}{2k}.$$

(b) If Γ is a punctured circle, then $f(\Gamma)$ is a line in $\mathbf{R}^2 - \{(0,0)\}$.
Proof This proof is similar to (a) and is left as an exercise. □

Example 7

Let f be an inversion in the circle $x^2 + y^2 = 1$. Let A be a subset of $\mathbf{R}^2$ given by

$$A = \{(x, y) \mid x^2 - x + y^2 = 0\}.$$

Find $f(A)$.

The center of the inversion is $(0, 0)$ and the radius is $\rho = 1$. The set A is a circle centered at $(1/2, 0)$ and radius $1/2$. In fact, the equation $(x - 1/2)^2 + y^2 = 1/4$ simplifies to the equation that defines A. From the above we conclude that $f(A)$ is a vertical line given by $x = 1$.

Example 8

Let f be an inversion in the circle $x^2 + y^2 = 1$. Let A be a subset of $\mathbf{R}^2$ given by

$$A = \{(x, y) \mid x = 1/2\}.$$

Find $f(A)$.

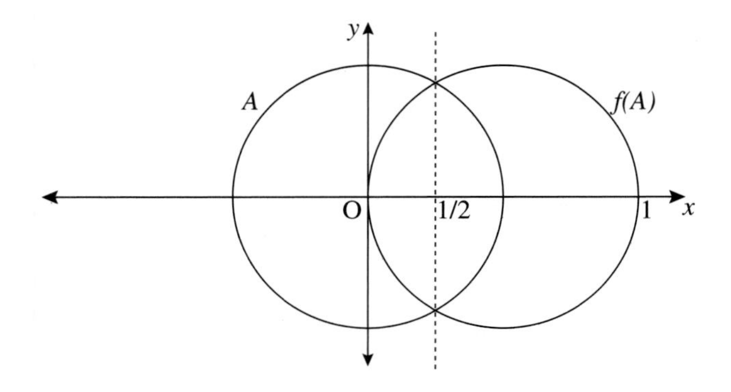

Figure 3.37

The set A is a vertical line (see Figure 3.37), and then $f(A)$ is a punctured circle of radius

$$\rho^2/2k = 1/[2(1/2)] = 1.$$

Therefore it is centered at $(1, 0)$. The equation in rectangular coordinates is $(x_1)^2 + y^2 = 1$ and in polar coordinates

$$r = 2\cos\theta.$$

Exercises

1. Let γ be a circle centered at O and radius ρ. Let P be a point such that $OP < \rho$. Show that there is an inversion that takes P to O and the circle γ to itself.
   Hint: Without loss of generality, suppose that $\rho = 1$, O is the origin of a rectangular coordinate system, and with respect to these coordinates $P = (a, 0)$.

2. (a) Repeat the procedure of Example 4 to find a formula for the reflection through line $y = mx, m < 0$.
   (b) Find a formula for the reflection through line $y = mx + b$.

3. Find the following isometries of $\mathbf{R}^2$.
   (a) A rotation that leaves the rectangle with vertices $(0,0), (0,2), (1,0), (1,2)$ invariant.
   (b) A reflection that leaves the square with vertices $(1,1), (1,-1), (-1,-1), (-1,1)$ invariant.
   (c) A rigid motion that leaves the triangle with vertices $(1,0), (3,0), (1,2)$ invariant.

4. Find ALL symmetries of the polygons below:
   (a) A square whose vertices are: $(0,0), (0,2), (2,2), (2,0)$.
   (b) A triangle with vertices
   $$(\frac{-\sqrt{3}}{2}, \frac{-1}{2}), \quad (\frac{\sqrt{3}}{2}, \frac{-1}{2}), \quad (0,1).$$
   (c) A triangle with vertices $(1,0), (1/2,1), (3/2,1)$.

5. Let $\mathcal{H}_O^r$ be a homothety with center O and ratio r. Find a formula for $\mathcal{H}_O^r$ in the following cases:
   (a) $O = (0,0)$ and $r = 2$.
   (b) $O = (1,0)$ and $r = 1/2$.
   (c) $O = (1,1)$ and $r = 3$.

6. Let f be an inversion in the circle $x^2 + y^2 = 1$. For each set A below, find $f(A)$ and say which points of A are fixed by f:
   (a) $A = \{(x,y) \in \mathbf{R}^2 \mid y = x, x < 0\}$.
   (b) $A = \{(x,y) \in \mathbf{R}^2 \mid x = 1/4\}$.

(c) $A = \{(x, y) \in \mathbf{R}^2 \mid y = 1\}$.
(d) $A = \{(x, y) \in \mathbf{R}^2 \mid c = 2\}$.
(e) $A = \{(x, y) \in \mathbf{R}^2 \mid y = 2x, x \neq 0\}$.
(f) $A = \{(x, y) \in \mathbf{R}^2 \mid x^2 - y + y^2 = 0\}$.

3.5 Appendix

We start by recalling a definition from set theory.

Definition 3.5.1 *Let X and Y be two nonempty sets. The* Cartesian product *$X \times Y$ is the set of ordered pairs*

$$\{(x, y) \mid x \in X,\ y \in Y\}.$$

Definition 3.5.2 *A* function *f from X to Y is a subset of $X \times Y$ ($f \subset X \times Y$) satisfying:*
(i) for each $x \in X$ there exists $y \in Y$ such that $(x, y) \in f$.
(ii) if $(x, y) = (x, y')$, then $y = y'$.

A function is usually thought of as a *rule* that assigns to each element of X one and only one element of Y. In Definition 3.5.2 this is made clear in (i) and (ii), by saying that each element of X is the first coordinate of only one ordered pair. But Definition 3.5.2 contains two other important ingredients of a function, namely, the sets X and Y. Therefore a function is not only a rule of assignment but a rule together with two sets.

Example 1

Given the following sets

$$X = \{-1, 1, 2, 3\}, \quad X' = \{2, 3, 4\}, \quad \text{and} \quad Y = \{1, 4, 9, 16, 25\},$$

consider the functions

$$f = \{(-1, 1), (1, 1), (2, 4), (3, 9)\}$$

and

$$g = \{(2, 4), (3, 9), (4, 16)\}.$$

We have $f \subset X \times Y$ while $g \subset X' \times Y$. Although the rule is the same (assigns the square of the number), f and g are two different functions, since they are two different sets.

A precise notation for a function should show its three ingredients, the rule and the two sets. Observe that

$$x \mapsto x^2 \qquad \text{or} \qquad f(x) = x^2$$

are not good notations for the functions of Example 1, since they do not show the sets involved and then they do not establish the difference between f and g.

The most common notation for a function is $f : X \to Y$. The set X is called the *domain* of f. It is the set that contains the elements that will be assigned to elements of Y. If element $y \in Y$ has been assigned to $x \in X$, then y is called the *image* of x and is denoted by $y = f(x)$. In this case, the point x is called the *preimage* of y. The set

$$f(X) = \{f(x) \mid x \in X\} \subset Y$$

is called the *image set* or *range* of f. Some authors do not give a name to set Y and some call it the *target set*. Functions are also called *mappings* or simply *maps*, and in geometry one often uses the name *transformation*.

Given two functions $f : X \to Y$ and $g : X \to Y$, if

$$f(x) = g(x), \quad \forall\, x \in X,$$

then $f = g$, since they have the same domain X, the same target set Y, and the same rule.

One of the simplest examples of functions is the *identity* function. In this text such a function is denoted by I_X, meaning a map $I : X \to X$ given by $I(x) = x$. When the set X is clear from the context, then we will simply denote it by I.

Definition 3.5.3 *Given $f : X \to Y$ and $g : Y \to Z$, the* composite *of f and g, denoted by $g \circ f$, consists of finding $y = f(x) \in Y$, followed by the image of y by g; that is, $g \circ f : X \to Z$ is given by*

$$(g \circ f)(x) = g(f(x)).$$

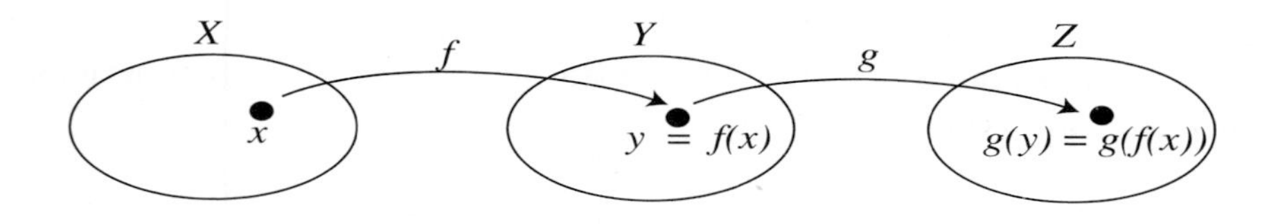

Figure 3.38 Composition of functions

Given $f : X \to Y$, $g : Y \to Z$, and $h : Z \to W$, we can find first $(g \circ f)(x)$ followed by h or find $f(x)$ and then its image by $h \circ g$. It turns out that the answer is the same, since

$$\begin{aligned}(h \circ (g \circ f))(x) &= h((g \circ f)(x)) \\ &= h(g(f(x))) = (h \circ g)(f(x)) \\ &= ((h \circ g) \circ f)(x), \quad \forall\, x \in X.\end{aligned}$$

It follows that

$$h \circ (g \circ f) = (h \circ g) \circ f.$$

This is known as the *associative* property of composition of functions. It allows us to simplify the notation and skip the parentheses; we simply write $h \circ g \circ f$.

If f is a function from a set X to itself, then the composite $f \circ f$ is denoted by f^2. The notation f^n means that f is composed with itself n times.

Definition 3.5.4 *Let $f : X \to Y$ be a function and let A be a proper subset of X ($A \subset X, A \neq X$). The* restriction *of f to A, denoted by $f_{|A}$, is defined as the function $f_{|A} : A \to Y$ whose rule is $f_{|A}(x) = f(x)$.*

Here is another example of the fact that, although the rule is the same, f and $f_{|A}$ are two different functions, since they have different domains. Now, if a function is defined on a subset A of X, such a function can be extended to a function from X to Y.

Definition 3.5.5 *Let $f : A \to Y$ be a function where A is a subset of X. An* extension *of f to X is a function $\tilde{f} : X \to Y$ such that $\tilde{f}(x) = f(x)$ for all $x \in A$.*

Example 2

Let $\mathbf{R}_+$ be the set of nonnegative real numbers and $f : \mathbf{R}_+ \to \mathbf{R}$ be given by $f(x) = \sqrt{x}$. Then f can be extended to $\tilde{f} : \mathbf{R} \to \mathbf{R}$ defined by $\tilde{f}(x) = \sqrt{|x|}$.

The extension of a function f is not unique. For example, the function of Example 2 can be extended to $\bar{f} : \mathbf{R} \to \mathbf{R}$ given by

$$\bar{f}(x) = \begin{cases} \sqrt{x}, & x \geq 0, \\ 0 & x < 0. \end{cases}$$

Let $f : X \to Y$ be a function and let us consider B a subset of Y. The set

$$f^{-1}(B) = \{x \in X \mid f(x) \in B\}$$

is called the *preimage* or *inverse image* of B. Notice that while the image of an element is only one element, the preimage of the subset $\{y\}$ may contain none or more than one element. In Example 1 above we have

$$f^{-1}(\{1\}) = \{-1, 1\} \qquad \text{and} \qquad f^{-1}(\{25\}) = \phi.$$

But there exist functions $f : X \to Y$ with the property that for each $y \in Y$, $f^{-1}(\{y\})$ has only one element.

Definition 3.5.6 *A function $f : X \to Y$ is said to be* injective *or* one-to-one *if, given $x_1, x_2 \in X$,*

$$f(x_1) = f(x_2) \quad \Rightarrow \quad x_1 = x_2.$$

This is equivalent to saying that

$$x_1 \neq x_2 \quad \Rightarrow \quad f(x_1) \neq f(x_2).$$

The definition above is also equivalent to saying that if $y \in f(X)$ then $f^{-1}(\{y\})$ contains only one element.

Definition 3.5.7 *A function $f : X \to Y$ is said to be* surjective *or* onto *if $f(X) = Y$; that is, for each $y \in Y$ there exists $x \in X$ such that $f(x) = y$ and hence $f^{-1}(\{y\}) \neq \phi$, $\forall\, y \in Y$.*

A function $f : X \to Y$ that is both injective and surjective is called *bijective* or *one-to-one correspondence* between X and Y.

If f is bijective, then for each $y \in Y$ the preimage $f^{-1}(\{y\})$ has only the element x such that $f(x) = y$. It is then possible to define another function $g : Y \to X$ by $g(y) = g(f(x)) = x$. Notice that $(g \circ f)(x) = x$, while

$$\begin{aligned}(f \circ g)(y) &= (f \circ g)(f(x)) \\ &= f(g(f(x))) \\ &= f((g \circ f)(x)) = f(x) = y,\end{aligned}$$

which we write as

$$(g \circ f) = I_X \qquad \text{and} \qquad (f \circ g) = I_Y.$$

This motivates the following definition.

Definition 3.5.8 *A function $f : X \to Y$ is said to be* invertible *if there exists a function $g : Y \to X$ such that*

$$(g \circ f) = I_X \qquad \text{and} \qquad (f \circ g) = I_Y.$$

Proposition 3.5.9 *A map $f : X \to Y$ is invertible if and only if it is bijective.*

Proof ($\Leftarrow$) This part follows from the considerations above.
($\Rightarrow$) First we show that f is injective. Suppose that $f(x_1) = f(x_2)$. Applying function g to both sides, we get

$$g(f(x_1)) = g(f(x_2)) \quad \Rightarrow \quad (g \circ f)(x_1) = (g \circ f)(x_2).$$

Since $g \circ f = I_X$, we conclude that

$$x_1 = (g \circ f)(x_1) = (g \circ f)(x_2) = x_2$$

and then f is injective.

Now we show that f is onto. Let $y \in Y$. We want to show that there exists $x \in X$ such that $f(x) = y$. Indeed, let x be given by $x = g(y)$. Applying function f to both sides, we conclude the proof, for we have

$$f(x) = f(g(y)) = (f \circ g)(y) = y,$$

where the last equality was implied by the fact that $f \circ g = I_Y$. □

Proposition 3.5.10 *Let $f : X \to Y$ be an invertible map. Then its inverse is unique.*

Proof Let h and g be functions from Y to X such that $(f \circ g) = I_Y$ and $(h \circ f) = I_X$. Then $(f \circ g)(y) = y$, and applying function h to both sides we obtain

$$h((f \circ g)(y)) = h(y).$$

On the other hand,

$$h((f \circ g)(y)) = (h \circ f)(g(y))$$

and, since $h \circ f = I_X$ and $g(y) \in X$, we get that

$$(h \circ f)(g(y)) = g(y)$$

and therefore

$$g(y) = h(y), \quad \forall\, y \in Y,$$

implying that $g = h$. □

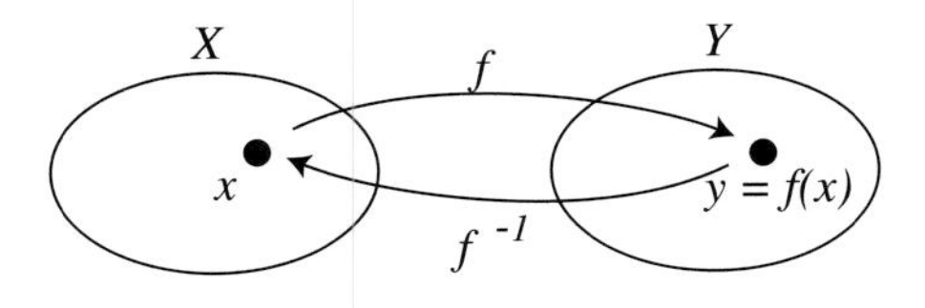

Figure 3.39

The only function with the property of Definition 3.5.8 is called the *inverse* of f and is denoted by f^{-1}. Observe that if f is invertible, then so is f^{-1}, and its inverse is $(f^{-1})^{-1} = f$.

Chapter 4

EUCLIDEAN 3-SPACE

4.1 Axiom System for 3-Dimensional Geometry

In this section we state postulates that will be used to study the geometry of 3-space. The undefined objects here are *point*, *line*, and *plane*. We will assume one undefined relation, namely, *points lie in planes*. In addition, all planes will be assumed to be *Euclidean planes*; that is, the axioms of plane Euclidean geometry, and consequently all its theorems, hold for every plane. The first axioms for 3-space are:

> **Sp_1** Any three points lie in at least one plane, and any three noncollinear points lie in exactly one plane.
>
> **Sp_2** There exist four distinct points with the property that no plane contains all four of them.

Definition 4.1.1 *We say that a line l* lies in a *plane $\mathcal{P}$ if the plane $\mathcal{P}$* contains *line l, that is, every point on l lies in $\mathcal{P}$.*

Recall that the first incidence axiom for the plane states that there exists only one line incident with two distinct points. It seems natural then to assume that if two points lie in a plane, then so does every point on the line through them. We state this fact in the next postulate.

> **Sp_3** If two points lie in a plane, then the line incident with these points lies in the same plane.

Definition 4.1.2 *A line is said to be* parallel *to a plane if they do not have any common point. Two planes are called* parallel planes *if they have no point in common.*

Proposition 4.1.3 *Let l and $\mathcal{P}$ be a line and plane, respectively. Then only one of the following holds:*
(i) l is parallel to $\mathcal{P}$.
(ii) l intersects $\mathcal{P}$ in only one point.
(iii) l lies in $\mathcal{P}$.

Proof If l is not parallel to $\mathcal{P}$, then there exists a point $P \in l \cap \mathcal{P}$. If P is the only point in their intersection, then we have *(ii)*. If not, then l lies in $\mathcal{P}$ by Axiom **Sp3**. □

Notice that if the intersection of two nonparallel planes contains three noncollinear points, then these planes are the same plane, by Axiom **Sp1**. If their intersection contains exactly two points, then it contains at least the line incident with them, by **Sp3**. However, we cannot state any conclusive result without the following axiom:

> **Sp4** The intersection of two planes contains at least two distinct points.

Proposition 4.1.4 *Let $\mathcal{P}_1$ and $\mathcal{P}_2$ be two planes. Then only one of the following holds:*
(i) $\mathcal{P}_1$ and $\mathcal{P}_2$ are parallel planes.
(ii) The intersection of $\mathcal{P}_1$ and $\mathcal{P}_2$ is a line.
(iii) $\mathcal{P}_1 = \mathcal{P}_2$.

Definition 4.1.5 *Four points $A, B, C,$ and D are called* coplanar *if there exists a plane that contains all four of them. Two lines are said to be* coplanar *if there exists a plane that contains both of them. Two noncoplanar lines are also called* skew-lines.

Observe that two skew-lines do not have any common point. Therefore, two distinct lines in space are said to be *parallel* if they are *coplanar* and have no point in common.

Proposition 4.1.6 *(a) If two distinct lines have a common point, then there exists only one plane containing them. Moreover, this common point is unique.*

(b) If l and m are two distinct lines in space, then only one of the following occurs:
(i) l and m are skew-lines.
(ii) l and m are concurrent lines.
(iii) l and m are parallel lines.

Proof *(a)* Let l_1 and l_2 denote the lines and $P \in l_1 \cap l_2$. Let Q_1 and Q_2 be points on l_1 and l_2, respectively, such that P, Q_1, and Q_2 are distinct points. It follows from Axiom $\mathbf{Sp_1}$ that there is a plane $\mathcal{P}$ containing P, Q_1, and Q_2. From Axiom $\mathbf{Sp_3}$ we have that $l_1 = \overleftrightarrow{PQ_1}$ and $l_2 = \overleftrightarrow{PQ_2}$ are contained in $\mathcal{P}$. Now, the incidence axiom $\mathbf{I_1}$ for the plane implies that l_1 and l_2 have only one common point and hence P, Q_1, and Q_2 are noncollinear points. It follows from Axiom $\mathbf{Sp_1}$ that the plane containing P, Q_1, and Q_2 and, hence containing l_1 and l_2, is unique.

The proof of *(b)* is left to the reader. □

Proposition 4.1.7 *The axioms above imply:*
(a) If A and B are two distinct points, then there exist at least two distinct planes containing them.
(b) For every line there exist at least two distinct planes containing it.
(c) There exist at least three distinct planes that do not have a common line.
(d) There exist two lines with the property that no plane contains both of them.
(e) For every line there exists at least one plane not containing it.
(f) For every plane there exists at least one line not contained in it.

Proof *(a)* It follows from $\mathbf{Sp_2}$ that there exist points P_1, P_2, P_3, and P_4 that are noncoplanar. From $\mathbf{Sp_1}$ we conclude that there is at least one plane $\mathcal{P}_i$ containing A, B, and P_i, for $i = 1, 2, 3, 4$. Notice that at least two of these planes are distinct, since P_1, P_2, P_3, and P_4 cannot lie in the same plane.
(b) Let A and B be two points on the line that we denote by l. It follows from part (a) that there exist at least two distinct planes containing A and B. Since A and B lie in each of these two planes, we have that A and B are in their intersection. Proposition 4.1.4(ii) implies that their intersection is a line which is incident with A and B. Therefore such a line is l, by Axiom $\mathbf{I_1}$.

(c) Let P_1, P_2, P_3, and P_4 be noncoplanar points, whose existence is guaranteed by Axiom $\mathbf{Sp_2}$. We claim that no three of these are collinear. In fact, suppose that P_1, P_2, and P_3 all lie on line l. It follows from $\mathbf{Sp_1}$ that there is a plane $\mathcal{P}$ containing P_1, P_2, and P_4. From $\mathbf{Sp_3}$ we get that l, and hence P_3, lie in $\mathcal{P}$, since P_1 and P_2 are in $\mathcal{P}$. Then $\mathcal{P}$ contains P_1, P_2, P_3, and P_4, contradicting that they are noncoplanar points.

Now, consider the four distinct planes that contain $\{P_1, P_2, P_3\}$, $\{P_1, P_2, P_4\}$,$\{P_1, P_3, P_4\}$, and $\{P_2, P_3, P_4\}$, respectively. We leave to the reader the verification that no three of these have a common line.

(d) Let P_1, P_2, P_3, and P_4 be noncoplanar points. From Axiom $\mathbf{Sp_1}$ we obtain that there is at least one plane $\mathcal{P}_1$ containing P_1, P_2, and P_3 and one plane $\mathcal{P}_2$ containing P_2, P_3, and P_4. Let l_1 be the line incident with P_1 and P_2 and l_2 incident with P_3 and P_4. If there exists a plane that contains l_1 and l_2, then such a plane contains contains points P_1, P_2, P_3, and P_4, which contradicts that such points are noncoplanar.

The proofs of *(e)* and *(f)* are left to the reader. □

In the study of the separation properties in space, planes play a role similar to that of lines in a plane. The results are closely analogous to those in the plane, and no new axioms are needed for their proofs.

Definition 4.1.8 *Let $\mathcal{P}$ be a plane and A and B two distinct points not in $\mathcal{P}$. We say that A and B are on the same side of $\mathcal{P}$ if A and B are on the same side of every line lying in $\mathcal{P}$ that is coplanar with $\overleftrightarrow{AB}$.*

It follows from the definitions above and Definition 1.2.7 that A and B are on the same side of $\mathcal{P}$ if and only if line segment $\overline{AB}$ does not intersect plane $\mathcal{P}$.

Definition 4.1.9 *Let $\mathcal{P}$ be a plane and A a point not in $\mathcal{P}$. The set of all points that are on the same side of $\mathcal{P}$ as A is called the (open)* half-space *bounded by $\mathcal{P}$ containing A. We denote it by $HS_{\mathcal{P},A}$.*

Observe that if A and B are on the same side of $\mathcal{P}$, then

$$HS_{\mathcal{P},A} = HS_{\mathcal{P},B}.$$

Theorem 4.1.10 (Space separation property) *Every plane determines exactly two disjoint half-spaces.*

Proof We leave the proof of this result as an exercise, since it is similar to the proof of the *plane separation property* of Chapter 1. □

Definition 4.1.11 *Let $\mathcal{P}_1$ and $\mathcal{P}_2$ be two intersecting planes and line $l = \mathcal{P}_1 \cap \mathcal{P}_2$. Let H_1 be one of the half-planes determined by l in $\mathcal{P}_1$. Consider a point P on plane $\mathcal{P}_2$ such that $P \notin l$, and its half-plane H_2 bounded by l and containing P. Then the set $\mathcal{D} = l \cup H_1 \cup H_2$ is called a* dihedral angle. *The line l is called the* edge *of a dihedral angle and the half-planes its* sides *or* faces.

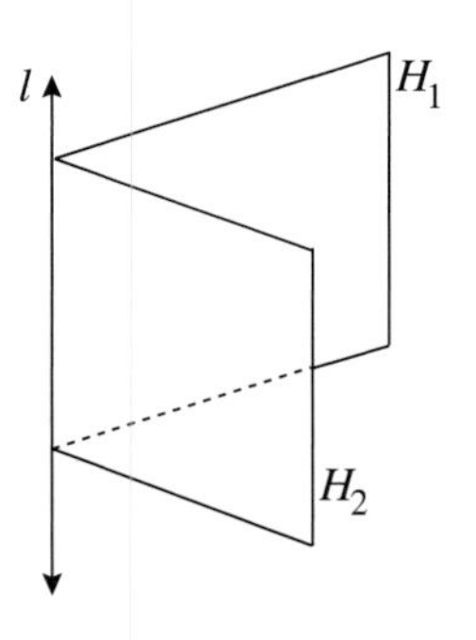

Figure 4.1

Definition 4.1.12 *Let $\mathcal{D} = l \cup H_1 \cup H_2$ be a dihedral angle and $\mathcal{P}_1$ and $\mathcal{P}_2$ the planes that contain H_1 and H_2, respectively. Let HS_i be the half-space determined by $\mathcal{P}_i$ that contains H_j, for $j \neq i$. The set*

$$\text{Int}(\mathcal{D}) = HS_1 \cap HS_2$$

is called the interior of the dihedral angle $\mathcal{D}$.

As in a plane, a subset $\mathcal{S}$ is said to be *convex* if for any two points $A, B \in \mathcal{S}$ the line segment $\overline{AB} \subset \mathcal{S}$.

Proposition 4.1.13 *(a) A half-space is convex.*
(b) The interior of a dihedral angle is convex.
(c) The closed half-space $\overline{HS_{\mathcal{P}}} = \mathcal{P} \cup HS_{\mathcal{P}}$ is a convex set.
(d) Let A and B be on different faces of a dihedral angle $\mathcal{D}$. Then any point P between A and B is in Int($\mathcal{D}$).

Proof *(a)* Let A and B be on the same side of a half-space $HS_{\mathcal{P}}$. Let l be a line in $\mathcal{P}$ that is coplanar with line $m = \overleftrightarrow{AB}$. Let $\mathcal{P}'$ be the plane detemined by l and m. In plane $\mathcal{P}'$, A and B are on the same half-plane $\mathcal{H}$ bounded by l, and since $\mathcal{H}$ is convex, we have that $\overline{AB} \in \mathcal{H}$. Since this is true for any arbitrary line l in $\mathcal{P}$ coplanar with m, we conclude that $\overline{AB} \in HS_{\mathcal{P}}$.
(b) It follows from the fact that the intersection of convex sets in space is convex.

We leave the proofs of *(c)* and *(d)* to the reader. □

Exercises

1. Prove Propositions 4.1.3, 4.1.4, 4.1.6(b), 4.1.7(e), and (f).

2. Prove Theorem 4.1.10

3. Let HS be a half-space determined by a plane $\mathcal{P}$. Show that HS contains at least four noncoplanar points.

4. Let A and B be points on different sides of a dihedral angle.
(a) Show that every point between A and B is in the interior of the dihedral angle.
(b) Let C be a point on the edge of the dihedral angle and $\mathcal{P}$ the plane containing A, B, and C. Show that every point in the intersection of $\mathcal{P}$ with the interior of the dihedral angle is an interior point of $\angle BCA$.

5. Show that if two parallel planes are intersected by a third plane, then the lines of intersection are parallel.

6. Show that if a line intersects one of two parallel planes, then it intersects the other.
Hint: Let $\mathcal{P}_1$ and $\mathcal{P}_2$ be the two parallel planes and l a line intersecting $\mathcal{P}_1$. Let Q be an arbitrary point in $\mathcal{P}_2$ and $\mathcal{P}'$ the plane determined by l and Q. Consider line $l' \in \mathcal{P}'$ through Q and parallel to l. Now we have that $\mathcal{P}'$ intersects $\mathcal{P}_2$ (why?) and the line $m = \mathcal{P}_2 \cap \mathcal{P}'$ intersects l (why?).

4.2 Perpendicular Lines and Planes

Before we make the first definition of this section we want observe the following: If a line l intersects plane $\mathcal{P}$ at a point P, then l and any line m lying in $\mathcal{P}$ and passing through P determine only one plane $\mathcal{P}'$. We then say that *l is perpendicular to m* if l and m are perpendicular lines of plane $\mathcal{P}'$.

Definition 4.2.1 *A line l intersecting a plane $\mathcal{P}$ at a point P is said to be* perpendicular to the plane $\mathcal{P}$ at P *if l is perpendicular to every line lying in $\mathcal{P}$ and passing through P. In this case, we also say that plane $\mathcal{P}$* is perpendicular to line l at P.

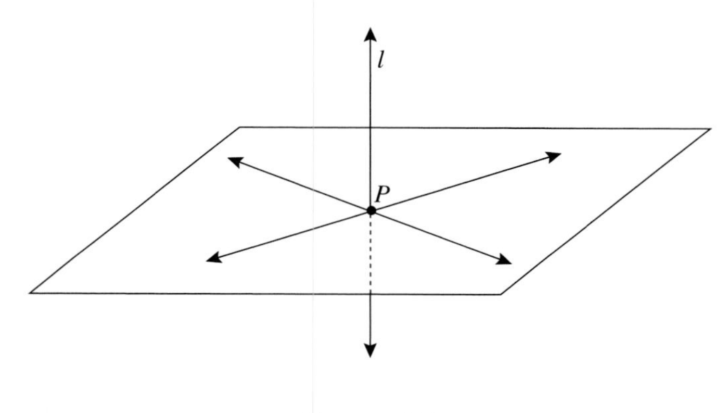

Figure 4.2

Lemma 4.2.2 *If a line l is perpendicular to two concurrent lines lying in a plane $\mathcal{P}$, then l is perpendicular to $\mathcal{P}$ at their point of intersection.*

Proof Let l_1 and l_2 be two lines on plane $\mathcal{P}$ passing through P and perpendicular to l. Let m be any other line lying in $\mathcal{P}$ and passing through P (see Figure 4.3). We will show that l is perpendicular to m. For that, let us consider points $P_1 \in l_1$ and $P_2 \in l_2$ such that P_1 and P_2 are on opposite sides of m. On l, we consider points A and B such that $AP = PB$. We then have

$$\triangle APP_1 \simeq \triangle BPP_1 \quad \text{and} \quad \triangle APP_2 \simeq \triangle BPP_2$$

by SAS. Therefore

$$AP_1 = BP_1 \quad \text{and} \quad AP_2 = BP_2,$$

which implies that

$$\triangle AP_1P_2 \simeq \triangle BP_1P_2,$$

by SSS. It follows that

$$\angle AP_1P_2 \simeq \angle BP_1P_2.$$

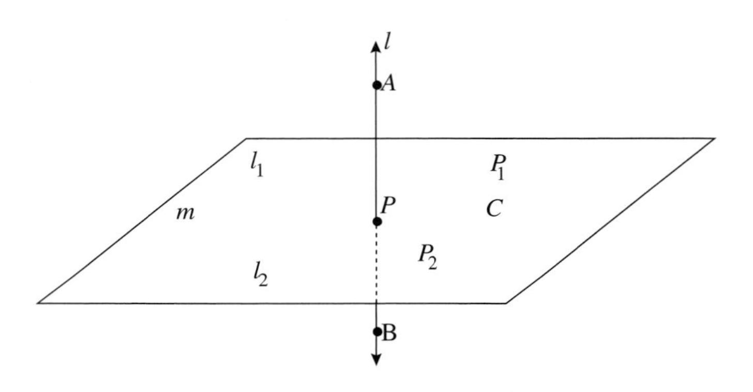

Figure 4.3

Now let C denote the point where line segment $\overline{P_1P_2}$ intersects line m. We claim that $\triangle AP_1C \simeq \triangle BP_1C$ by SAS. In fact, we have a common side $\overline{P_1C}$ and $AP_1 = BP_1$. Notice that

$$\angle AP_1C = \angle AP_1P_2 \quad \text{and} \quad \angle BP_1C = \angle BP_1P_2$$

and hence

$$\triangle AP_1C \simeq \triangle BP_1C.$$

It follows that $AC = BC$ and hence $\triangle CAB$ is isosceles. Since P is the midpoint of $\overline{AB}$, we conclude that $\overline{AB}$ is perpendicular to $\overline{PC}$, which means l is perpendicular to m. □

Definition 4.2.3 *A plane $\mathcal{P}$ is said to be a* perpendicular bisector *of a line segment $\overline{AB}$ if $\mathcal{P}$ is perpendicular to $\overline{AB}$ at its midpoint.*

Proposition 4.2.4 *For every line l and for every point P on l there exists a unique plane which is perpendicular to l at P.*

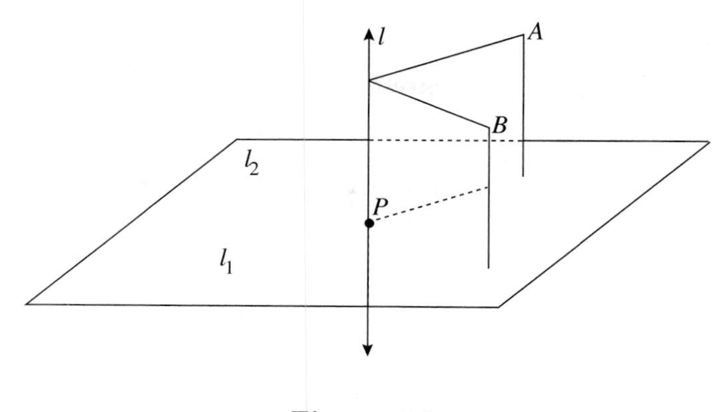

Figure 4.4

Proof We show first the existence of the perpendicular plane. Let A be a point not lying on l. On the plane $\mathcal{P}_1$ determined by A and l there exists a unique line l_1 that is perpendicular to l and passes through P. Considering another point B not lying in $\mathcal{P}_1$, we obtain a plane $\mathcal{P}_2$ determined by B and l. On this plane we consider line l_2 that is perpendicular to l through P. Observe that l_1 and l_2 are concurrent at P. Let $\mathcal{P}$ be the plane that contains lines l_1 and l_2. Since l is perpendicular to l_1 and l_2, it is perpendicular to plane $\mathcal{P}$ by Lemma 4.2.2. Therefore $\mathcal{P}$ is perpendicular to l at P.

To show the uniqueness, we suppose that plane $\mathcal{P}'$ is also perpendicular to l at P. Let $\mathcal{P}''$ be any plane containing line l and let $m = \mathcal{P} \cap \mathcal{P}''$ and $m' = \mathcal{P}' \cap \mathcal{P}''$. Notice that l is perpendicular to both lines m and m'. If $\mathcal{P} \neq \mathcal{P}'$, then $m \neq m'$. This implies that there are two lines m and m' in $\mathcal{P}''$ that are perpendicular to l at P, which is clearly a contradiction. Therefore $\mathcal{P} = \mathcal{P}'$. □

Corollary 4.2.5 *Every line segment has a unique perpendicular plane bisector.*

Corollary 4.2.6 *Let l and m be two perpendicular lines and let P denote their intersection point. Then m lies in the plane perpendicular to l at P.*

Proof Let $\mathcal{P}$ be the plane perpendicular to l at P. Let n be a line contained in $\mathcal{P}$ and through P. Then l is perpendicular to l. If m is not in $\mathcal{P}$, then m and n determine a plane $\mathcal{P}' \neq \mathcal{P}$; furthermore, $\mathcal{P}'$ is also perpendicular to l at Q, by Lemma 4.2.2. But this contradicts the uniqueness of Proposition 4.2.4 □

Theorem 4.2.7 *For every plane $\mathcal{P}$ and for every point P in $\mathcal{P}$ there exists a unique line which is perpendicular to $\mathcal{P}$ at P.*

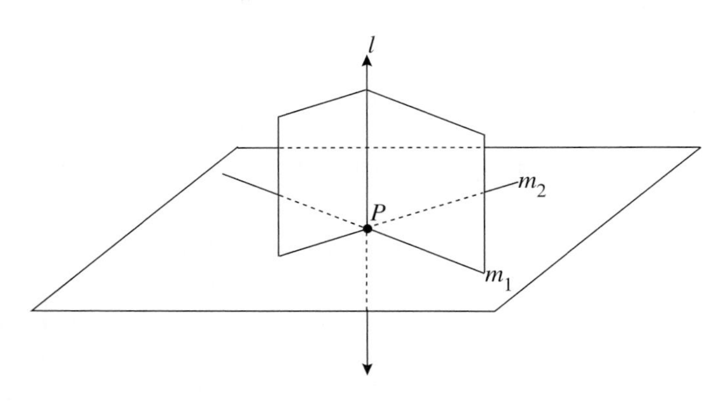

Figure 4.5

Proof Let m_1 be a line lying in $\mathcal{P}$ and passing through P. Let $\mathcal{P}_1$ be the plane perpendicular to m_1 at P. Similarly, we consider another line $m_2 \subset \mathcal{P}$ and a plane $\mathcal{P}_2$ perpendicular to m_2 at P. Let $l = \mathcal{P}_1 \cap \mathcal{P}_2$ (see Figure 4.5). Then l is perpendicular to m_1 and m_2 and hence perpendicular to $\mathcal{P}$ by Lemma 4.2.2. Now suppose that l' is another line perpendicular to $\mathcal{P}$ at P and $\mathcal{P}'$ the plane determined by l and l'. Let $m = \mathcal{P} \cap \mathcal{P}'$. We then obtain two lines, l and l', in $\mathcal{P}'$ which are perpendicular to m at P. This contradiction implies that $l = l'$. □

Lemma 4.2.8 *Any two distinct lines perpendicular to the same plane are parallel.*

Proof Let l_1 and l_2 be perpendicular to the plane $\mathcal{P}$ at points P and Q, respectively (see Figure 4.6). We will show first that they are coplanar. For that, consider the midpoint M of $\overline{PQ}$ and on the perpendicular bisector of $\overline{PQ}$ points A and B such that $AM = MB$. It is easy to see that $\triangle APM \simeq \triangle BPM$ and hence $PA = PB$. Let X be an arbitrary point of l_1. Since l_1 is perpendicular to both $\overline{PA}$ and $\overline{PB}$, we obtain

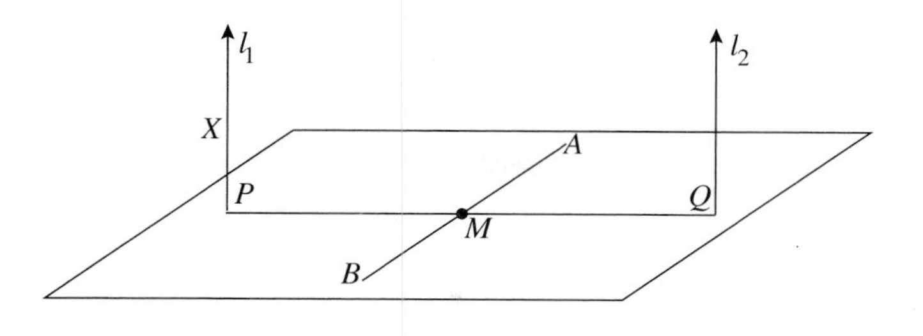

Figure 4.6

that $\triangle XPA \simeq \triangle XPB$, which in turn implies that $AX = BX$. Therefore $\triangle XAB$ is isosceles and $\overline{XM}$ is perpendicular to $\overline{AB}$. This shows that every point X on l_1 is in the perpendicular bisector plane of $\overline{AB}$. Similarly one shows that every point on l_2 is in the unique perpendicular bisector plane. Therefore lines l_1 and l_2 lie in the same plane.

Now to conclude that l_1 and l_2 are parallel, we observe that line $\overleftrightarrow{PQ}$ is also in the perpendicular bisector plane of $\overline{AB}$. Therefore lines l_1, l_2, and $\overleftrightarrow{PQ}$ are coplanar, and the alternate interior angle theorem implies that l_1 and l_2 are parallel, since they have a common perpendicular line $\overleftrightarrow{PQ}$. □

Theorem 4.2.9 *For every plane $\mathcal{P}$ and for every point P not in $\mathcal{P}$ there exists a unique line passing through P that is perpendicular to $\mathcal{P}$.*

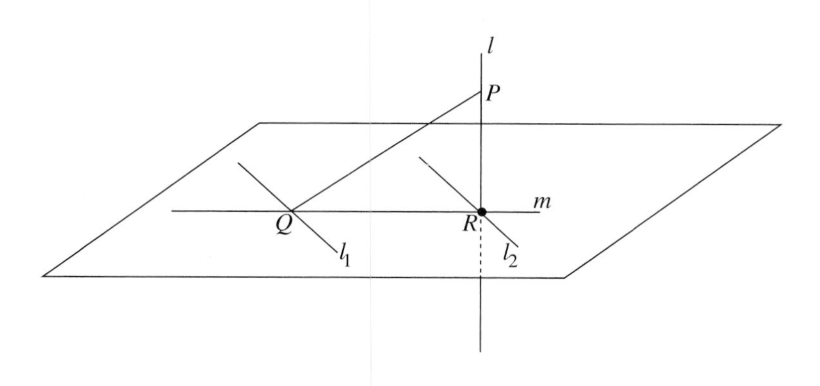

Figure 4.7

Proof Let l_1 be any line in $\mathcal{P}$. In the plane containing l_1 and P we drop a perpendicular from P to l_1 whose foot we denote by Q. Let m be the line in $\mathcal{P}$ that passes through Q and is also perpendicular to l_1. Now in the plane containing m and P, let l be the perpendicular dropped from P to m. We claim that line l is perpendicular to $\mathcal{P}$. In fact, let $R = l \cap m$ and l_2 a line through R and perpendicular to m. Note that l_2 is parallel to l_1. We will show that l is also perpendicular to l_2. For that we consider plane $\mathcal{P}'$ containing points P, Q, and R. Let l_2' be the line perpendicular to $\mathcal{P}'$ at R. Observe that l_1 is perpendicular to $\mathcal{P}'$ at Q and hence l_1 and l_2' are parallel, by Lemma 4.2.8. Since the plane that contains l_1 and l_2' is $\mathcal{P}$, we conclude that $l_2' = l_2$, because parallel lines in a Euclidean plane are unique. Therefore l_2 is perpendicular to any line in $\mathcal{P}'$ passing through R, in particular, line l.

The uniqueness part of the proof is left to the reader. □

Definition 4.2.10 *Let $\mathcal{D} = l \cup H_1 \cup H_2$ be a dihedral angle and $\mathcal{P}$ be a plane perpendicular to the edge l. The angle given by the intersection $\mathcal{P} \cap \mathcal{D}$ is called the* plane angle *of the dihedral angle.*

Lemma 4.2.11 *All plane angles of the same dihedral angle are congruent.*

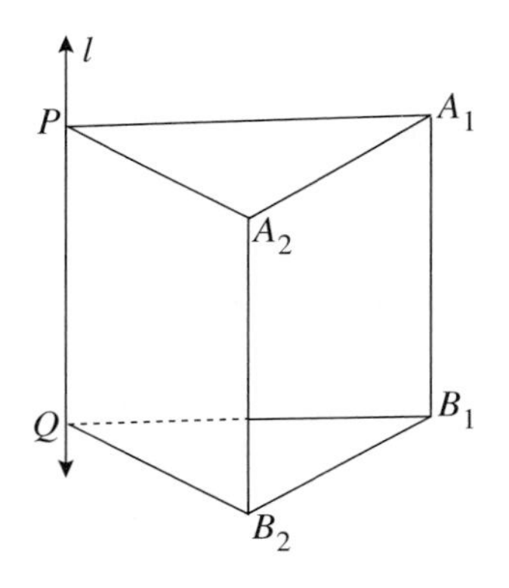

Figure 4.8

Proof Let $\angle P$ and $\angle Q$ be two plane angles of the dihedral angle $\mathcal{D} = l \cup H_1 \cup H_2$. Let $A_i, B_i \in H_i$ such that

$$PA_1 = QB_1 \quad \text{and} \quad PA_2 = QB_2.$$

Notice that $\overline{PA_1}$ and $\overline{QB_1}$ are coplanar and both perpendicular to the edge l. Therefore $\overline{PA_1}$ is parallel to $\overline{QB_1}$. This implies that $\square PA_1B_1Q$ is a parallelogram and hence $PQ = A_1B_1$. Similarly, one obtains $PQ = A_2B_2$, which in turn implies $A_1B_1 = A_2B_2$. We then conclude that $\triangle PA_1A_2 \simeq \triangle QB_1B_2$ by SSS, implying that

$$\angle A_1PA_2 \simeq \angle B_1QB_2.$$

□

Using Lemma 4.2.11, we can now define the measure of a dihedral angle.

Definition 4.2.12 *The measure of a dihedral angle $\mathcal{D}$ is the measure of all plane angles of $\mathcal{D}$. A dihedral angle is a* right angle *if its plane angles are right angles.*

Definition 4.2.13 *Two planes are said to be* perpendicular *if their union contains a right dihedral angle.*

Proposition 4.2.14 *Let l be a line perpendicular to plane $\mathcal{P}$. Then any plane $\mathcal{P}'$ containing l is perpendicular to the plane $\mathcal{P}$.*

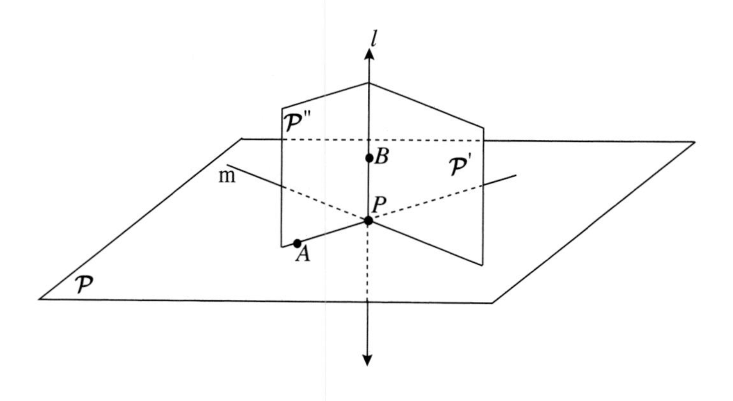

Figure 4.9

Proof Let $m = \mathcal{P} \cap \mathcal{P}'$ and $P = m \cap l$. Observe that line m is the edge of all four dihedral angles determined by $\mathcal{P}$ and $\mathcal{P}'$. Let $\mathcal{P}''$ be a plane perpendicular to m at P. Since plane $\mathcal{P}''$ contains line l, we have that $l = \mathcal{P}' \cap \mathcal{P}''$. Let $A \in \mathcal{P} \cap \mathcal{P}''$ and $B \in l$. Since $\mathcal{P}''$ is perpendicular to the edge m, we obtain that $\angle APB$ is a plane angle.

Since $\angle APB$ is a right angle, we conclude that $\mathcal{P}$ and $\mathcal{P}'$ are perpendicular planes. □

Exercises

1. Show that if a line is perpendicular to one of two parallel planes, it is perpendicular to the other.

2. Show that any two planes perpendicular to the same line are parallel.

3. Show that if a plane is perpendicular to one of two parallel lines, then it is perpendicular to the other.

4. Show that if two lines are each parallel to a third line, then they are parallel to each other.

5. Let $\mathcal{P}_1$ and $\mathcal{P}_2$ be two perpendicular planes and $l = \mathcal{P}_1 \cap \mathcal{P}_2$. Show that the line in $\mathcal{P}_1$ perpendicular to l is perpendicular to $\mathcal{P}_2$.

4.3 Rigid Motions in 3-Space

A map T from 3-space to itself with the property that

$$T(A)\,T(B) = AB$$

for all points A and B is called a *rigid motion.*

Proposition 4.3.1 *Let T be a rigid motion of 3-space, with $T(A) = A'$ and $T(B) = B'$. Then:*
(a) T is injective.
(b) $T(\overline{AB}) = \overline{A'\,B'}$.
(c) T preserves angle measure.
(d) T is onto.

The content of Proposition 4.3.1 is analogous to that of Proposition 3.1.3. The reader is asked to prove it in Exercise 1 of this section. Its proof consists of repeating the same arguments used in Chapter 3.

Recall that a point P is said to be a *fixed point* of a transformation T if $T(P) = P$. The transformation that fixes all points of space is called the identity map and is denoted by I.

Proposition 4.3.2 *Let T and S be two rigid motions of 3-space.*
(a) If T fixes four noncoplanar points, then $T = I$.
(b) If $T(A_1) = S(A_1)$, $T(A_2) = S(A_2)$, $T(A_3) = S(A_3)$, $T(A_4) = S(A_4)$, and A_1, A_2, A_3, A_4 are noncoplanar points, then $T = S$.

Proof *(a)* Let A_1, A_2, A_3, A_4 be the noncoplanar points fixed by T. Notice that no three of these are collinear, and thus let us consider plane $\mathcal{P}$ determined by A_1, A_2, A_3. It follows from Exercise 2 that T restricted to plane $\mathcal{P}$ is a rigid motion of plane $\mathcal{P}$ that fixes three non-collinear points of $\mathcal{P}$ (observe that $T(\mathcal{P})$ and $\mathcal{P}$ have three noncollinear common points). From the result of Chapter 3 we conclude that

$$T(P) = P, \quad \forall\, P \in \mathcal{P}.$$

Now suppose that for point $Q \notin \mathcal{P}$, we have $T(Q) = Q' \neq Q$; it follows that $Q \neq A_4$. Let us consider l the only line through Q that is perpendicular to plane $\mathcal{P}$ at point P. Since $T(P) = P$ and T is a rigid motion, we have $PQ = PQ'$, and the line through Q' and P is perpendicular to plane $\mathcal{P}$ at P, for T preserves angle measure. By the uniqueness of the perpendicular line, we conclude that $\overleftrightarrow{PQ'} = l$. Moreover, P is the midpoint of segment $\overline{QQ'}$. Consider now $\triangle A_4QQ'$. Since A_4 is a point fixed by T, we have

$$A_4Q' = T(A_4)T(Q) = A_4Q,$$

which implies that $\triangle A_4QQ'$ is isosceles. Therefore $\overleftrightarrow{PA_4} \perp l$, which in turn implies that $A_4 \in \mathcal{P}$ by Corollary 4.2.6; this fact contradicts that A_1, A_2, A_3, A_4 are noncoplanar.
(b) Since $S^{-1} \circ T$ is a rigid motion and fixes A_1, A_2, A_3, and A_4, we have that $S^{-1} \circ T = I$ and hence $S = T$. □

The fundamental types of rigid motions of space are:

Reflection through a plane $\mathcal{P}$

Let P be an arbitrary point of the space. If $P \in \mathcal{P}$, the reflection of P is itself. If $P \notin \mathcal{P}$, consider m be a line perpendicular to $\mathcal{P}$ and through a point P and let $O = l \cap \mathcal{P}$. The *reflection of* P in $\mathcal{P}$ is the point $P' \in m$ such that P and P' are on opposite sides of $\mathcal{P}$ and $OP = OP'$ (see Figure 4.10). It follows that all points of $\mathcal{P}$ are fixed by the reflection.

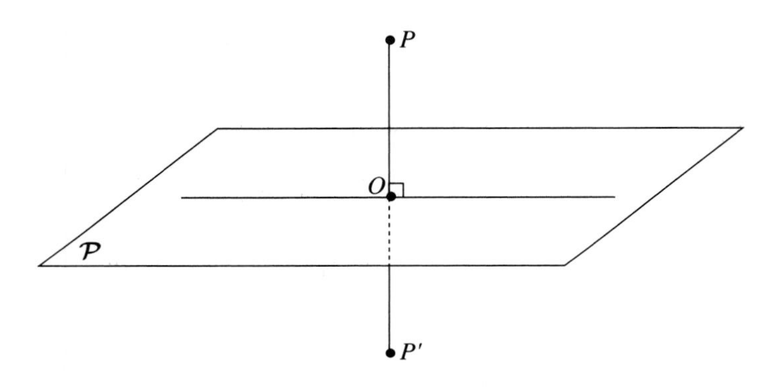

Figure 4.10

Rotation about a line l through an angle θ

Let P be an arbitrary point of the space. The image of P by rotation

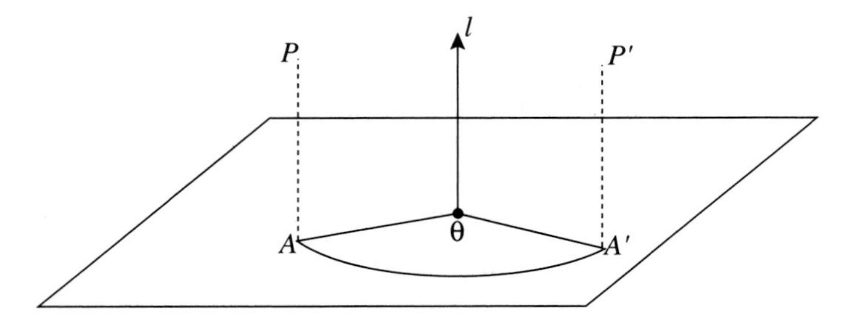

Figure 4.11

about l through an angle θ is obtained as follows:

(i) Consider the plane $\mathcal{P}$ through point P that is perpendicular to the axis of rotation l.
(ii) Let O be given by $O = l \cap \mathcal{P}$.
(iii) Let R be a rotation of plane $\mathcal{P}$ about O through the angle θ.

The image of P is the point $P' = R(P)$. Observe that all points on l are fixed points of this transformation.

Translations

As in the plane, a *translation* in space is a motion that moves all points the same length in the direction of a directed line segment $\overset{\mapsto}{AB}$.

Proposition 4.3.3 *Let S_1 and S_2 be reflections through planes $\mathcal{P}_1$ and $\mathcal{P}_2$, respectively.*
(a) If $l = \mathcal{P}_1 \cap \mathcal{P}_2$, then $S_2 \circ S_1$ is a rotation about line l.
(b) If $\mathcal{P}_1$ and $\mathcal{P}_2$ are parallel planes, then $S_2 \circ S_1$ is a translation in the direction of a line perpendicular to both planes.

Proof (a) First notice that $S_2 \circ S_1(Q) = Q$ for all points $Q \in l$. Choose a point $P \notin l$ and let $\mathcal{P}'$ be the plane perpendicular to l that contains P. Let $O = \mathcal{P}' \cap l$, $l_1 = \mathcal{P}_1 \cap \mathcal{P}'$, and $l_2 = \mathcal{P}_2 \cap \mathcal{P}'$. We have $S_2(S_1(O)) = O$. Notice that the image of a point in $\mathcal{P}'$ by S_1 coincides with the image of the same point by $\bar{S}_1$, the reflection of $\mathcal{P}'$ through line l_1 (see Exercise 3). The same observation is true for $\bar{S}_2$, the reflection of $\mathcal{P}'$ through line l_2.

Recall from Chapter 3 that $\bar{S}_2 \circ \bar{S}_1$ is a rotation of $\mathcal{P}'$ about O that we denote by $\bar{R}$. Now let R be a rotation of space about l. Then $R(Q) = Q = S_2 \circ S_1(Q)$ for points on l and $R(A) = \bar{R}(A)$ for all points $A \in \mathcal{P}'$, which implies

$$\bar{R}(A) = \bar{S}_2 \circ \bar{S}_1(A) = S_2 \circ S_1(A) \quad \forall\, A \in \mathcal{P}'.$$

Therefore, $R = S_2 \circ S_1$, by Proposition 4.3.2.
(b) Let l be a line that is perpendicular to both planes and let $A_1 = l \cap \mathcal{P}_1$ and $A_2 = l \cap \mathcal{P}_2$. Let m_i be the line in plane $\mathcal{P}_i$ that is perpendicular to l at A_i. We then have that l, m_1, and m_2 determine a plane, say $\mathcal{P}'$, and $m_1 \parallel m_2$. Using Exercise 3, we conclude that $S_i(P) = \bar{S}_i(P)$ for point $P \in \mathcal{P}'$. It follows from Chapter 3 that $\bar{S}_2 \circ \bar{S}_1$ is translation of

$\mathcal{P}'$ along line l. Since any other line l' that is perpendicular to $\mathcal{P}_1$ and $\mathcal{P}_2$ is parallell to l, we conclude that $S_2 \circ S_1$ coincides with translation along line l. □

Exercises

1. Review the proof of Proposition 1.3 of Chapter 3. Use similar arguments to prove Proposition 4.3.1.

2. Let T be a rigid motion of 3-space. Show the following:
   (a) If l is a line, then so is $T(l)$.
   (b) If $\mathcal{P}$ is a plane, then so is $T(\mathcal{P})$.
   Hint: Consider points P and Q such that $P \in \mathcal{P}$ and $Q \in l$, the line perpendicular to $\mathcal{P}$ at P. Show first that $T(\mathcal{P}) \subset \mathcal{P}'$, where $\mathcal{P}'$ is the plane perpendicular to line $l' = \overleftrightarrow{P'Q'}$, $P' = T(P)$, and $Q' = T(Q)$. Then show that every point in $\mathcal{P}'$ is the image of a point in $\mathcal{P}$.

3. Let S be the reflection of 3-space in plane $\mathcal{P}$, $l \subset \mathcal{P}$, and $\mathcal{P}'$ a plane perpendicular to l. Let $m = \mathcal{P} \cap \mathcal{P}'$ and consider $\bar{S}$, the reflection of $\mathcal{P}'$ through m. Show that $S(P) = \bar{S}(P)$ for all points $P \in \mathcal{P}'$.
   Hint: Use Exercise 2.

4. Show that every rotation in 3-space is the composition of two reflections. Describe these two reflections.
   Hint: Use Propositions 4.3.2, 4.3.3, and the similar result of Chapter 3 for a plane rotation.

5. Show that every translation in 3-space is the composition of two reflections. Describe these two reflections.
   Hint: Use Propositions 4.3.2, 4.3.3, and the similar result of Chapter 3 for a plane translation.

6. Let T be a rigid motion of 3-space that is not the identity.
   (a) Prove that if T fixes points A and B, then it fixes all points on line $\overleftrightarrow{AB}$.
   (b) Prove that if all fixed points of T are collinear, then T is a rotation about the line that contains the fixed points.

(c) Prove that if T fixes three noncollinear points A, B, and C, then T is the reflection through the plane determined by points A, B and C.
(d) Prove that if T has only one fixed point, then T is the composition of a reflection with a rotation.

4.4 Vectors in 3-Space

In the next chapter we will study rigid motions of higher-dimensional spaces. In order to better describe them we will study in this section vectors in 3-space. We start this section reviewing vectors as they are often used in calculus, physics, and mechanics. A *vector* in a plane

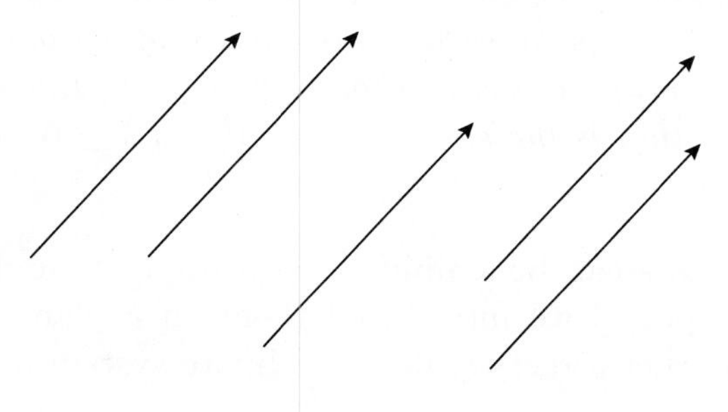

Figure 4.12 Equivalent vectors

or 3-space is represented geometrically by a directed line segment or an arrow. The direction of the arrow is the direction of the vector, while its length describes the magnitude of the vector and is usually called the *norm of the vector.* Directed line segments having the same length and same direction are called *equivalent.* A vector is defined as an *equivalence class* of this equivalence relation. As vectors, we regard them as *equal*, even though the line segments may be located in different positions. Therefore if $\overset{\longmapsto}{AB}$ and $\overset{\longmapsto}{CD}$ are equivalent and $\overset{\longmapsto}{AB}$ represents a vector v and $\overset{\longmapsto}{CD}$ represents a vector u, we write $v = u$. The vector of norm zero is called the *zero vector* and will be denoted by $\mathbf{0}$.

Definition 4.4.1 *Let v and u be two vectors and k a real number. We define:*
(i) $v + u$, called the sum *of u and v, is represented by the diagonal of the parallelogram determined by v and u.*

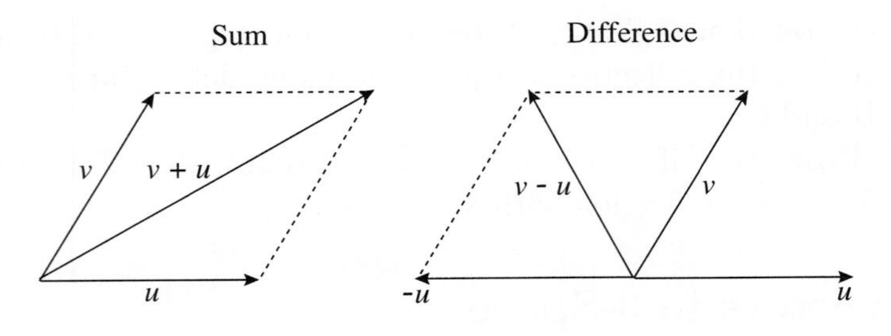

Figure 4.13

(ii) $-v$, called the negative *of v, is the vector having the same norm, but oppositely directed.*
(iii) $v - u = v + (-u)$ is called the subtraction *of u from v.*
(iv) the product *kv is the vector whose length is $|k|$ times the length of v and whose direction is the same as that of v if $k \geq 0$ and opposite to that of v if $k < 0$.*

The study of vectors can be simplified by using rectangular coordinate systems. In Chapter 3 we introduced them on a plane. We describe now how to construct a rectangular coordinate system in 3-space:

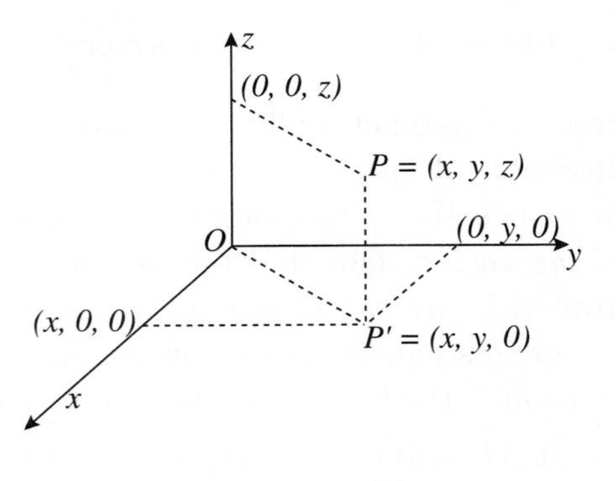

Figure 4.14

Choose a plane $\mathcal{P}$ and coordinatize it as in Chapter 2. We then have chosen a point O that we called origin of $\mathcal{P}$ and two perpendicular axes that we labeled x and y. Using Theorem 4.2.7, we consider the unique line through O that is perpendicular to $\mathcal{P}$. Now we have three mutually perpendicular lines passing through point O. We call the lines *coordinate axes* and point O the *origin*. Label the third line as z-axis and then select a positive direction for each axis using the "right-hand rule." This means that if you point the thumb of your right hand in the positive x-direction, and your first finger in the positive y-direction, then the rest of your fingers of your right hand curl toward the positive z-direction. Choose the same unit of length for each coordinate axis. Each pair of coordinate axes determines a plane called the *coordinate plane*. There are three such planes, referred as the xy-plane, xz-plane, and yz-plane. The xy-plane is then the original plane $\mathcal{P}$.

If P is a point in 3-space, using Theorem 4.2.9 we consider the unique line l through P that is perpendicular to the xy-plane. Let P' be the point of intersection of l with the xy-plane; in this plane P' is written in a unique way as the pair of real numbers (x, y); we then write P' in space as the triple $(x, y, 0)$. Now, on the plane that contains the z-axis and line l, we drop a perpendicular from P to the z-axis; let Z denote the foot of this perpendicular and z the real number corresponding to point Z. Notice that $\square ZPP'O$ is a rectangle in the plane containing P, P', and O. Therefore $z = OZ = P'P$, and rectangle $\square ZPP'O$ assigns in a unique way the third coordinate of P. The *coordinates* of P are defined to be (x, y, z), identifying the 3-space with the set

$$\mathbf{R}^3 = \{(x, y, z) \mid x, y, z \in \mathbf{R}\}.$$

Using a rectangular coordinate system, we locate a vector v in $\mathbf{R}^3$ so that its initial point is at the origin of a rectangular coordinate system. Then the *coordinates of its terminal point* are called the *components* of v, and we write

$$v = (x, y, z)$$

where $P = (x, y, z)$ and $\overset{\mapsto}{OP}$ represents v.

Suppose now that a vector v is not positioned with its initial point at the origin (see Figure 4.15); that is, v is represented by $\overset{\mapsto}{P_1P_2}$. Let $P_1 = (x_1, y_1, z_1)$ and $P_2 = (x_2, y_2, z_2)$. From Definition 4.4.1 we get

$$\overset{\mapsto}{OP_2} - \overset{\mapsto}{OP_1} = \overset{\mapsto}{P_1P_2}.$$

Therefore the components of v are

$$v = (x_2 - x_1,\ y_2 - y_1,\ z_2 - z_1).$$

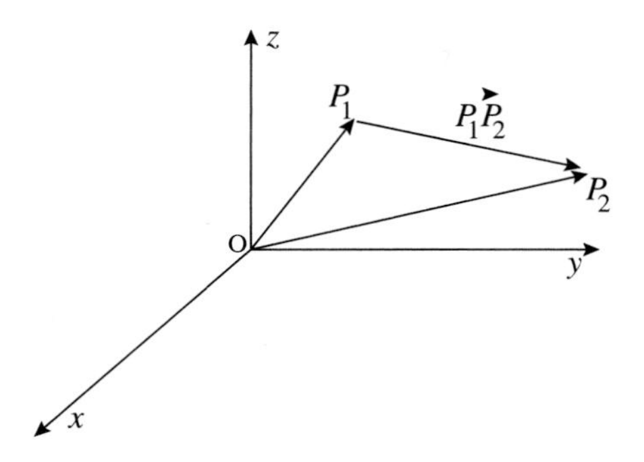

Figure 4.15

Proceeding as in Chapter 3, applying the Pythagorean theorem twice, we obtain that the length of a segment $\overline{PQ}$ is given by

$$P_1P_2 = \sqrt{(x_1 - x_2)^2 + (y_1 - y_2)^2 + (z_1 - z_2)^2}.$$

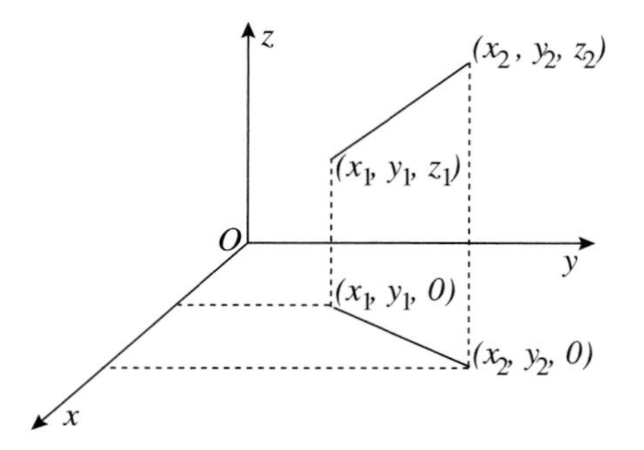

Figure 4.16

Let $||v||$ denote the *length* or the *norm* of a vector $v = (x, y, z)$. Since $||v||$ is the length of any directed line segment that represents v, we obtain

$$||v|| = \sqrt{x^2 + y^2 + z^2}.$$

Observe that $||v|| = 0$ if and only if $v = (0, 0, 0)$.

The dot product

Definition 4.4.2 *Let* $v = (x_1, y_1, z_1)$ *and* $u = (x_2, y_2, z_2)$ *be vectors in* $\mathbf{R}^3$. *The* dot product *or* Euclidean inner product *is defined by*

$$v \cdot u = x_1 x_2 \ + \ y_1 y_2 \ + \ z_1 z_2.$$

Proposition 4.4.3 *The dot product satisfies the following properties:*
(a) $v \cdot u = u \cdot v$ *for all vectors* v, u *in* $\mathbf{R}^3$.
(b) $(w + v) \cdot u = w \cdot u \ + \ v \cdot u$ *for all vectors* w, v, u *in* $\mathbf{R}^3$.
(c) $v \cdot v = ||v||^2$ *for all vectors* v *in* $\mathbf{R}^3$.
(d) Let us assume that vectors v *and* u *are positioned so that their initial points coincide. Let* θ *denote the measure of the angle between* v *and* u. *If* $v \neq 0$ *and* $u \neq 0$, *then*

$$v \cdot u = ||v|| \ ||u|| \cos\theta.$$

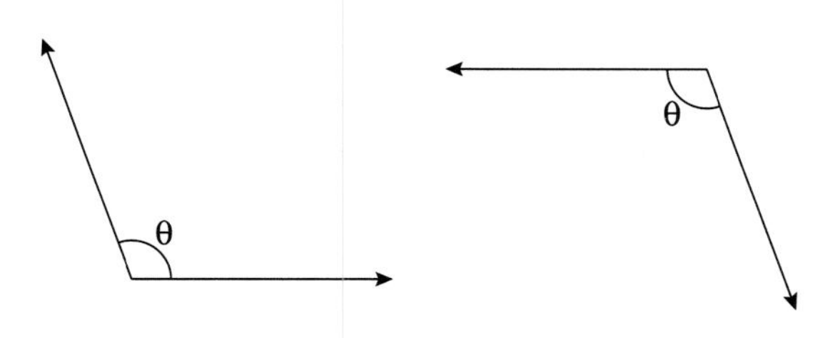

Figure 4.17

(e) Let θ *be the angle between* v *and* u. *Then*

$$\theta < 90^\circ \quad \text{if and only if} \quad u \cdot v > 0,$$

$$\theta > 90^\circ \quad \text{if and only if} \quad u \cdot v < 0,$$

$$\theta = 90^\circ \quad \text{if and only if} \quad u \cdot v = 0.$$

Proof The proofs of *(a)*, *(b)*, and *(c)* follow immediately from the definition. For *(d)*, let $u = (x_1, y_1, z_1)$ and $v = (x_2, y_2, z_2)$. Let us

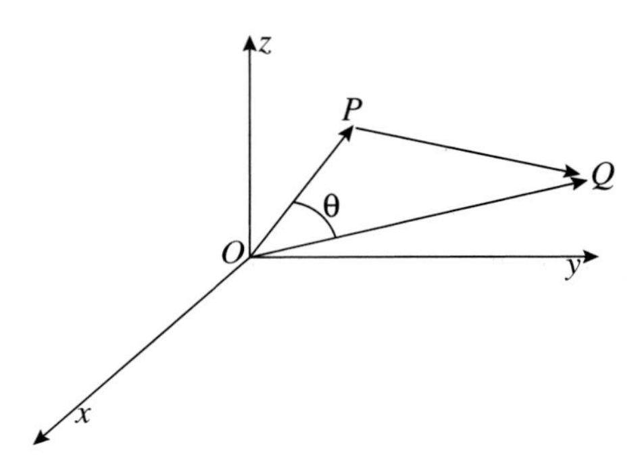

Figure 4.18

consider points P and Q (see Figure 4.18) such that $\overset{\longmapsto}{OP}$ and $\overset{\longmapsto}{OQ}$ represent u and v respectively. From the law of cosines we have

$$||\overset{\longmapsto}{PQ}||^2 = ||v||^2 + ||u||^2 - 2||u||\,||v||\cos\theta.$$

Since $\overset{\longmapsto}{PQ} = v - u = ((x_2 - x_1), (y_2 - y_1), (z_2 - z_1))$, substituting above we get

$$(x_2 - x_1)^2 + (y_2 - y_1)^2 + (z_2 - z_1)^2$$
$$= x_2^2 + y_2^2 + z_2^2 + x_1^2 + y_1^2 + z_1^2 - 2||u||\,||v||\cos\theta.$$

Expanding the left-hand side of the equation above, we obtain after simplifying

$$x_1x_2 + y_1y_2 + z_1z_2 = ||u||\,||v||\cos\theta.$$

Now *(e)* is an immediate consequence of *(d)* and the definition of cosine.

□

Definition 4.4.4 *Two vectors v and u are said to be* orthogonal *if their dot product is zero.*

Lines and planes in 3-space

We shall show now how to obtain equations of lines and planes in 3-space.

Let P_0, P, and Q be three distinct points on a line l. Then the directed line segments satisfy the relation

$$\overrightarrow{P_0P} = c\,\overrightarrow{P_0Q} = d\,\overrightarrow{PQ}, \quad \text{for some} \quad c, d \in \mathbf{R},$$

that is, $\overrightarrow{P_0P}, \overrightarrow{P_0Q}$, and $\overrightarrow{PQ}$ are all multiples of the same vector. We then make the following definition.

Definition 4.4.5 *We say that line l is in the direction of vector v if, given any two distinct points P and Q on l, then $\overrightarrow{PQ} = cv$, for some real number c.*

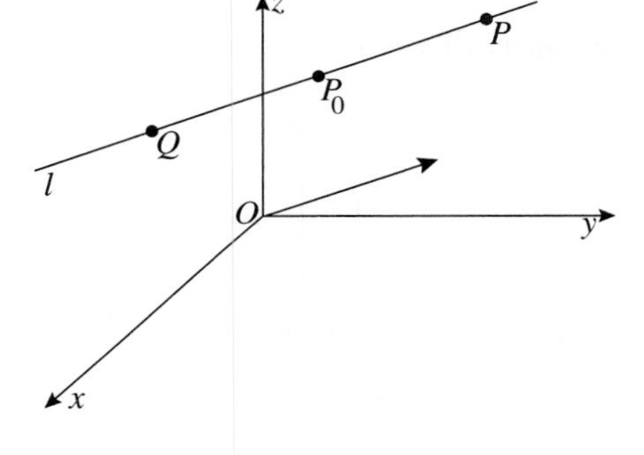

Figure 4.19

Parametric equations for a line

Let $P = (x, y, z)$ be a point on the line l through $P_0 = (x_0, y_0, z_0)$ and in the direction of vector $v = (a, b, c)$. Then x, y, and z satisfy

$$\overrightarrow{P_0P} = (x - x_0, y - y_0, z - z_0) = c\,(a, b, c), \quad \text{for some} \quad c \in \mathbf{R}.$$

It follows that the vector equation

$$(x - x_0, y - y_0, z - z_0) = t\,(a, b, c), \quad t \in \mathbf{R},$$

describes all points P on line l. The vector equation above gives rise to

$$x = x_0 + ta, \quad y = y_0 + tb, \quad z = z_0 + tc, \qquad t \in \mathbf{R},$$

which are the *parametric equations* for line l.

Definition 4.4.6 *We say that vector n is a* normal vector *of plane $\mathcal{P}$, if it is parallel to a line perpendicular to $\mathcal{P}$.*

Equation of a plane

Let $\mathcal{P}$ be a plane through the point $P_0 = (x_0, y_0, z_0)$ and $n = (a, b, c)$ a *normal vector*. Then, if P is any other point in $\mathcal{P}$, the vector $\overrightarrow{PP_0}= (x - x_0, y - y_0, z - z_0)$ is orthogonal to n, and therefore the equation

$$\overrightarrow{PP_0} \cdot n = (x - x_0, y - y_0, z - z_0) \cdot (a, b, c) = 0$$

describes all points P on the plane $\mathcal{P}$. Multiplying out and collecting the terms, we have

$$ax + by + cz + d = 0$$

where a, b, and c are not all zero.

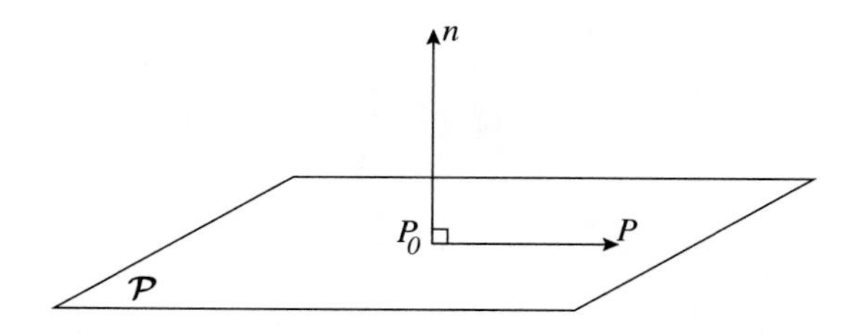

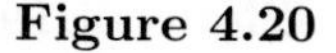

Figure 4.20

The vector product

Definition 4.4.7 *The* vector product *of vectors $u = (u_1, u_2, u_3)$ and $v = (v_1, v_2, v_3)$ (in that order) is the unique vector denoted by $u \wedge v$ such that*

$$(u \wedge v) \cdot w = \det(w, u, v) \quad \text{for all } w = (w_1, w_2, w_3) \in \mathbf{R}^3.$$

Here $\det(u, v, w)$ means the *determinant* of the matrix

$$\begin{pmatrix} w_1 & w_2 & w_3 \\ u_1 & u_2 & u_3 \\ v_1 & v_2 & v_3 \end{pmatrix},$$

which is given by

$$\det(u, v, w) = w_1(u_2v_3 - v_2u_3) \; - \; w_2(u_1v_3 - v_1u_3) \; + \; w_3(u_1v_2 - v_1u_2).$$

It is immediate from the definition that

$$u \wedge v = (u_2v_3 - v_2u_3, \; u_1v_3 - v_1u_3, \; u_1v_2 - v_1u_2).$$

It is common to denote the vector product by $u \times v$ and call it *the cross product.*

Proposition 4.4.8 *The vector product has the following properties:*
(a) $u \wedge v = -v \wedge u$ *(anticommutativity).*
(b) $(au + bw) \wedge v = au \wedge v + bw \wedge v$, *for all* $u, v \in bfR^3$ *and all real numbers* a, b.
(c) $u \wedge v = 0$ *if and only if* $u = cv$, *for some real number* c.
(d) $(u \wedge v) \cdot u = (u \wedge v) \cdot v = 0$.

It follows from Proposition 4.4.8(d) that the vector $u \wedge v \neq 0$ is orthogonal to u and v. If u and v are represented by line segments with the same initial point P, Proposition 4.4.8(d) implies that $u \wedge v$ is parallel to the line that is perpendicular to the plane containing u and v at P. Therefore $u \wedge v$ is normal to the plane determined by u and v. This fact implies the following proposition.

Proposition 4.4.9 *The equation of the plane passing through* 3 *non-collinear points* P_1, P_2, *and* P_3 *is given by*

$$\overrightarrow{PP_1} \wedge \overrightarrow{PP_2} \cdot \overrightarrow{PP_3} = 0.$$

Proof If P is a point in the plane passing through P_1, P_2, and P_3, then the vectors $\overrightarrow{PP_i}$, $i = 1, 2, 3$, are in such a plane and the vector $\overrightarrow{PP_1} \wedge \overrightarrow{PP_2}$ is normal to the plane. □

In the rest of this section we shall describe in rectangular coordinates the main properties of rigid motions of 3-space. In the following propositions, P' will denote the image of the point P by a transformation T.

Proposition 4.4.10 *Let $T : \mathbf{R}^3 \to \mathbf{R}^3$ be a rigid motion. Then:*
(a) $\| \overrightarrow{P' Q'} \| = \| \overrightarrow{P Q} \|, \quad \forall\, P, Q \in \mathbf{R}^3$.
(b) For all points A, B, C of $\mathbf{R}^3$ we have

$$\overrightarrow{A'B'} \cdot \overrightarrow{A'C'} = \overrightarrow{AB} \cdot \overrightarrow{AC}.$$

Proof *(a)* The conclusion follows from the definition of the norm of a vector, that is,

$$P'Q' = \| \overrightarrow{P'Q'} \| \quad \text{and} \quad P'Q' = PQ = \| \overrightarrow{PQ} \|.$$

(b) Recall that we proved in Proposition 4.3.1(c) that rigid motions of the 3-space preserve angle measure. Then, applying part(a) of this proposition and part (d) of Proposition 4.4.3, we finish this proof. □

In $\mathbf{R}^3$, we define the *distance* between two points P and Q, denoted by $||P - Q||$, as the length of line segment $\overline{PQ}$. Then we have the formula

$$||P - Q|| = \| \overrightarrow{OP} - \overrightarrow{OQ} \| = \sqrt{(x_1 - x_2)^2 + (y_1 - y_2)^2 - (z_1 - z_2)^2},$$

which is called *Euclidean* distance because it was obtained using the Pythagorean theorem, which is strictly Euclidean. The word *isometry* means "same distance." A rigid motion is then an isometry that preserves Euclidean distance.

In the next chapter we will classify all rigid motions of 3-space by classifying first the ones with a fixed point (see also Exercise 6 of Section 4.3). Notice that if a transformation has a fixed point O, the rectangular system can be placed such that point O is the origin of the system. For rigid motions with a fixed point we have the following result.

Proposition 4.4.11 *Let $T : \mathbf{R}^3 \to \mathbf{R}^3$ be a rigid motion such that $T((0, 0, 0)) = (0, 0, 0)$. Let v be a vector with components $v = (x, y, z)$. Placing the initial point of v at $(0, 0, 0)$, we define $T(v) = T((x, y, z))$. Then for all vectors u, v we have,*

$$T(v) \cdot T(u) = v \cdot u$$

and in particular, $||T(v)|| = ||v||$.

Proof Let $P = v = (x_1, y_1, z_1)$, $Q = u = (x_2, y_2, z_2)$. Then $v = \overset{\longmapsto}{OP}$ and $u = \overset{\longmapsto}{OQ}$, where $O = (0, 0, 0)$. By Proposition 4.4.10 we have

$$\overset{\longmapsto}{O'P'} \cdot \overset{\longmapsto}{O'Q'} = \overset{\longmapsto}{OP} \cdot \overset{\longmapsto}{OQ} = v \cdot u.$$

Since $T(O) = O$, $T(v) = T(P) = P'$, and $T(u) = T(Q) = Q'$, we have

$$\overset{\longmapsto}{OP'} \cdot \overset{\longmapsto}{OQ'} = T(v) \cdot T(u)$$

and hence $T(v) \cdot T(u) = v \cdot u$. □

Using rectangular coordinates, we now write a formula for each of the types of rigid motion of space.

Reflection through a plane $\mathcal{P}$

Placing a rectangular system such that plane $\mathcal{P}$ is the xy-plane, we describe the reflection through $\mathcal{P}$ by

$$T : \mathbf{R}^3 \to \mathbf{R}^3$$

$$(x, y, z) \mapsto (x, y, -z).$$

Rotation about a line l through an angle θ

Let us place a rectangular system such that line l is the z-axis. Then for $P = (x, y, z)$, let $A = (x, y, 0)$ and $A' = R(A) = (x', y', 0)$ (see Figure 4.21). Therefore

$$T : \mathbf{R}^3 \to \mathbf{R}^3$$

is given by

$$T((x, y, z)) = \begin{pmatrix} x' \\ y' \\ z' \end{pmatrix} = \begin{pmatrix} \cos\theta & -\sin\theta & 0 \\ \sin\theta & \cos\theta & 0 \\ 0 & 0 & 1 \end{pmatrix} \begin{pmatrix} x \\ y \\ z \end{pmatrix}$$

that is, such a rotation is written as

$$T((x, y, z)) = (x \cos\theta - y \sin\theta, x \sin\theta + y \cos\theta, z).$$

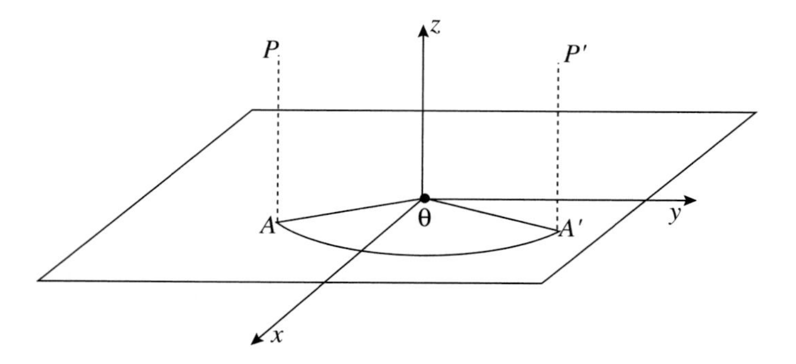

Figure 4.21

Translations

Since a *translation* moves all points the same distance in the same direction, the direction and the distance are very well characterized by a vector $\vec{v}$, and using a rectangular system, we can write $v = (a, b, c)$. Therefore a translation is written as

$$T : \mathbf{R}^3 \to \mathbf{R}^3$$

$$(x, y, z) \mapsto (x + a, y + b, z + c).$$

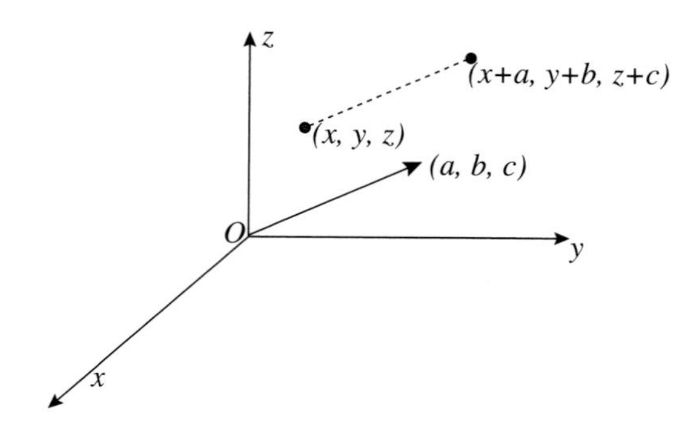

Figure 4.22

Exercises

1. Find an equation of the plane through the given point and with the given normal vector:
(a) $P = (1, 4, 5)$, $n = (7, 1, 4)$.
(b) $P = (1, 0, 1)$, $n = (1, 1, 1)$.
(c) $P = (0, 1, 2)$, $n = (-1, 0, 1)$.

2. Find an equation of the plane through the given point and parallel to the given plane:
(a) $P = (5, 3, 2)$, $x + 2y - 3z + 4 = 0$.
(b) $P = (2, 3, 4)$, $z = 3$.
(c) $P = (1, 0, -1)$, $z = 2x + y$.

3. Find an equation of the plane through the given point and containing the given line:
(a) $P = (1, 0, 1)$, $x = -1 + 2t$, $y = 2 - 3t$, $z = 3 - t$.
(b) $P = (-1, 0, 1)$, $x = 1 - t$, $y = 3t$, $z = 1 + t$.
(c) $P = (1, 1, 1)$, $x = 2t$, $y = 2$, $z = 1 + t$.

4. Define the angle between two planes as the the smallest dihedral angle determined by them. Show that the measure of the angle between two planes equals the measure of the angle betweeen their normal vectors.

5. Determine whether the planes are parallel, perpendicular, or neither. If neither, determine the angle between them.
(a) $x + z = 1$ and $y + z = 1$.
(b) $8x + 6y - 2z = 1$ and $z = 4x + 3y$.
(c) $x + 4y - 3z = 1$ and $-3x + 6y + 7z = 0$.

6. Consider a point $P = (x_0, y_0, z_0)$ not in the plane with equation $ax + by + cz + d = 0$. Let l be a line through P that is perpendicular to the given plane and O the point where l intersects the plane. Show that

$$PO = \frac{|ax_0 + by_0 + cz_0 - d|}{\sqrt{a^2 + b^2 + c^2}}.$$

PO is called the *distance* from point P to plane $\mathcal{P}$.

Hint: Observe that the vector $\overset{\mapsto}{OP}$ is parallel to $n = (a, b, c)$; then use the dot product.

7. Determine the angle of intersection of plane $3x + 4y + 7z + 8 = 0$ and line $x - 2 = 3t,\ \ y - 3 = 5t,\ \ z - 5 = 9t$.

8. Show that the distance D between the parallel planes $ax + by + cz + d_1 = 0$ and $ax + by + cz + d_2 = 0$ is

$$D = \frac{|d_1 - d_2|}{\sqrt{a^2 + b^2 + c^2}}.$$

9. The vector projection of vector $v = \overset{\longmapsto}{OP}$ onto vector $u = \overset{\longmapsto}{OQ}$ is defined as vector $w = \overset{\longmapsto}{OR}$ such that $\overset{\longleftrightarrow}{PR} \perp \overset{\longleftrightarrow}{OR}$.
(a) Show that the vector projection of vector v onto vector u is given by

$$\frac{v \cdot u}{||u||^2} u.$$

(b) Show that the vector

$$w = v - \frac{v \cdot u}{||u||^2} u$$

is orthogonal to u.

10. Show the following identity for the vector product:

$$(A \wedge B) \cdot (C \wedge D) = (A \cdot C)(B \cdot D) - (A \cdot D)(B \cdot C).$$

11. Given the vectors $v \neq 0$ and w, show that there exists a vector u such that $u \wedge v = w$ if and only if v is perpendicular to w. Is this vector u uniquely determined?

12. Given two nonparallel planes $a_i x + b_i y + c_i z + d_i = 0,\ i = 1, 2$, show that their line of intersection may be parametrized as

$$x - x_0 = v_1 t, \quad y - y_0 = v_2 t, \quad z - z_0 = v_3 t,$$

where (x_0, y_0, z_0) belongs to the intersection and

$$v = (v_1, v_2, v_3) = n_1 \wedge n_2, \quad \text{for} \quad n_i = (a_i, b_i, c_i).$$

13. Find a formula for the reflection through the plane $x = 5$.
Hint: Let L be the translation that maps plane $x = 5$ to the yz-plane. Let S be a reflection through yz-plane; then consider the composition $L^{-1} \circ S \circ L$.

14. Find a formula for the rotation about line $x = 1, \quad y = 1$ through the angle $\theta = 45^\circ$.
Hint: Compose again with a suitable translation.

15. (a) Following the notion of homothety defined in Chapter 3, define a homothety as a map from $\mathbf{R}^3$ to $\mathbf{R}^3$.
(b) Find a formula for the homothety of center $(0, 0, 0)$ and ratio $r = -2$.
(c) Find a formula for the homothety of center $(1, 0, 1)$ and ratio $r = 2$.

Chapter 5

EUCLIDEAN n-SPACE

5.1 The n-Space

Most of the geometrical ideas of a plane or 3-space can be extended to any dimension; that is, an ordered n-tuple of real numbers is thought of as a point of an "n-dimensional space." We first make the following definition.

Definition 5.1.1 *Let n be a positive integer. The set*

$$\mathbf{R}^n = \{(x_1, \dots, x_n) \mid x_i \in \mathbf{R},\ \forall i = 1, \dots, n\}$$

*is called n-*space*. The elements $(x_1, \dots, x_n)$ of $\mathbf{R}^n$ are called* vectors.

On $\mathbf{R}^n$ we define the following operations:

Vector addition
For $v = (x_1, \dots, x_n)$ and $u = (y_1, \dots, y_n)$ in $\mathbf{R}^n$,

$$v + u = (x_1 + y_1, \dots, x_n + y_n)$$

Scalar multiplication
For $v = (x_1, \dots, x_n)$ and $\lambda \in \mathbf{R}$,

$$\lambda v = (\lambda x_1, \dots, \lambda x_n).$$

The result below contains the properties of these two operations. Its proof is straightforward.

Theorem 5.1.2 *Let* $\mathbf{0}$ *denote the zero vector* $(0, \ldots, 0)$. *Then for all vectors* u, v, w *and all real numbers* λ, μ *we have:*
(a) $v + u = u + v$
(b) $(v + u) + w = v + (u + w)$
(c) $\mathbf{0} = \mathbf{0} + v$
(d) $v + (-v) = \mathbf{0}$
(e) $\lambda(\mu v) = (\lambda\mu)v$
(f) $\lambda(v + u) = \lambda v + \lambda u$
(g) $(\lambda + \mu)v = \lambda v + \mu v$
(h) $1v = v$.

Definition 5.1.3 *Let* $v = (x_1, \ldots, x_n)$ *and* $u = (y_1, \ldots, y_n)$ *be any vectors in* $\mathbf{R}^n$. *The* Euclidean inner product $v \cdot u$ *is defined by*

$$v \cdot u = x_1 y_1 + \cdots + x_n y_n.$$

The Euclidean norm or Euclidean length of $v = (x_1, \ldots, x_n)$ *is defined by*

$$||v|| = \sqrt{v \cdot v} = \sqrt{x_1^2 + \cdots + x_n^2}.$$

Observe that the Euclidean inner product is just a generalization of the dot product of $\mathbf{R}^2$ and $\mathbf{R}^3$. Further, $||v||$ is the length of line segment $\overline{\mathbf{0}v}$, obtained from the Pythagorean theorem. The Euclidean inner product satisfies the same properties of the dot product that we list below.

Proposition 5.1.4 *For all vectors* v, u *of* $\mathbf{R}^n$ *we have*
(a) $v \cdot u = u \cdot v$
(b) $(w + v) \cdot u = w \cdot u + v \cdot u$
(c) $v \cdot v = ||v||^2$
(d) $v \cdot u = ||v|| ||u|| \cos\theta$, *where* θ *is the angle determined by* v *and* u *that satisfies* $0 \leq \theta \leq 180°$.

Definition 5.1.5 *The* Euclidean distance *between two points* $P = (x_1, \ldots, x_n)$ *and* $Q = (y_1, \ldots, y_n)$ *is defined by*

$$d(P, Q) = ||P - Q|| = \sqrt{(x_1 - y_1)^2 + \cdots + (x_n - y_n)^2}.$$

The set $\mathbf{R}^n$ *endowed with this distance is called* Euclidean n-space.

We finish this section by generalizing to $\mathbf{R}^n$ the notion of rigid motion.

Definition 5.1.6 *A map $T : \mathbf{R}^n \to \mathbf{R}^n$ is called a* rigid motion *or a* Euclidean isometry *of $\mathbf{R}^n$ if*

$$||T(P) - T(Q)|| = ||P - Q|| \quad \forall\, P, Q \in \mathbf{R}^n.$$

5.2 Basis and Change of Coordinates

In the previous section an ordered n-tuple $(x_1, \ldots, x_n)$ of real numbers was called a vector whose components are the real numbers $x_1, \ldots, x_n$. Now consider the particular vectors

$$e_1 = (1, 0, \ldots, 0), \quad e_2 = (0, 1, 0, \ldots, 0), \quad \text{and} \quad e_n = (0, \ldots, 0, 1).$$

Then the components $x_1, \ldots, x_n$ of a vector v are the numbers that make true the vector equation

$$v = (x_1, \ldots, x_n) = x_1 e_1 + \cdots + x_n e_n.$$

However, many problems in geometry become easier to solve if we assign to $v = (x_1, \ldots, x_n)$ another n-tuple of real numbers that plays a similar role with respect to a special set of vectors that the n-tuple $(x_1, \ldots, x_n)$ does with respect to the set $\{e_1, \ldots, e_n\}$. First we introduce some important concepts of linear algebra.

Definition 5.2.1 *Let $v_1, \ldots, v_k$ be vectors in $\mathbf{R}^n$. A vector v is called a* linear combination *of the vectors $v_1, \ldots, v_k$ if there exist real numbers $c_1, \ldots, c_k$ such that*

$$v = c_1 v_1 + \cdots + c_k v_k$$

Example 1

Consider the vectors $v_1 = (2, 1)$ and $v_2 = (0, 1)$. Show that $v = (1, -1)$ is a linear combination of v_1 and v_2.

We have to show that there are real numbers c_1 and c_2 such that

$$(1, -1) = c_1(2, 1) + c_2(0, 1)$$

which is equivalent to saying that the system of linear equations

$$2c_1 = 1, \qquad c_1 + c_2 = -1$$

has a solution. It is easy to see that $c_1 = 1/2$ and $c_2 = -3/2$ is a solution for the system.

Example 2

Consider the vectors $v_1 = (1, 2, -1)$ and $v_2 = (0, 1, 1)$. Show that $v = (0, 1, 0)$ is *not* a linear combination of v_1 and v_2.

As in Example 1 we equate the components and obtain the system

$$c_1 = 0, \qquad 2c_1 + c_2 = 1, \qquad -c_1 + c_2 = 0,$$

which is clearly inconsistent.

Example 3

Consider the vectors $v_1 = (1, 1), v_2 = (1, 0), v_3 = (1, -1)$. Show that $v = (0, 1)$ is a linear combination of v_1, v_2, and v_3.

A linear combination $(0, 1) = c_1(1, 1) + c_2(1, 0) + c_3(1, -1)$ gives rise to the system

$$c_1 + c_2 + c_3 = 0, \qquad c_1 - c_3 = 0.$$

Observe that this system has infinitely many solutions. Therefore v is not uniquely expressed as a linear combination of v_1, v_2, and v_3.

Definition 5.2.2 *A set of vectors $\beta = \{v_1, \ldots, v_k\}$ is called a* basis *of $\mathbf{R}^n$ if each vector v of $\mathbf{R}^n$ can be uniquely expressed as a linear combination of the vectors in β.*

Example 4

Consider $\beta = \{e_1, e_2, \ldots, e_n\}$, where

$$e_1 = (1, 0, \ldots, 0), \quad e_2 = (0, 1, 0, \ldots, 0), \quad \text{and} \quad e_n = (0, \ldots, 0, 1).$$

It is clear that each vector $v = (x_1, \ldots, x_n)$ is written in a unique way as a linear combination of the vectors $e_1, e_2, \ldots, e_n$. This is called the *standard basis* of $\mathbf{R}^n$.

Definition 5.2.3 *A set of vectors* $\{v_1, \ldots, v_k\}$ *of* $\mathbf{R}^n$ *is said to be* linearly independent *if whenever* $c_1v_1 + \cdots + c_kv_k = \mathbf{0}$, *where* $\mathbf{0}$ *denotes the zero vector, then necessarily* $c_1 = 0, \ldots, c_n = 0$. *If* $v_1, \ldots, v_k$ *are not linearly independent, we say that* $v_1, \ldots, v_k$ *are* linearly dependent.

Lemma 5.2.4 *Let* $\{v_1, \ldots, v_k\}$ *be a set of linearly independent vectors of* $\mathbf{R}^n$. *Then*
(a) $k \leq n$.
(b) If v *is a linear combination of* $v_1, \ldots, v_k$, *there exist unique real numbers such that* $v = c_1v_1 + \cdots + c_kv_k$.

Proof Consider the linear combination

$$\mathbf{0} = (0, \ldots, 0) = c_1v_1 + \cdots + c_kv_k.$$

Writing $v_i = (x_1^i, \cdots, x_n^i)$ and equating the components, we obtain

$$c_1x_1^1 + \cdots + c_kx_1^k = 0$$
$$\vdots$$
$$c_1x_n^1 + \cdots + c_kx_n^k = 0,$$

which is a homogeneous system of linear equations that has k unknowns and n equations. Since this system has only the trivial solution, we conclude that $k \leq n$.

Now to prove *(b)*, suppose there exist real numbers $b_1, \ldots, b_k$ such that

$$v = b_1v_1 + \cdots + b_kv_k.$$

Since we also have $v = c_1v_1 + \cdots + c_kv_k$, subtracting the second equation from the first we get

$$\mathbf{0} = (b_1 - c_1)v_1 + \cdots + (b_k - c_k)v_k.$$

But $v_1, \ldots, v_k$ are linearly independent, and then we conclude that

$$b_1 - c_1 = 0, \ldots, b_k - c_k = 0,$$

and hence $b_i = c_i$ for all $i = 1, \ldots, k$. □

Proposition 5.2.5 *Let* $\beta = \{v_1, \ldots, v_k\}$ *be a basis of* $\mathbf{R}^n$. *Then* $v_1, \ldots, v_k$ *are linearly independent and* $k = n$.

Proof Consider the linear combination

$$\mathbf{0} = (0, \ldots, 0) = c_1 v_1 + \cdots + c_k v_k.$$

Of course the zero vector can also be written as

$$\mathbf{0} = (0, \ldots, 0) = 0v_1 + \cdots + 0v_k.$$

Since β is a basis, the zero vector is written in a unique way as a linear combination of $v_1, \ldots, v_k$. Therefore $c_1 = 0, \ldots, c_k = 0$ and $v_1, \ldots, v_k$ are linearly independent. It follows from Lemma 5.2.4(a) that $k \leq n$. We will show that $n \leq k$.

Again because β is a basis, each vector of the standard basis is written as a linear combination of $v_1, \ldots, v_k$. We have then

$$e_i = b_1^i v_1 + \cdots + b_k^i v_k.$$

Now we consider a linear combination $\mathbf{0} = c_1 e_1 + \cdots + c_n e_n$. Writing each vector e_i in terms of the basis β, we obtain

$$(c_1 b_1^1 + \cdots + c_n b_1^n) v_1 + \cdots + (c_1 b_k^1 + \cdots + c_n b_k^n) v_k = \mathbf{0}.$$

Since $v_1, \ldots, v_k$ are linearly independent, we have

$$\begin{aligned} c_1 b_1^1 + \cdots + c_n b_1^n &= 0 \\ &\vdots \\ c_1 b_k^1 + \cdots + c_n b_k^n &= 0, \end{aligned}$$

which is a homogeneous system of linear equations with n unknowns $c_1, \ldots, c_n$ and k equations. But we know that the only solution is $c_1 = 0, \ldots, c_n = 0$ because $e_1, \ldots, e_n$ are also linearly independent. Therefore $n \leq k$. □

Proposition 5.2.6 *Let $\beta = \{v_1, \ldots, v_n\}$ be a set of linearly independent vectors of $\mathbf{R}^n$. Then β is a basis of $\mathbf{R}^n$.*

Proof If each vector v can be expressed as a linear combination of $v_1, \ldots, v_n$, since β is linearly independent, Lemma 5.2.4(b) implies that each v is expressed in a unique way as a linear combination of the vectors in β. This implies that β is a basis. Therefore, if β is not a basis, then there exists $v \in \mathbf{R}^n$ that is not a linear combination of $v_1, \ldots, v_n$. We will show that this implies that the set $\gamma = \{v_1, \ldots, v_n, v\}$ is also linearly independent, which contradicts Lemma 5.2.4(a), since γ contains $n+1$ vectors. In order to show that γ is linearly independent we let $c_1, \ldots, c_n, c$ be such that

$$c_1 v_1 \ + \cdots + \ c_n v_n \ + \ cv = \mathbf{0}.$$

If $c \neq 0$, then we can write

$$v = \frac{-c_1}{c} v_1 \ + \cdots + \ \frac{-c_n}{c} v_n,$$

which contradicts our assumption that v is not a linear combination of $v_1, \ldots, v_n$. Then $c = 0$ and, substituting above, we get

$$c_1 v_1 \ + c \ldots + \ c_n v_n = \mathbf{0}.$$

Since $v_1, \ldots, v_n$ are linearly independent, the equation above implies

$$c_1 = 0, \ldots, c_n = 0.$$

□

Definition 5.2.7 *An* ordered basis *of $\mathbf{R}^n$ is a basis of $\mathbf{R}^n$ endowed with a special order; that is, it is a sequence of n linearly independent vectors of $\mathbf{R}^n$.*

Let $\beta = \{v_1, \ldots, v_n\}$ be an ordered basis of $\mathbf{R}^n$ and

$$v = c_1 v_1 \ + \cdots + \ c_n v_n.$$

Each vector v determines a unique n-tuple of real numbers $(c_1, \ldots, c_n)$ whose entries are the coefficients of its linear combination of the vectors of β. We call such an n-tuple *coordinates of* v relative to β.

From now on, the n-tuple $(c_1, \ldots, c_n)$ will denote the coordinates relative to the standard basis of $\mathbf{R}^n$. We will use the notation $(c_1, \ldots, c_n)_\beta$ for coordinates relative to some ordered basis β. The column

$$[v]_\beta = (c_1, \ldots, c_n)_\beta^T = \begin{pmatrix} c_1 \\ \vdots \\ c_n \end{pmatrix}_\beta$$

is called the *coordinate column* of v relative to basis β. Vectors in $\mathbf{R}^n$ are usually expressed by its standard coordinates. When the basis is changed, then a relationship between the standard and new coordinates is needed. Such a relationship is given by a matrix usually called *change of basis matrix.* For instance, let $\mathcal{S} = \{e_1, \ldots, e_n\}$ denote the standard basis and consider the basis $\beta = \{v_1, \ldots, v_n\}$. Now we write each vector of $\mathcal{S}$ as linear combination of the vectors of β.

$$e_1 = a_{11}v_1 \ + \cdots + \ a_{n1}v_n$$

$$\vdots$$

$$e_n = a_{n1}v_1 \ + \cdots + \ a_{nn}v_n.$$

The change of basis matrix from $\mathcal{S}$ to β, also called the *transition matrix* and denoted by $[I]^{\mathcal{S}}_{\beta}$, is the matrix whose entries are a_{ij}. Similarly, the change of basis matrix from β to $\mathcal{S}$, $[I]^{\beta}_{\mathcal{S}} = (b_{ij})$, has entries b_{ij} obtained from the linear combinations

$$v_1 = b_{11}e_1 \ + \cdots + \ b_{n1}e_n$$

$$\vdots$$

$$v_n = b_{n1}e_1 \ + \cdots + \ b_{nn}e_n.$$

The change of basis matrices relate the two coordinate columns by the equations

$$[v]_{\beta} = [I]^{\mathcal{S}}_{\beta}\,[v],$$

$$[v] = [I]^{\beta}_{\mathcal{S}}\,[v]_{\beta}.$$

Example 5

Consider the basis $\beta = \{(1,1,0), (1,-1,0), (0,0,1)\}$. Find $[I]^{\beta}_{\mathcal{S}}$ and $[I]^{\mathcal{S}}_{\beta}$.

It follows easily from the above that

$$[I]^{\beta}_{\mathcal{S}} = \begin{pmatrix} 1 & 1 & 0 \\ 1 & -1 & 0 \\ 0 & 0 & 1 \end{pmatrix}.$$

For the second matrix we write

$$(1,0,0) = a(1,1,0) + b(1,-1,0) + c(0,0,1),$$

$$(0,1,0) = d(1,1,0) + e(1,-1,0) + f(0,0,1),$$

$$(1,0,0) = g(1,1,0) + h(1,-1,0) + i(0,0,1).$$

Equating the components and solving the system, we get $a = b = d = 1/2$, $e = -1/2$, $c = f = g = h = 0$, and $i = 1$. Then we have

$$[I]_{\beta}^{\mathcal{S}} = \begin{pmatrix} 1/2 & 1/2 & 0 \\ 1/2 & -1/2 & 0 \\ 0 & 0 & 1 \end{pmatrix}.$$

It is a standard fact in linear algebra (whose proof we omit) that $[I]_{\mathcal{S}}^{\beta}$ is an invertible matrix and the inverse is the matrix $[I]_{\beta}^{\mathcal{S}}$. It follows that the determinant of any change of basis matrix A, denoted by $\det(A)$, is nonzero.

In geometry we are particularly interested in the relationship among coordinates relative to bases that share important geometric properties with the standard basis. The vectors of the standard basis are orthogonal to each other and are all unit vectors.

Definition 5.2.8 *A set of vectors $S = \{v_1, \ldots, v_k\}$ is called* orthogonal *if $v_i \cdot v_j = 0$ for $i \neq j$. In addition, if all vectors are unit vectors, that is, if $v_i \cdot v_i = 1$ for all $i = 1, \ldots, k$, then S is called an* orthonormal set.

Proposition 5.2.9 *Let $\beta = \{v_1, \ldots, v_n\}$ be a set of nonzero orthogonal vectors of $\mathbf{R}^n$. Then β is a basis of $\mathbf{R}^n$.*

Proof We will show that an orthogonal set with nonzero vectors is linearly independent, and then the result follows Proposition 5.2.6. Let us then consider $c_1, \ldots, c_n$ such that

$$c_1 v_1 + \cdots + c_n v_n = \mathbf{0}.$$

For each v_i we have

$$(c_1 v_1 + \cdots + c_n v_n) \cdot v_i = \mathbf{0} \cdot v_i = 0.$$

From the properties of the dot product we obtain

$$c_1\, v_1 \cdot v_i + \cdots + c_i\, v_i \cdot v_i + \cdots + c_n\, v_n \cdot v_i = 0.$$

But $v_i \cdot v_j = 0$ if $i \neq j$ and hence we are left with

$$c_i\, v_i \cdot v_i = c_i ||v_i||^2 = 0,$$

and since $||v_i||^2 \neq 0$, we conclude that $c_i = 0$. Therefore

$$c_1 = 0, \ldots, c_n = 0,$$

and β is linearly independent. □

An orthogonal set of $\mathbf{R}^n$ is called an *orthogonal* basis. In particular if $\{v_1, \ldots, v_n\}$ is an orthonormal set, then it is called an *orthonormal* basis.

Proposition 5.2.10 *Let $\beta = \{v_1, \ldots, v_n\}$ be an orthonormal basis of $\mathbf{R}^n$ and $[v]_\beta = (c_1, \ldots, c_n)$ the coordinates of v relative to basis β. Then*

$$c_i = v \cdot v_i \quad \text{for all} \quad i = 1, \ldots, n,$$

that is,

$$v = v \cdot v_1\, v_1 + v \cdot v_2\, v_2 + \cdots + v \cdot v_n\, v_n.$$

Proof We have that v is written as a linear combination of the vectors of basis β as

$$v = c_1 v_1 \; + \cdots + \; c_n v_n.$$

Taking the dot product of v with vector v_i, we have

$$v \cdot v_i = (c_1 v_1 \; + \cdots + \; c_n v_n) \cdot v_i.$$

Using the properties of the dot product, we obtain

$$v \cdot v_i = c_1\, v_1 \cdot v_i + \cdots + c_i\, v_i \cdot v_i + \cdots + c_n\, v_n \cdot v_i.$$

Since β is orthonormal, we have that

$$v_j \cdot v_i = 0, \;\; j \neq i \quad \text{and} \quad v_i \cdot v_i = 1,$$

which, substituted in the line above, gives

$$v \cdot v_i = c_i \; v_i \cdot v_i = c_i.$$

Now we substitute the c_i's in the expression of v and the proof is complete. □

Example 6

Let $\beta = \{v_1, v_2, v_3\}$, where

$$v_1 = \frac{\sqrt{2}}{2}(1, -1, 0), \quad v_2 = \frac{\sqrt{2}}{2}(1, 1, 0), \quad v_3 = (0, 0, -1).$$

The set β is an orthonormal basis (the reader should verify this fact).

Let the vector $(1, 2, -1)$ be written as a linear combination of the vectors in β by

$$(1, 2, -1) = c_1 v_1 + c_2 v_2 + c_3 v_3.$$

The coordinates $(c_1, c_2, c_3)_\beta$ could be found as in the previous examples, namely, equating the components and solving the resulting linear system of equations. But using Proposition 5.2.10 we obtain

$$\begin{aligned} c_1 &= (1, 2, -1) \cdot v_1 = (1, 2, -1) \cdot \frac{\sqrt{2}}{2}(1, -1, 0) = -\frac{\sqrt{2}}{2} \\ c_2 &= (1, 2, -1) \cdot v_2 = (1, 2, -1) \cdot \frac{\sqrt{2}}{2}(1, 1, 0) = -\frac{\sqrt{3}}{2} \\ c_3 &= (1, 2, -1) \cdot v_3 = (1, 2, -1) \cdot (0, 0, -1) = 1. \end{aligned}$$

Now let us study the change of basis matrix where $\mathcal{S}$ still denotes the standard basis and $\beta = \{v_1, \ldots, v_n\}$ is an orthonormal basis. Writing each vector v_i as

$$v_i = (x_{1i}, \ldots, x_{ni}),$$

we have

$$[I]_{\mathcal{S}}^{\beta} = \begin{pmatrix} x_{11} & x_{12} & \ldots & x_{1n} \\ x_{21} & x_{22} & \ldots & x_{2n} \\ \vdots & \vdots & & \vdots \\ x_{n1} & x_{n2} & \ldots & x_{nn} \end{pmatrix}.$$

For simplicity, let us denote the transition matrix by P. Let P^T be the transpose of P; the multiplication $P^T P$ gives

$$\begin{pmatrix} x_{11} & x_{21} & \dots & x_{n1} \\ x_{12} & x_{22} & \dots & x_{n2} \\ \vdots & \vdots & & \vdots \\ x_{1n} & x_{2n} & \dots & x_{nn} \end{pmatrix} \begin{pmatrix} x_{11} & x_{12} & \dots & x_{1n} \\ x_{21} & x_{22} & \dots & x_{2n} \\ \vdots & \vdots & & \vdots \\ x_{n1} & x_{n2} & \dots & x_{nn} \end{pmatrix} = \begin{pmatrix} z_{11} & z_{12} & \dots & z_{1n} \\ z_{21} & z_{22} & \dots & z_{2n} \\ \vdots & \vdots & & \vdots \\ z_{n1} & z_{n2} & \dots & z_{nn} \end{pmatrix}$$

Notice that

$$z_{11} = x_{11}x_{11} + x_{21}x_{21} + \cdots + x_{n1}x_{n1} = v_1 \cdot v_1 = 1,$$

$$z_{12} = x_{11}x_{12} + x_{21}x_{22} + \cdots + x_{n1}x_{n2} = v_1 \cdot v_2 = 0,$$

for $||v_1|| = 1$ and $v_1 \cdot v_2 = 0$. In general we have

$$x_{1i}^2 + x_{2i}^2 + \cdots + x_{ni}^2 = v_i \cdot v_i = 1,$$

$$x_{1i}x_{1j} + x_{2i}x_{2j} + \cdots + x_{ni}x_{nj} = v_i \cdot v_j = 0, \quad i \neq j,$$

for $\{v_1, \dots, v_n\}$ is an orthonormal basis. Therefore

$$P^T P = \begin{pmatrix} 1 & 0 & \dots & 0 \\ 0 & 1 & \dots & 0 \\ \vdots & \vdots & & \vdots \\ 0 & 0 & \dots & 1 \end{pmatrix}.$$

This implies that $P^T = P^{-1}$, and we conclude that the two transition matrices are related by

$$[I]_\beta^\mathcal{S} = \left([I]_\mathcal{S}^\beta\right)^{-1} = \left([I]_\mathcal{S}^\beta\right)^T.$$

Example 7

Let us consider the basis β of Example 6. We then have that

$$[I]_\beta^\mathcal{S} = \begin{pmatrix} \frac{\sqrt{2}}{2} & \frac{\sqrt{2}}{2} & 0 \\ -\frac{\sqrt{2}}{2} & \frac{\sqrt{2}}{2} & 0 \\ 0 & 0 & -1 \end{pmatrix}.$$

From the above we get that

$$[I]_{\mathcal{S}}^{\beta} = ([I]_{\beta}^{\mathcal{S}})^T = \begin{pmatrix} \frac{\sqrt{2}}{2} & -\frac{\sqrt{2}}{2} & 0 \\ \frac{\sqrt{2}}{2} & \frac{\sqrt{2}}{2} & 0 \\ 0 & 0 & -1 \end{pmatrix}.$$

Definition 5.2.11 *A matrix A is called an* orthogonal matrix *if $A^{-1} = A^T$.*

Lemma 5.2.12 *(a) Let $\mathcal{S}$ denote the standard basis and β be an orthonormal basis of $\mathbf{R}^n$. Then*

$$\det\left([I]_{\mathcal{S}}^{\beta}\right) = \det\left([I]_{\beta}^{\mathcal{S}}\right) = \pm 1.$$

(b) The determinant of any orthogonal matrix is ± 1.

Proof Since $\left([I]_{\mathcal{S}}^{\beta}\right)\left([I]_{\mathcal{S}}^{\beta}\right)^{-1} = I$, we get

$$\det\left[[I]_{\mathcal{S}}^{\beta}\right]\det\left[([I]_{\mathcal{S}}^{\beta})^{-1}\right] = 1.$$

But we also have that $[I]_{\beta}^{\mathcal{S}} = \left([I]_{\mathcal{S}}^{\beta}\right)^{-1} = \left([I]_{\mathcal{S}}^{\beta}\right)^T$, and any matrix and its transpose have the same determinant. Therefore

$$\left(\det\left([I]_{\mathcal{S}}^{\beta}\right)\right)^2 = \left(\det\left([I]_{\beta}^{\mathcal{S}}\right)\right)^2 = 1,$$

which is the result of part (a) of the lemma. Part (b) has the same proof. □

Exercises

1. Investigate the linear independence of the following vectors
   (a) $(1, 1, 1, 0), (1, -1, 2, 1), (0, 2, -1, -1)$.
   (b) $(1, 1, 1, 0, -1), (1, -1, 2, -1, 0), (0, 2, -1, -1, 0), (-1, -1, 3, 0, 0)$.

2. (i) Find the coordinates of $v = (2, 0, 1, 4)$ in the following ordered bases for $\mathbf{R}^4$:
   (a) $\beta = \{(1, 1, 1, 0), (1, -1, 0, 0), (0, 0, 1, -1), (0, 0, 1, 1)\}$.

(b) $\beta' = \{(0,1,1,),(1,0,1,-1),(1,-1,0,1),(1,1,-1,0)\}$.
(ii) (a) Find the transition matrix P from β to β'.
(b) Find the coordinates of v in the basis β' using the transition matrix of part (a).

3. Let $\mathcal{S}$ denote the standard ordered basis of $\mathbf{R}^3$ and $\beta = \{v_1, v_2, v_3\}$ be an orthonormal basis of $\mathbf{R}^3$. Let

$$[I]_{\mathcal{S}}^{\beta} = \begin{pmatrix} a & 0 & c \\ 0 & 1 & 0 \\ b & 0 & d \end{pmatrix}$$

be the change of basis matrix. Find all possible values for a, b, c, and d.

4. Find an orthogonal matrix whose first row is $(1/3, 2/3, 2/3)$.

5.3 Linear Transformations

Definition 5.3.1 *A mapping $T : \mathbf{R}^n \to \mathbf{R}^n$ is called a* linear transformation *if*
(i) $T(u+v) = T(u) + T(v)$ for all vectors u and v in $\mathbf{R}^n$.
(ii) $T(cv) = cT(v)$ for all vectors v in $\mathbf{R}^n$ and all real numbers c.

Example 1
Let $T : \mathbf{R}^4 \to \mathbf{R}^4$ such that

$$T((x_1, x_2, x_3, x_4)) = (x_1 - x_2, x_1 + x_3 + x_4, x_3 - x_2, x_2 + x_4).$$

We verify now that T is a linear transformation.
(i) Let $v = (x_1, x_2, x_3, x_4)$ and $u = (y_1, y_2, y_3, y_4)$. Then

$$v + u = (x_1 + y_1, x_2 + y_2, x_3 + y_3, x_4 + y_4)$$

and hence

$$\begin{aligned} T(v+u) &= ((x_1 + y_1) - (x_2 + y_2),\ (x_1 + y_1) + (x_3 + y_3) + (x_4 + y_4), \\ &\quad (x_3 + y_3) - (x_2 + y_2),\ (x_2 + y_2) + (x_4 + y_4)) \\ &= (x_1 - x_2, x_1 + x_3 + x_4, x_3 - x_2, x_2 + x_4) \\ &\quad + (y_1 - y_2, y_1 + y_3 + y_4, y_3 - y_2, y_2 + y_4) \\ &= T(v) + T(u). \end{aligned}$$

(ii) If c is a real number, $cv = (cx_1, cx_2, cx_3, cx_4)$. Then,

$$\begin{aligned} T(cv) &= T((cx_1,\ cx_2,\ cx_3,\ cx_4)) \\ &= (cx_1 - cx_2, cx_1 + cx_3 + cx_4, cx_3 - cx_2, cx_2 + cx_4) \\ &= c\,(x_1 - x_2, x_1 + x_3 + x_4, x_3 - x_2, x_2 + x_4) \\ &= cT(v). \end{aligned}$$

We now give more examples of linear transformations; the verification that they are linear is left as an exercise.

Example 2
Let $T : \mathbf{R}^2 \to \mathbf{R}^2$ be a counterclockwise rotation about the origin through an angle θ. From Chapter 3 we get that if $v = (x, y)$, then

$$T(v) = \begin{pmatrix} x' \\ y' \end{pmatrix} = \begin{pmatrix} \cos\theta & -\sin\theta \\ \sin\theta & \cos\theta \end{pmatrix} \begin{pmatrix} x \\ y \end{pmatrix}.$$

Example 3
Let $T : \mathbf{R}^3 \to \mathbf{R}^3$ be a rotation through an angle θ about the x-axis. Therefore T restricted to the yz-plane is a rotation of $\mathbf{R}^2$. If $v = (x, y, z)$, from Example 2 of Section 3.4 we conclude that

$$T(v) = \begin{pmatrix} x' \\ y' \\ z' \end{pmatrix} = \begin{pmatrix} 1 & 0 & 0 \\ 0 & \cos\theta & -\sin\theta \\ 0 & \sin\theta & \cos\theta \end{pmatrix} \begin{pmatrix} x \\ y \\ z \end{pmatrix}.$$

Example 4
Let $T : \mathbf{R}^3 \to \mathbf{R}^3$ be a reflection through the yz-plane. If $v = (x, y, z)$, then

$$T(v) = T((x, y, z)) = (-x, y, z).$$

Proposition 5.3.2 *(a) Let $T : \mathbf{R}^n \to \mathbf{R}^n$ be a linear map and $\mathbf{0}$ denote the origin of $\mathbf{R}^n$. Then $T(\mathbf{0}) = \mathbf{0}$.*
(b) Let $\{v_1, \ldots, v_n\}$ be a basis of $\mathbf{R}^n$. Then a mapping $T : \mathbf{R}^n \to \mathbf{R}^n$ is linear if and only if

$$T(\sum_{i=1}^{n} c_i v_i) = \sum_{i=1}^{n} c_i T(v_i)$$

for all real numbers $c_1, \ldots, c_n$.
(c) Let $\{v_1, \ldots, v_n\}$ *be a basis of* $\mathbf{R}^n$. *To each vector* v_i *assign a vector* w_i. *Then there exists a unique linear transformation* $T : \mathbf{R}^n \to \mathbf{R}^n$ *such that* $T(v_i) = w_i$. *Such a transformation on an arbitrary vector* $v = \sum_{i=1}^n c_i v_i$ *is given by*

$$T(v) = \sum_{i=1}^{n} c_i w_i$$

Proof (a) Consider the real number 0. Then

$$T(\mathbf{0}) = T(0\mathbf{0}) = 0T(\mathbf{0}) = \mathbf{0}.$$

(b) It is easy to see that if T is linear, then T preserves linear combinations, and thus

$$T(\sum_{i=1}^{n} c_i v_i) = \sum_{i=1}^{n} c_i T(v_i)$$

for all sets of vectors $\{v_1, ..., v_n\}$ of $\mathbf{R}^n$.

Conversely, since $\{v_1, ..., v_n\}$ is a basis, given $u, v \in \mathbf{R}^n$ we write

$$v = a_1 v_1 + \cdots + a_n v_n,$$

$$u = b_1 v_1 + \cdots + b_n v_n.$$

By hypothesis, we have

$$T(v) = \sum_{i=1}^{n} a_i T(v_i) \quad \text{and} \quad T(u) = \sum_{i=1}^{n} b_i T(v_i).$$

Then $v + u = \sum_{i=1}^n c_i v_i$, where $c_i = a_i + b_i$, and we have

$$\begin{aligned} T(v+u) &= T(\sum_{i=1}^{n} c_i v_i) = \sum_{i=1}^{n} c_i T(v_i) \\ &= \sum_{i=1}^{n} (a_i + b_i) T(v_i) \\ &= \sum_{i=1}^{n} a_i T(v_i) + \sum_{i=1}^{n} b_i T(v_i) \\ &= T(v) + T(u). \end{aligned}$$

Now if c is a real number, $cv = \sum_{i=1}^{n} ca_i v_i$, and then

$$\begin{aligned} T(cv) &= T(\sum_{i=1}^{n} ca_i v_i) = \sum_{i=1}^{n} ca_i T(v_i) \\ &= c\sum_{i=1}^{n} a_i T(v_i) = cT(v). \end{aligned}$$

(c) It follows from (b) that T is linear. We now show that this T is the unique linear transformation satisfying $T(v_i) = w_i$. Suppose that there exists a linear transformation S such that $S(v_i) = w_i$. Given $v = \sum_{i=1}^{n} c_i v_i$, since S is linear,

$$S(v) = \sum_{i=1}^{n} c_i S(v_i) = \sum_{i=1}^{n} c_i w_i.$$

Then $T(v) = S(v)$ for all $v \in \mathbf{R}^n$; that is, $T = S$. □

It follows from part (c) of the proposition above that if $\{v_1, \ldots, v_n\}$ is basis, then a linear transformation T is completely and uniquely determined by the vectors $T(v_1), \ldots, T(v_n)$. Another important feature of linear transformations is that they can be represented by matrices. Some properties of a linear transformation can then be studied by studying the matrix that represents the transformation.

In order to find such a matrix we choose two ordered bases, one for the domain and another for the target space of the transformation. More precisely, if $\beta = \{v_1, \ldots, v_n\}$ and $\gamma = \{w_1, \ldots, w_n\}$ are two ordered bases of $\mathbf{R}^n$, then we write

$$\begin{aligned} T(v_1) &= a_{11}w_1 + a_{21}w_2 + \cdots + a_{n1}w_n, \\ &\vdots \\ T(v_n) &= a_{1n}w_1 + a_{2n}w_2 + \cdots + a_{nn}w_n, \end{aligned}$$

which yields the matrix

$$[T]_\gamma^\beta = \begin{pmatrix} a_{11} & \cdots & a_{1n} \\ \vdots & & \vdots \\ a_{n1} & \cdots & a_{nn} \end{pmatrix}.$$

If the same basis is chosen for the domain and the target space, we simplify the notation, writing only $[T]_\beta$. In the case that both bases are the standard basis $\mathcal{S}$ of $\mathbf{R}^n$, we call $[T]_\mathcal{S}$ the *standard matrix* of T.

Example 5

Let $T : \mathbf{R}^3 \to \mathbf{R}^3$ be given by

$$T(x, y, z) = (2x - 3y + z, -x - y + 2z, y - 2z).$$

The standard matrix of T is easy to find, for we have

$$\begin{array}{lclcl} T((1,0,0)) & = & (2,-3,1) & = & 2(1,0,0) - 3(0,1,0) + 1(0,0,1), \\ T((0,1,0)) & = & (-1,-1,2) & = & -1(1,0,0) - 1(0,1,0) + 2(0,0,1), \\ T((0,0,1)) & = & (0,1,2) & = & 0(1,0,0) + 1(0,1,0) + 2(0,0,1), \end{array}$$

and then the matrix is given by

$$\begin{pmatrix} 2 & -1 & 0 \\ -3 & -1 & 1 \\ 1 & 2 & -2 \end{pmatrix}.$$

Example 6

Let $T : \mathbf{R}^2 \to \mathbf{R}^2$ be given by $T(x, y) = (x + y, -2x + 4y)$. Here, we want to find the matrix that represents T with respect to basis $\beta = \{v_1, v_2\}$, where $v_1 = (1, 1)$ and $v_2 = (1, 2)$. We compute

$$T(v_1) = T((1,1)) = (2,2) = 2(1,1) + 0(1,2),$$

$$T(v_2) = T((1,2)) = (3,6) = 0(1,1) + 3(1,2).$$

Therefore

$$[T]_\beta = \begin{pmatrix} 2 & 0 \\ 0 & 3 \end{pmatrix}.$$

We notice that the matrix of T with respect to basis β is diagonal, while the standard matrix (find it) is not so simple. This suggests that for this particular linear map, the basis β is more suitable, since the linear map is represented by a simple matrix.

In choosing a convenient basis for which the matrix that represents a linear map is as simple as possible, we should be careful and write the coordinate of the vectors with respect to the new basis. As we saw before, this is done by finding the change of basis matrix. Then we have

$$[T(v)]_\beta = [T]_\beta \cdot [v]_\beta = [T]_\beta \cdot [I]_\beta^{\mathcal{S}}[v].$$

Example 7

Let $T : \mathbf{R}^3 \to \mathbf{R}^3$ be given by $T(x, y, z) = (3x - 2y, -2x + 3y, 5z)$. The standard matrix of T is

$$\begin{pmatrix} 3 & -2 & 0 \\ -2 & 3 & 0 \\ 0 & 0 & 5 \end{pmatrix}$$

but with respect to the basis $\beta = \{(1, 1, 0), (-1, 1, 0), (0, 0, 1)\}$ the matrix representing T is

$$[T]_\beta = \begin{pmatrix} 1 & 0 & 0 \\ 0 & 5 & 0 \\ 0 & 0 & 5 \end{pmatrix}$$

(the reader should verify this). We find the change of basis matrix $[I]_\beta^{\mathcal{S}}$. We have

$$(1, 0, 0) = a(1, 1, 0) + b(-1, 1, 0) + c(0, 0, 1),$$
$$(0, 1, 0) = d(1, 1, 0) + e(-1, 1, 0) + f(0, 0, 1),$$
$$(1, 0, 0) = g(1, 1, 0) + h(-1, 1, 0) + i(0, 0, 1).$$

Equating the components and solving the system, we get $a = d = e = 1/2$, $b = -1/2$, $c = f = g = h = 0$, and $i = 1$. Thus

$$[I]_\beta^{\mathcal{S}} = \begin{pmatrix} 1/2 & 1/2 & 0 \\ -1/2 & 1/2 & 0 \\ 0 & 0 & 1 \end{pmatrix}.$$

Now when we say that the image of point (x, y, z) has coordinates (a, b, c) with respect to basis β, we write $(a, b, c)_\beta$, and the numbers a, b, c are obtained by the multiplication of matrices below.

$$\begin{pmatrix} a \\ b \\ c \end{pmatrix}_\beta = \begin{pmatrix} 1 & 0 & 0 \\ 0 & 5 & 0 \\ 0 & 0 & 5 \end{pmatrix} \begin{pmatrix} 1/2 & 1/2 & 0 \\ -1/2 & 1/2 & 0 \\ 0 & 0 & 1 \end{pmatrix} \begin{pmatrix} x \\ y \\ z \end{pmatrix}.$$

Therefore

$$a = \frac{1}{2}(x+y), \qquad b = \frac{5}{2}(-x+y), \qquad z = 5z,$$

which means

$$T(x,y,z) = a(1,1,0) + b(-1,1,0) + c(0,0,1).$$

Notice that if we substitute a, b, c above, we get the original formula for T, which is $T(x,y,z) = (3x-2y, -2x+3y, 5z)$.

The following theorem (whose proof we also omit here) shows a relationship between matrices that represent the same linear transformation with respect to two different bases.

Theorem 5.3.3 *Let $T : \mathbf{R}^n \to \mathbf{R}^n$ be a linear transformation and β and γ two ordered bases of $\mathbf{R}^n$. Then*

$$[T]_\beta = [I]_\beta^\gamma \, [T]_\gamma \, [I]_\gamma^\beta.$$

Recall that $\left([I]_\beta^\gamma\right)^{-1} = [I]_\gamma^\beta$ and hence $\det\left(\left([I]_\beta^\gamma\right)^{-1}\right) \cdot \det\left([I]_\gamma^\beta\right) = 1$. This implies that

$$\det([T]_\beta) = \det([T]_\gamma).$$

We then define the *determinant* of a linear transformation $T : \mathbf{R}^n \to \mathbf{R}^n$, denoted by $\det(T)$, as the determinant of the matrix $[T]_\beta$, where β is any basis of $\mathbf{R}^n$.

Definition 5.3.4 *Let $T : \mathbf{R}^n \to \mathbf{R}^n$ be a linear transformation such that* $\det(T) \neq 0$. *Then T is said to be* orientation-preserving *if* $\det(T) > 0$; *if* $\det(T) < 0$, *then we say that T is* orientation-reversing.

Exercises

1. Let $\mathcal{S}$ denote the standard ordered basis of $\mathbf{R}^2$, $\beta = \{(1,1),(1,0)\}$. Consider $T : \mathbf{R}^2 \to \mathbf{R}^2$ defined by $T(x,y) = (x-y, 2x+y)$.
   (a) Find $[T]_\mathcal{S}$, that is, the standard matrix of T.
   (b) Find $[T]_\beta$.
   (c) Find $[T]_\mathcal{S}^\beta$.
   (d) Find $[T]_\beta^\mathcal{S}$.

2. Let $\mathcal{S}$ denote the standard ordered basis of $\mathbf{R}^3$ and

$$\beta = \{(1,1,1),(1,1,0),(0,0,1,)\}.$$

Let $S : \mathbf{R}^3 \to \mathbf{R}^3$ be the linear map represented by

$$[S]^{\beta}_{\mathcal{S}} = \begin{pmatrix} 1 & 0 & 0 \\ 0 & 1 & 0 \\ 0 & 0 & -1 \end{pmatrix}.$$

(a) Find a formula for S.
(b) Find $[S]_\beta$.
(c) Find $[S]^{\mathcal{S}}_{\beta}$.

5.4 Orthogonal Transformations

The linear transformations of Examples 2, 3, and 4 of the previous section are isometries either of the plane or of the 3-space. Of course we cannot expect that every isometry of Euclidean space is a linear transformation, since there are isometries with no fixed points and hence they cannot fix the origin. However, we will show that every Euclidean isometry is the composition of a special type of linear transformation with a translation of $\mathbf{R}^n$. We now define this special type of linear map.

Definition 5.4.1 *A linear map $T : \mathbf{R}^n \to \mathbf{R}^n$ such that*

$$T(v) \cdot T(u) = v \cdot u, \quad \forall\, u, v\, \in \mathbf{R}^n,$$

is called an orthogonal transformation *of $\mathbf{R}^n$.*

Proposition 5.4.2 *(a) An orthogonal transformation is a rigid motion.*
(b) An orthogonal transformation maps orthonormal bases onto orthonormal bases.
(c) Let $T : \mathbf{R}^n \to \mathbf{R}^n$ be an orthogonal transformation. Then

$$\det(T) = \pm 1.$$

Proof (a) Let x and y be any two points of $\mathbf{R}^n$.

$$\begin{aligned} ||T(x) - T(y)||^2 &= (T(x) - T(y)) \cdot (T(x) - T(y)) \\ &= T(x) \cdot T(x) - 2T(x) \cdot T(y) + T(y) \cdot T(y) \\ &= x \cdot x - 2x \cdot y + y \cdot y \\ &= (x - y) \cdot (x - y) \\ &= ||x - y||^2. \end{aligned}$$

(b) Let $\gamma = \{v_1, \ldots, v_n\}$ be an orthonormal basis. Since T is an orthogonal map, we have

$$T(v_i) \cdot T(v_i) = v_i \cdot v_i = 1 \quad \text{and} \quad T(v_i) \cdot T(v_j) = v_i \cdot v_j = 0, \;\; i \neq j.$$

It follows that the set $\beta = \{T(v_1), \ldots, T(v_n)\}$ is an orthonormal basis.
(c) Consider the basis $\beta = \{T(e_1), \ldots, T(e_n)\}$, where $\{e_1, \ldots, e_n\}$ is the standard basis. The entries of the standard matrix of T are given by

$$T(e_i) = a_{1i}e_1 + \cdots + a_{ni}e_n, \quad i = 1, \ldots, n.$$

Notice that the a_{ij}'s are also the entries of the transition matrix $[I]_{\mathcal{S}}^{\beta}$. Since $\mathcal{S}$ and β are both orthonormal bases, we conclude that $[I]_{\mathcal{S}}^{\beta}$ is an orthogonal matrix, and therefore $\det(T) = \pm 1$. □

Definition 5.4.3 *A* translation *of* $\mathbf{R}^n$ *is the map* $L_v : \mathbf{R}^n \to \mathbf{R}^n$ *such that* $L_v(x) = x + v$.

We point out here that translations preserve the dot product between oriented line segments. In fact, let $u' = L_v(u) = u + v$ and $w' = L_v(w) = w + v$, where u and w are points of $\mathbf{R}^n$. Notice that the vectors u (thought of as an oriented line segment $\overset{\mapsto}{\mathbf{0}u}$) and $\overset{\mapsto}{u'v} = u' - v$ have the same components; likewise the vectors w and $\overset{\mapsto}{w'v} = w' - v$. Therefore

$$u \cdot w = (u' - v) \cdot (w' - v).$$

However, a translation is not an orthogonal map, for

$$u' \cdot w' = (u + v) \cdot (w + v) \neq u \cdot w.$$

We are ready now for the main result of this section.

Theorem 5.4.4 *A map $T : \mathbf{R}^n \to \mathbf{R}^n$ is a rigid motion of $\mathbf{R}^n$ if and only if it is a composition of a translation and an orthogonal transformation.*

Proof Since both an orthogonal transformation and a translation are rigid motions, so is the composition of them.

Conversely, let T be a rigid motion and $v = T(\mathbf{0})$. Let us define $F : \mathbf{R}^n \to \mathbf{R}^n$ by $F(x) = T(x) - v$. Then $F(\mathbf{0}) = \mathbf{0}$; furthermore,

$$L_v \circ F\ (x) = L_v(T(x) - v) = T(x) - v + v = T(x)$$

which implies that T is the transformation F followed by a translation. Therefore, if we prove that F is an orthogonal transformation, the proof will be complete. First, observe that F is also a rigid motion, since $F = T \circ L_{-v}$ is the composition of the rigid motions. Moreover, F preserves the dot product of $\mathbf{R}^n$. To see this, we calculate

$$\begin{aligned} -2F(x) \cdot F(y) &= ||F(x) - F(y)||^2 - ||F(x)||^2 - ||F(y)||^2 \\ &= ||F(x) - F(y)||^2 - ||F(x) - F(\mathbf{0})||^2 \\ &\quad -||F(y) - F(\mathbf{0})||^2 \\ &= ||x - y||^2 - ||x - \mathbf{0}||^2 - ||y - \mathbf{0}||^2 \\ &= -2x \cdot y. \end{aligned}$$

Therefore if we show that F is a linear transformation we will have shown that F is an orthogonal transformation.

In order to show that F is a linear map, we consider an orthonormal basis $\{v_1, \ldots, v_n\}$ of $\mathbf{R}^n$. It follows from Proposition 5.2.10 that each vector $x \in \mathbf{R}^n$ can be written as

$$x = (x \cdot v_1)\, v_1 + \cdots + (x \cdot v_n)v_n.$$

Since F preserves the dot product, the set $\{F(v_1), \ldots, F(v_n)\}$ is an orthonormal set, hence it is an orthonormal basis of $\mathbf{R}^n$ by Proposition 5.2.9. Therefore $F(x)$ is written as

$$F(x) = (F(x) \cdot F(v_1))\ F(v_1)\ + \cdots +\ (F(x) \cdot F(v_n))\ F(v_n).$$

Using again the fact that F preserves the dot product, we obtain

$$F(x) = (x \cdot v_1)\ F(v_1)\ + \cdots +\ (x \cdot v_n)\ F(v_n),$$

which implies that F is linear by Proposition 5.3.2(b). □

Corollary 5.4.5 *A rigid motion of* $\mathbf{R}^n$ *that fixes the origin is an orthogonal transformation, and in particular it is a linear transformation.*

We can then use Proposition 5.3.2(c) as a method for finding a formula for rigid motions that fix the origin, since they are linear transformations. Proposition 5.3.2(c) says that once we know how a linear transformation T acts on the vectors of a basis, T is completely determined.

Example 1

Let $T : \mathbf{R}^3 \to \mathbf{R}^3$ be the reflection through the plane $x - y + z = 0$.
(a) Find the image of $(1, 2, 2)$.
(b) Find a formula for T.

Since the reflection fixes all points of the plane, and plane $x - y + z = 0$ passes through $(0, 0, 0)$, we have that the origin is fixed by T. Thus T is a linear map, by Corollary 5.4.5. We choose two linearly independent vectors in the plane. For instance, let us consider vectors $(1, 1, 0)$ and $(0, 1, 1)$. We then have $T((1, 1, 0)) = (1, 1, 0)$ and $T((0, 1, 1)) = (0, 1, 1)$. Choose a vector that is orthogonal to the given plane, for instance $(1, -1, 1)$. Because T is a reflection through this plane,

$$T((1, -1, 1)) = -(1, -1, 1) = (-1, 1, -1).$$

(a) Since $\beta = \{(1, 1, 0), (0, 1, 1), (1, -1, 1)\}$ is a basis of $\mathbf{R}^3$, vector $(1, 2, 2)$ can be written as

$$(1, 2, 2) = a(1, 1, 0) \; + \; b(0, 1, 1) \; + \; c(1, -1, 1),$$

which gives rise to the system

$$1 = a + c, \quad 2 = a + b - c, \quad \text{and} \quad 2 = b + c.$$

The unique solution of the system is

$$a = \frac{2}{3}, \quad b = \frac{5}{3}, \quad \text{and} \quad c = \frac{1}{3}.$$

Therefore

$$(1, 2, 2) = \frac{2}{3}\,(1, 1, 0) \; + \; \frac{5}{3}\,(0, 1, 1) \; + \; \frac{1}{3}\,(1, -1, 1),$$

and its image is

$$\begin{aligned} T(1,2,2) &= \frac{2}{3}\,T(1,1,0) + \frac{5}{3}\,T(0,1,1) + \frac{1}{3}\,T(1,-1,1) \\ &= \frac{2}{3}\,(1,1,0) + \frac{5}{3}\,(0,1,1) + \frac{1}{3}\,(-1,1,-1) \\ &= (\frac{1}{3},\frac{8}{3},\frac{2}{3}). \end{aligned}$$

(b) For a vector of coordinates (x, y, z) we repeat the procedure, writing

$$(x,y,z) = a(1,1,0) + b(0,1,1) + c(1,-1,1)$$

and then we have

$$x = a + c, \quad y = a + b - c, \quad \text{and} \quad z = b + c.$$

Solving for a, b, c, we get

$$a = \frac{1}{3}(2x + y - z), \quad b = \frac{1}{3}(-x + y + 2z), \quad \text{and} \quad c = \frac{1}{3}(x - y + z).$$

Applying Proposition 5.3.2(d), we obtain

$$\begin{aligned} T((x,y,z)) &= aT((1,1,0)) + bT((0,1,1)) + cT((1,-1,1)) \\ &= a(1,1,0) + b(0,1,1) + c(-1,1,-1). \end{aligned}$$

The formula is then obtained by substituting the values of a, b, and c, that is,

$$\begin{aligned} T((x,y,z)) &= \frac{1}{3}(2x + y - z, 2x + y - z, 0) \\ &\quad + \frac{1}{3}(0, -x + y + 2z, -x + y + 2z) \\ &\quad + \frac{1}{3}(-x + y - z, x - y + z, -x + y - z) \\ &= \frac{1}{3}(x + 2y - 2z, 2x + y + 2z, -2x + 2y + z). \end{aligned}$$

The method above gives the following formula for reflections.

Proposition 5.4.6 *Let $\mathcal{P}$ be a plane through the origin and n a vector normal to it. The reflection through $\mathcal{P}$ is given by*

$$S(v) = v - \frac{2(v \cdot n)}{||n||^2}\, n$$

Proof Let T be the reflection through $\mathcal{P}$. We will show that the map S defined above is T. For that, consider the basis $\beta = \{v_1, v_2, n\}$ such that v_1 and v_2 are linearly independent vectors of $\mathcal{P}$. Recall that $T(v_i) = v_i, i = 1, 2$, and $T(n) = -n$. From the definition of S, we obtain that $S(v_1) = v_1$ and $S(v_2) = v_2$, since $v_1 \cdot n = 0 = v_2 \cdot n$. Further,

$$S(n) = n - \frac{2(n \cdot n)}{||n||^2} n = n - 2\frac{||n||^2}{||n||^2} n = -n,$$

and hence $S(n) = T(n)$. Thus we have that S and T agree on all vectors of basis β. Therefore $S = T$. □

Corollary 5.4.7 *Let S be the reflection of $\mathbf{R}^3$ through a plane $\mathcal{P}$. Then* $\det(S) = -1$.

Proof We find the matrix that represents S with respect to the basis $\beta = \{v_1, v_2, n\}$ of the proof of Proposition 5.4.6. We have

$$[S]_\beta = \begin{pmatrix} 1 & 0 & 0 \\ 0 & 1 & 0 \\ 0 & 0 & -1 \end{pmatrix},$$

which implies $\det(S) = -1$ □

Applying the formula of Proposition 5.4.6 to the plane of Example 1 above, we obtain:

$$\begin{aligned} S(x, y, z) &= (x, y, z) - 2\frac{(x, y, z) \cdot (1, -1, 1)}{||(1, -1, 1)||^2}(1, -1, 1) \\ &= (x, y, z) - \frac{2x - 2y + 2z}{3}(1, -1, 1) \\ &= \frac{1}{3}(x + 2y - 2z, 2x + y + 2z, -2x + 2y + z). \end{aligned}$$

Example 2

Let $T : \mathbf{R}^3 \to \mathbf{R}^3$ be the reflection through the plane $x - y + z + 1 = 0$. Find a formula for T.

Now the plane $x - y + z + 1 = 0$ does not go through $(0, 0, 0)$ but it is parallel to $x - y + z = 0$. The vector $u = (0, 1, 0)$ is in the plane. Therefore the translation $L_v : \mathbf{R}^3 \to \mathbf{R}^3$, where $v = -u$, maps point

$(0,1,0)$ to $(0,0,0)$, and thus, it maps plane $x-y+z+1=0$ onto plane $x-y+z=0$. The map L_v given by

$$L_v(x,y,z) = (x, y-1, z).$$

In Example 1, we found a formula for the reflection through plane $x-y+z=0$; let us now denote such a transformation by S. Then, the reflection we are looking for in this example is given by $T = L_v^{-1} \circ S \circ L_v$. Since

$$L_v^{-1}(x,y,z) = L_u(x,y,z) = (x, y+1, z),$$

we have

$$\begin{aligned} T(x,y,z) &= (L_v^{-1} \circ S \circ L_v)(x,y,z) = L_v^{-1}(S(x,y-1,z)) \\ &= L_v^{-1}(\frac{1}{3}(x+2y-2-2z, 2x+y-1+2z, \\ &\quad -2x+2y-2+z)) \\ &= \frac{1}{3}(x+2y-2-2z, 2x+y+2z, -2x+2y-2+z). \end{aligned}$$

We point out here the composition stated by Theorem 5.4.4. The rigid motion T can be written as $T = F \circ L_w$, where

$$F(x,y,z) = \frac{1}{3}(x+2y-2z, 2x+y+2z, -2x+2y+z)$$

and $w = (-2, 0, -2)$.

We finish this section showing how to use Theorem 5.3.3 for finding formulas for rotations.

Example 3

Let $T : \mathbf{R}^3 \to \mathbf{R}^3$ be a rotation about the line $\{t(1,1,1) \mid t \in \mathbf{R}\}$ through an angle θ. Find a formula for T.

Observe that line $\{t(1,1,1) \mid t \in \mathbf{R}\}$ goes through the origin, and since it is fixed by the rotation T, we conclude that T is a linear transformation, by Corollary 5.4.5.

In this problem we use the matrix rotation. Just observe that in Examples 2 and 3 of the previous section, we had the standard coordinates of $\mathbf{R}^2$ and $\mathbf{R}^3$; that is, coordinates with respect to the standard orthonormal basis.

For this example, we choose an orthonormal basis of $\mathbf{R}^3$ adapted to the axis of rotation. Recall that the equation of the plane through $(0,0,0)$ orthogonal to the axis is

$$x + y + z = 0.$$

First, choose two orthogonal vectors in the plane $x + y + z = 0$. We start with $w_1 = (1,-1,0)$; then we need another vector w_2 in the plane that is orthogonal to $w_1 = (1,-1,0)$. Since w_2 must be orthogonal to the vectors normal to plane $x+y+z = 0$, we select $w_3 = (1,1,1)$ (which is orthogonal to plane $x + y + z = 0$) and take

$$w_2 = w_3 \wedge w_1 = (1,1,-2).$$

The set $\{(1,-1,0),(1,1,-2),(1,1,1)\}$ is an orthogonal basis of $\mathbf{R}^3$. Normalizing each vector, we obtain an orthonormal basis $\beta = \{v_1, v_2, v_3\}$, where

$$v_1 = \frac{\sqrt{2}}{2}(1,-1,0), \quad v_2 = \frac{\sqrt{6}}{6}(1,1,-2) \quad \text{and} \quad v_3 = \frac{\sqrt{3}}{3}(1,1,1)\}.$$

Since $T(v_1)$ and $T(v_2)$ lie in the plane $x + y + z = 0$, we have

$$\begin{aligned} T(v_1) &= \cos\theta\, v_1 + \sin\theta\, v_2, \\ T(v_2) &= -\sin\theta\, v_1 + \cos\theta\, v_2. \end{aligned}$$

Further, $T(v_3) = v_3$, and thus the matrix that represents T with respect to basis β is

$$[T]_\beta = \begin{pmatrix} \cos\theta & -\sin\theta & 0 \\ \sin\theta & \cos\theta & 0 \\ 0 & 0 & 1 \end{pmatrix}.$$

Using Theorem 5.3.3, we have

$$T(x,y,z) = [T]_{\mathcal{S}}(x,y,z)^T = [I]_{\mathcal{S}}^{\beta} \cdot [T]_\beta \cdot [I]_{\beta}^{\mathcal{S}}(x,y,z)^T.$$

The change of basis matrix from basis β to the the standard basis is

$$[I]_{\mathcal{S}}^{\beta} = \begin{pmatrix} \sqrt{2}/2 & \sqrt{6}/6 & \sqrt{3}/3 \\ -\sqrt{2}/2 & \sqrt{6}/6 & \sqrt{3}/3 \\ 0 & -2\sqrt{6}/6 & \sqrt{3}/3 \end{pmatrix}.$$

Recall that the change of basis matrix between orthonormal bases is an orthogonal matrix. Since β is orthonormal, $[I]_\beta^\mathcal{S}$ is orthogonal and hence $[I]_\beta^\mathcal{S} = \left([I]_\mathcal{S}^\beta\right)^T$. Therefore

$$[I]_\beta^\mathcal{S} = \begin{pmatrix} \sqrt{2}/2 & -\sqrt{2}/2 & 0 \\ \sqrt{6}/6 & \sqrt{6}/6 & -2\sqrt{6}/6 \\ \sqrt{3}/3 & \sqrt{3}/3 & \sqrt{3}/3 \end{pmatrix}.$$

Multiplying the three matrices, we obtain $[T]_\mathcal{S}$ given by

$$\begin{pmatrix} \frac{1}{3}(1+2\cos\theta) & \frac{1}{3}(1-\cos\theta-\sqrt{3}\sin\theta) & \frac{1}{3}(1-\cos\theta+\sqrt{3}\sin\theta) \\ \frac{1}{3}(1-\cos\theta+\sqrt{3}\sin\theta) & \frac{1}{3}(1+2\cos\theta) & \frac{1}{3}(1-\cos\theta-\sqrt{3}\sin\theta) \\ \frac{1}{3}(1-\cos\theta-\sqrt{3}\sin\theta) & \frac{1}{3}(1-\cos\theta+\sqrt{3}\sin\theta) & \frac{1}{3}(1+2\cos\theta) \end{pmatrix}$$

and then

$$T(x,y,z) = [T]_\mathcal{S}\begin{pmatrix} x \\ y \\ z \end{pmatrix}.$$

In particular, if $\theta = 90°$, we have

$$T(x,y,z) = \begin{pmatrix} \frac{1}{3} & \frac{1}{3}(1-\sqrt{3}) & \frac{1}{3}(1+\sqrt{3} \\ \frac{1}{3}(1+\sqrt{3}) & \frac{1}{3} & \frac{1}{3}(1-\sqrt{3}) \\ \frac{1}{3}(1-\sqrt{3}) & \frac{1}{3}(1+\sqrt{3}) & \frac{1}{3} \end{pmatrix}\begin{pmatrix} x \\ y \\ z \end{pmatrix}.$$

Example 4

Let $T : \mathbf{R}^3 \to \mathbf{R}^3$ be a rotation about the line $\{t(1,1,1)+(1,0,-1) \mid t \in \mathbf{R}\}$ through the angle $90°$. Find a formula for T.

Now this rotation is not a linear transformation, since the axis of rotation is not passing through $(0,0,0)$. However, lines $\{t(1,1,1) + (1,0,-1)\}$ and $\{t(1,1,1)\}$ are parallel, and the second goes through the origin. Let L denote a translation that carries point $(1,0,-1)$ to $(0,0,0)$. Such a translation will carry line $\{t(1,1,1) + (1,0,-1)\}$ to $\{t(1,1,1)\}$ and is given by

$$L(x,y,z) = (x-1, 0, z+1).$$

Let R be a rotation about line $\{t(1,1,1)\}$ through the angle $90°$. Then the transformation T is given by

$$T = L^{-1} \circ R \circ L.$$

Since $L^{-1}(x,y,z) = (x+1, 0, z-1)$, we have that

$$T(x,y,z) = R(x-1, y, z+1) + (1,0,-1).$$

Using the formula obtained for R in Example 2, we get that $T(x,y,z)$ is given by

$$\begin{pmatrix} \frac{1}{3} & \frac{1}{3}(1-\sqrt{3}) & \frac{1}{3}(1+\sqrt{3}) \\ \frac{1}{3}(1+\sqrt{3}) & \frac{1}{3} & \frac{1}{3}(1-\sqrt{3}) \\ \frac{1}{3}(1-\sqrt{3}) & \frac{1}{3}(1+\sqrt{3}) & \frac{1}{3} \end{pmatrix} \begin{pmatrix} x-1 \\ y \\ z+1 \end{pmatrix} + \begin{pmatrix} 1 \\ 0 \\ -1 \end{pmatrix}.$$

Exercises

1. Let $\mathcal{P}_1$ and $\mathcal{P}_2$ be two nonparallel planes in $\mathbf{R}^3$ and S_1 and S_2 reflections through $\mathcal{P}_1$ and $\mathcal{P}_2$, respectively. Suppose that line $l = \mathcal{P}_1 \cap \mathcal{P}_2$. Show, using the methods of this chapter, that $S_2 \circ S_1$ is a rotation about line l.

The next four problems involve some *solids* of $\mathbf{R}^3$. The reader not familiar with them is referred to the last section of Chapter 6 for their description.

2. Let P be the pyramid whose vertices are

$$(-\sqrt{3}/2, -1/2, 0), (\sqrt{3}/2, -1/2, 0), (0,1,0), (0,0,1).$$

Write a nonidentity rotation about the z-axis that leaves P invariant.

3. Let P be the pyramid whose vertices are

$$(0, -1/2, -\sqrt{3}/2), (0, -1/2, \sqrt{3}/2), (0,1,0), (1,0,0).$$

Write a nonidentity rotation that leaves P invariant.

4. Let P be the pyramid with a square base whose vertices are

$$(0,-2,0), (1,0,1), (1,0,-1), (-1,0,1), (-1,0,-1).$$

Write a nonidentity rotation that leaves P invariant.

5. Let P be the prism whose vertices are
$$(0,0,-1), \quad (0,2,-1), \quad (0,0,2), \quad (0,2,2)$$
$$(1,0,-1), \quad (1,2,-1), \quad (1,0,2), \quad (1,2,2).$$
Find a reflection of $\mathbf{R}^3$ that leaves P invariant.

6. Let $T : \mathbf{R}^3 \to \mathbf{R}^3$ be the reflection through plane $y - z = 0$. Find $T(x, y, z)$.

7. Let $T : \mathbf{R}^3 \to \mathbf{R}^3$ be the reflection through the plane containing vectors $(1, 1, 0)$ and $(0, 1, 1)$. Find $T(x, y, z)$.

8. Let $T : \mathbf{R}^3 \to \mathbf{R}^3$ be a rotation about line $\{t(1, 0, 1) \mid t \in \mathbf{R}\}$ and through the angle $\theta = 180°$. Find $T(x, y, z)$.

9. Let $T : \mathbf{R}^3 \to \mathbf{R}^3$ be the reflection through plane $x + z - 1 = 0$. Find $T(x, y, z)$.

10. Let $T : \mathbf{R}^3 \to \mathbf{R}^3$ be a rotation about line $\{t(1, -1, 1)+(1, 0, 1) \mid t \in \mathbf{R}\}$ and through the angle $\theta = 180°$. Find $T(x, y, z)$.

11. Let W be a plane given by $x - y + z = 0$. Let R be a rotation of W centered at the origin through an angle θ.

12. For each transformation below determine if it is a reflection of $\mathbf{R}^3$; if so, find the plane of reflection.
(a) $T(x, y, z) = 1/3(x - 2y - 2z, -2x + y - 2z, -2x - 2y + z)$.
(b) $T(x, y, z) = 2/3(2x - y - 2z, -x + y - 2z, -2x - 2y + z)$.
(c) $T(x, y, z) = 1/5(-4x + 3z - 1, 5y + 1, 4z - 3x)$.

13. For each transformation below determine if it is a rotation of $\mathbf{R}^3$; if so, find the axis of rotation.
(a) $T(x, y, z) = (-x, y, -z)$.
(b) $T(x, yz) = (-x - 1, y + 1, 1 - z)$.
(c) $T(x, y, z) = -\sqrt{2}/2(x + y, x - y, (2\sqrt{2}/2)z)$.
(d) $T(x, y, z) = \sqrt{2}/2(-x + y, -x - y, (2\sqrt{2}/2)z)$.

14. Let S be the reflection of W through line $\{t(1, 1, 0) \mid t \in \mathbf{R}\}$.
(a) Extend R to a rotation of $\mathbf{R}^3$.
(b) Extend S to a reflection of $\mathbf{R}^3$.
(c) Extend S to a rotation of $\mathbf{R}^3$.

15. Prove that no orthogonal operator can both a rotation and a reflection.

16. Give an example of an orthogonal operator that is neither a reflection nor a rotation.

17. Prove that every a finite reflection group G of the plane has a fixed point. (This result is the first step of Exercise 23 of Section 3.1.) The proof is below; fill in the details.
(1) Let $G = \{T_1, \ldots, T_n\}$. Recall that each $T_j \in G$ is the composition of an orthogonal transformation with a translation; that is, $T_j = F_j + v_j$, where F_j is an orthogonal linear transformation. Suppose that T_j is not a translation, which implies that $F_j \neq 0$. Then consider the point
$$y = \frac{1}{n}\Big(T_1(x) + \cdots + T_n(x)\Big).$$
Use the linearity of F_j to conclude that y is a fixed point of T_j.
(2) We claim that there exists $T_j \in G$ for which $F_{j0} \neq 0$. In fact, suppose in search of a contradiction, that all T_{j0} are translations. Since compositions of translations are translations, G would not contain any reflections. Find the contradiction and conclude that there exists $T_{j0} \in G$ that is not a translation.
(3) Suppose now that there exists $L \in G$ that is a translation. Then $T = L \circ T_{j0} \in G$, where T_{j0} is the map of step (2). It follows that T is not a translation either, and then $T(y) = y$. On the other hand,
$$T(y) = L(T_j(y)) = L(y),$$
implying that $L(y) = y$. But this contradicts that L is a translation (why?).
(4) Conclude that
$$T_j(y) = y, \quad \forall j = 1, \ldots, n.$$

5.5 The Classification of Orthogonal Transformations of Euclidean 2- and 3-Space

Let $T : \mathbf{R}^2 \to \mathbf{R}^2$ be a reflection through a line l through the origin. Observe that if a vector v is on the line, then $T(v) = v$, while if v is orthogonal to the line, $T(v) = -1v$. In other words, some vectors are mapped by T to multiples of themselves. Consider now a rotation of $\mathbf{R}^2$ through an angle θ. It is clear that if $\theta \neq 180°, 360°$, then no vector is taken by T to a multiple of itself.

Definition 5.5.1 *Let* $T : \mathbf{R}^n \to \mathbf{R}^n$ *be a linear transformation. A* nonzero *vector* $v \in \mathbf{R}^n$ *is called an* eigenvector *of* T *if* $T(v)$ *is a multiple of* v*, that is, there exists a real number* λ *such that*

$$T(v) = \lambda v.$$

The number λ *is called an* eigenvalue *of* T *and the vector* v *is said to be an eigenvector* corresponding *to the eigenvalue* λ*.*

Lemma 5.5.2 *Let* $T : \mathbf{R}^n \to \mathbf{R}^n$ *be an orthogonal transformation.*
(a) If λ *is an eigenvalue of* T*, then* $|\lambda| = 1$*.*
(b) If v *and* w *are eigenvectors of* T *corresponding to distinct eigenvalues, then* v *and* w *are orthogonal.*

Proof Let v be an eigenvector of T corresponding to the eigenvalue λ. Then

$$||T(v)||^2 = T(v) \cdot T(v) = \lambda v \cdot \lambda v = \lambda^2 v \cdot v = \lambda^2 ||v||^2.$$

Since T preserves the norm of vectors, we conclude that $\lambda^2 = 1$ and hence $|\lambda| = 1$.

For (b), let v and w be eigenvectors corresponding to the eigenvalues 1 and -1, respectively. Recall that T preserves the dot product and hence

$$v \cdot w = T(v) \cdot T(w) = v \cdot -w = -v \cdot w,$$

which implies $v \cdot w = 0$. □

Procedure for finding eigenvalues and eigenvectors

Given $T : \mathbf{R}^n \to \mathbf{R}^n$, we want to find $x_1, \ldots, x_n$ and λ so that

$$T((x_1, \ldots, x_n)) = \lambda\,(x_1, \ldots, x_n).$$

Let $A = (a_{ij})$ denote the standard matrix of T, that is, $a_{ij} = T(e_i) \cdot e_j$, where $\{e_1, \ldots, e_n\}$ is the standard basis of $\mathbf{R}^n$. The equation above is equivalent to the matrix equation

$$A \begin{pmatrix} x_1 \\ \vdots \\ x_n \end{pmatrix} = \lambda \begin{pmatrix} x_1 \\ \vdots \\ x_n \end{pmatrix} = \lambda \begin{pmatrix} \lambda & \ldots & 0 \\ \vdots & & \vdots \\ 0 & \ldots & \lambda \end{pmatrix} \begin{pmatrix} x_1 \\ \vdots \\ x_n \end{pmatrix} = \lambda I \begin{pmatrix} x_1 \\ \vdots \\ x_n \end{pmatrix}$$

which gives rise to the homogeneous system

$$\begin{aligned} (a_{11} - \lambda)x_1 \ + \cdots + a_{1n}x_n &= 0, \\ &\vdots \\ a_{n1}x_1 \ + \cdots + (a_{nn} - \lambda)x_n &= 0. \end{aligned}$$

If T has an eigenvalue λ with corresponding eigenvector $v = (x_1, \ldots, x_n)$, then $(x_1, \ldots, x_n)$ is a nontrivial solution of the system above, for an eigenvector is a nonzero vector. Therefore the determinant of the coefficient matrix $A - \lambda I$ must be zero.

We summarize the procedure in the following steps:

1. Compute the standard matrix of T.
2. Compute $\det(A - \lambda I)$, where A and I are the standard and the identity matrix, respectively.
3. Find the real solutions (if they exist) of the equation $\det(A - \lambda I) = 0$. They are precisely the eigenvalues of T.
4. For each eigenvalue of T find the corresponding eigenvectors by finding the solutions for the corresponding system.

Example 1

Let $T : \mathbf{R}^3 \to \mathbf{R}^3$ be given by $T((x, y, z)) = (3x - 2y, -2x + 3y, 5z)$.
Step 1: The standard matrix of T is

$$A = \begin{pmatrix} 3 & -2 & 0 \\ -2 & 3 & 0 \\ 0 & 0 & 5 \end{pmatrix}$$

Step 2: $\det(A - \lambda I) = (\lambda - 1)(\lambda - 5)^2$.
Step 3: $\det(A - \lambda I) = (\lambda - 1)(\lambda - 5)^2 = 0$ implies

$$\lambda = 1, \qquad \lambda = 5.$$

Step 4: An eigenvector is a nontrivial solution of

$$\begin{pmatrix} 3-\lambda & -2 & 0 \\ -2 & 3-\lambda & 0 \\ 0 & 0 & 5-\lambda \end{pmatrix} \begin{pmatrix} x \\ y \\ z \end{pmatrix} = \begin{pmatrix} 0 \\ 0 \\ 0 \end{pmatrix}.$$

For $\lambda = 1$ we have

$$\begin{pmatrix} 2 & -2 & 0 \\ -2 & 2 & 0 \\ 0 & 0 & 4 \end{pmatrix} \begin{pmatrix} x \\ y \\ z \end{pmatrix} = \begin{pmatrix} 0 \\ 0 \\ 0 \end{pmatrix},$$

implying that $z = 0$ and $x = y$. Therefore, $(1, 1, 0)$ is an eigenvector of T corresponding to the eigenvalue $\lambda = 1$.

For $\lambda = 5$ we have

$$\begin{pmatrix} -2 & -2 & 0 \\ -2 & -2 & 0 \\ 0 & 0 & 0 \end{pmatrix} \begin{pmatrix} x \\ y \\ z \end{pmatrix} = \begin{pmatrix} 0 \\ 0 \\ 0 \end{pmatrix}$$

implying that z can be any number and $x = -y$. Therefore $(1, -1, 0)$ and $(0, 0, 1)$ are two linearly independent eigenvectors of T corresponding to the eigenvalue $\lambda = 5$.

Proposition 5.5.3 *Let* $T : \mathbf{R}^2 \to \mathbf{R}^2$ *be an orthogonal transformation. Then* T *is either a rotation or a reflection.*

Proof Let (a, b) denote $T((1, 0))$. Since T preserves length, we have $a^2 + b^2 = 1$. Let us choose θ, $0 \leq \theta \leq 180°$, such that

$$a = \cos\theta \quad \text{and} \quad b = \sin\theta.$$

Since T preserves length and orthogonality, $T(0, 1) = \pm(-b, a)$.

If $T(0, 1) = (-b, a)$, we have

$$\begin{aligned} T((1,0)) &= (a, b) \\ &= a(1,0) + b(0,1) \\ &= \cos\theta(1,0) + \sin\theta(0,1), \\ T((0,1)) &= (-b, a) \\ &= -b(1,0) + a(0,1) \\ &= -\sin\theta(1,0) + \cos\theta(0,1). \end{aligned}$$

Thus the standard matrix of T is

$$A = \begin{pmatrix} \cos\theta & -\sin\theta \\ \sin\theta & \cos\theta \end{pmatrix},$$

which implies that T is a counterclockwise rotation of the plane about the origin through the angle θ.

If $T(0,1) = (b, -a)$, the standard matrix of T (the reader should verify this) is

$$A = \begin{pmatrix} \cos\theta & \sin\theta \\ \sin\theta & -cos\theta \end{pmatrix}.$$

In this case the equation $\det(A - \lambda I) = 0$ is

$$\lambda^2 - \cos^2\theta - \sin^2\theta = \lambda^2 - 1 = (\lambda - 1)(\lambda + 1) = 0$$

and hence 1 and -1 are eigenvalues of T. Let v and w be eigenvectors corresponding to the eigenvalues 1 and -1, respectively. It follows from Lemma 5.5.2(b) that v and w are orthogonal. Now consider the reflection S through the line $l = \{av \mid a \in \mathbf{R}\}$. We have $S(v) = v$ and $S(w) = -w$. Therefore S and T coincide on the basis $\{v, w\}$, which gives $S = T$ by Proposition 5.3.2(c). □

Lemma 5.5.4 *Let $T : \mathbf{R}^3 \to \mathbf{R}^3$ be an orthogonal transformation with eigenvalues* 1 *and* -1. *Then:*
(a) If v and w are linearly independent eigenvectors of T corresponding to the eigenvalue 1, *then T is the reflection through the plane determined by v and w.*
(b) If v, w are linearly independent eigenvectors of T corresponding to the eigenvalue -1, then T is the rotation about the line orthogonal to the plane determined by v and w through the angle $180°$.

Proof *(a)* Let S be the reflection through the plane $\mathcal{P}$ determined by v, w. Then

$$S(v) = v, \quad S(w) = w \quad \text{and} \quad S(n) = -n,$$

where n is a normal vector to $\mathcal{P}$. Now if v and w are eigenvectors of T corresponding to the eigenvalue 1, Lemma 5.5.2(b) implies that n is also an eigenvector of T corresponding to the eigenvalue -1. Then the actions of S and T coincide on a basis of $\mathbf{R}^3$, namely the basis $\{v, w, n\}$,

and then $S = T$ by Proposition 5.3.2(c).
(b) Now consider R a rotation about the line orthogonal to the plane determined by v and w through the angle $180°$. We have

$$R(v) = -v, \quad R(w) = -w, \quad \text{and} \quad R(n) = n,$$

where n again denotes a normal vector to the plane determined by v and w. Then the action of R coincides with the action of T on the basis $\{v, w, n\}$, which implies that $R = T$.

Theorem 5.5.5 *Let $T : \mathbf{R}^3 \to \mathbf{R}^3$ be an orthogonal transformation. Then T can be decomposed into the composition of a rotation and at most one reflection.*

Proof Let A be the standard matrix of T. Consider the algebraic equation $\det(A - \lambda I) = 0$. This is a equation of degree 3 and hence has a real solution; that is, T has an eigenvalue λ. Since T is an orthogonal transformation, $\lambda = \pm 1$.

Suppose $\lambda = 1$. Let v be an eigenvector of T corresponding to the eigenvalue 1. Let W be the plane orthogonal to v through the origin. The restriction of T to W is an orthogonal transformation of the plane W. If T acts on the points of W as a rotation, then T is a rotation of $\mathbf{R}^3$. If not, T restricted to W is a reflection and hence has eigenvalues 1 and -1. Then by Lemma 5.5.4(a) T is the reflection through the plane determined by v and w, where w is an eigenvector corresponding to the eigenvalue 1.

Suppose $\lambda = -1$. Let v now denote an eigenvector of T corresponding to the eigenvalue -1 and W be the plane orthogonal to v through the origin. If the restriction of T to W is a reflection, then there exist vectors $w, n \in W$, that are eigenvectors of T corresponding to the eigenvalues -1 and 1, respectively. By Lemma 5.5.4(b), T is a rotation about the line $\{tn \,|\, t \in \mathbf{R}\}$. Suppose then that T restricted to W is a rotation. The matrix representing T with respect to the basis $\{v, w, n\}$ is

$$\begin{pmatrix} -1 & 0 & 0 \\ 0 & \cos\theta & -\sin\theta \\ 0 & \sin\theta & \cos\theta \end{pmatrix} = \begin{pmatrix} 1 & 0 & 0 \\ 0 & \cos\theta & -\sin\theta \\ 0 & \sin\theta & \cos\theta \end{pmatrix} \begin{pmatrix} -1 & 0 & 0 \\ 0 & 1 & 0 \\ 0 & 0 & 1 \end{pmatrix}$$

which describes a reflection through the plane W determined by w and n followed by a rotation about the line through the origin and parallel to v. □

Corollary 5.5.6 *Let T be an orthogonal transformation of $\mathbf{R}^3$. Then T is a rotation if and only if* $\det(T) = 1$.

Proof Let us suppose that T is a rotation and $\beta = \{v_1, v_2, v_3\}$ is an orthonormal basis, where v_1 is on the axis of rotation. Then the matrix that represents T with repect to β is

$$\begin{pmatrix} 1 & 0 & 0 \\ 0 & \cos\theta & -\sin\theta \\ 0 & \sin\theta & \cos\theta \end{pmatrix}$$

and therefore has determinant 1.

Conversely, suppose that $\det(T) = 1$. It follows from Theorem 5.5.5 that T is the composition of a rotation with at most one reflection. Suppose then $T = R \circ S$, where R is a rotation and S a reflection. Since reflections have determinant -1, from the properties of the determinant we get that $\det(T) = \det(R)\det(S) = -1$, contradicting that $\det(T) = 1$. It follows that $T = R$, that is, T is a rotation. □

Proposition 5.5.7 *(a) Let $\mathcal{P}_1$ and $\mathcal{P}_1$ be two nonparallel planes through the origin of $\mathbf{R}^3$ and S_1 and S_2 reflections through $\mathcal{P}_1$ and $\mathcal{P}_2$, respectively. Let l be given by $l = \mathcal{P}_1 \cap \mathcal{P}_2$. Then $S_2 \circ S_1$ is a rotation about line l.*
(b) Every rotation about a line through the origin is the composition of two reflections.

Proof (a) Let v denote a unit vector lying on line l. Let n_i, $i = 1, 2$, be a unit vector orthogonal to plane $\mathcal{P}_i$. The set $\beta = \{v, n_1, n_2\}$ is basis of $\mathbf{R}^3$. Now we compute the matrix $[S_2 \circ S_1]_\beta$. Since v is fixed by S_1 and S_2, we have

$$S_2 \circ S_1(v) = S_2(S_1(v)) = v.$$

Using the formula in Proposition 5.4.6, we get

$$\begin{aligned} S_2 \circ S_1(n_1) &= S_2(-n_1) \\ &= -S_2(n_1) \end{aligned}$$

$$
\begin{aligned}
&= -n_1 + 2(n_1 \cdot n_2)n_2, \\
S_2 \circ S_1(n_2) &= S_2(n_2 - 2(n_2 \cdot n_1)n_1) \\
&= -n_2 - 2(n_2 \cdot n_1)S_2(n_1) \\
&= -n_2 - 2(n_2 \cdot n_1)[n_1 - 2(n_2 \cdot n_1)n_2] \\
&= -2(n_2 \cdot n_1)n_1 \; + \; [-1 \; + \; 4(n_2 \cdot n_1)^2]n_2.
\end{aligned}
$$

Therefore,

$$
[S_2 \circ S_1]_\beta = \begin{pmatrix} 1 & 0 & 0 \\ 0 & -1 & -2(n_1 \cdot n_2) \\ 0 & 2(n_1 \cdot n_2) & -1 + 4(n_2 \cdot n_1)^2 \end{pmatrix},
$$

which has determinant 1, implying that $S_2 \circ S_1$ is a rotation. Since the vector v is fixed by $S_2 \circ S_1$, we conclude that line l is the axis of rotation.
(b) Let R be a rotation and l its axis passing through $(0,0,0)$. Let $\mathcal{P}$ be a plane orthogonal to l at the origin. By the definition of rotation, if $v \in \mathcal{P}$, $R(v) \in \mathcal{P}$ and R restricted to $\mathcal{P}$ is a rotation of plane $\mathcal{P}$ centered at $(0,0,0)$. Let $\bar{S}_1$ and $\bar{S}_2$ be reflections of $\mathcal{P}$ through lines l_1 and l_2, respectively. We have that

$$
R(w) = \bar{S}_2 \circ \bar{S}_1(w), \;\; \forall w \in \mathcal{P}.
$$

Let $\mathcal{P}_1$ be the plane determined by lines l_1 and l, and $\mathcal{P}_2$ by l_2 and l. Denoting by S_1 and S_2 the reflections of $\mathbf{R}^3$ through planes $\mathcal{P}_1$ and $\mathcal{P}_2$, we get from part (a) that $S_2 \circ S_1$ is a rotation about line l. Since $S_2 \circ S_1(v) = R(v)$ for v lying on l and also for vectors in $\mathcal{P}$, we conclude that $R = S_2 \circ S_1$, for they have the same action on the basis, say, $\{v, w_1, w_2\}$, with $w_1, w_2 \in \mathcal{P}$. □

Proposition 5.5.8 *Let $T : \mathbf{R}^3 \to \mathbf{R}^3$ be an orthogonal transformation. Then:*
(a) If $\det(T) = 1$, *then* 1 *is an eigenvalue of T and T is a rotation about the line $\{tv \,|\, t \in \mathbf{R}\}$, where v is an eigenvector corresponding to the eigenvalue* 1.
(b) If $\det(T) = -1$ *and*
(i) 1 *is an eigenvalue of T, then T is the reflection through the plane determined by the eigenvectors corresponding to the eigenvalue* 1;
(ii) -1 *is the only eigenvalue of T, then T is the composition of a rotation and one reflection.*

Proposition 5.5.9 *Let T be a rigid motion of 3-space. Then T can be expressed as the composition of at most four reflections.*

Proof If T is an orthogonal transformation, Theorem 5.5.5 combined with Proposition 5.5.7(b) implies that T is the composition of at most three reflections. If T is not an orthogonal transformation, we let $v = T(\mathbf{0})$ and consider the plane $\mathcal{P}$ that is the perpendicular bisector of $\overline{v\mathbf{0}}$. Let S be the reflection through $\mathcal{P}$. Now $S \circ T$ is a rigid motion; moreover,

$$S(T(\mathbf{0}) = S(v) = \mathbf{0},$$

which implies that $S \circ T$ is an orthogonal transformation, which we denote by F. Then F is the composition of at most three reflections. Since $S^{-1} = S$, we conclude

$$T = S \circ F$$

and hence T is the composition of at most four reflections. □

Exercises

1. Prove Proposition 5.5.8.

2. Consider the following matrices

$$A = \begin{pmatrix} 1 & 0 & 0 \\ 0 & \cos\theta & -\sin\theta \\ 0 & \sin\theta & \cos\theta \end{pmatrix} \quad \text{and} \quad B = \begin{pmatrix} \cos\phi & -\sin\phi & 0 \\ \sin\phi & \cos\phi & 0 \\ 0 & 0 & 1 \end{pmatrix}.$$

   Let T and S be the linear transformations represented by A and B with respect to the standard basis of $\mathbf{R}^3$.
   (a) Prove that $T \circ S$ is a rotation.
   (b) Find its axis in the case that $\theta = 90°$ and $\phi = 45°$.

5.6 Orientation

In Chapter 4, when coordinatizing 3-space, we chose an orientation for it by choosing a positive side for each axis x, y, and z satisfying the right-hand rule. This notion of orientation can be translated as

$$e_3 = e_1 \wedge e_2,$$

where $\{e_1, e_2, e_3\}$ is the standard basis of $\mathbf{R}^3$ and $\wedge$ denotes the vector product. It follows that $e_3 \cdot (e_1 \wedge e_2) = 1$. From this fact we define the orientation of bases in $\mathbf{R}^3$.

Definition 5.6.1 *An ordered basis $\beta = \{v_1, v_2, v_3\}$ is said to be* positively oriented *or* right-handed *if $v_3 \cdot (v_1 \wedge v_2) > 0$. When it is negative, the basis is said to be* negatively oriented *or* left-handed.

Recall that if the vectors v_j's of a basis β of $\mathbf{R}^3$ have coordinates $v_j = (a_{1j}, a_{2j}, a_{3j})$, the transition matrix $[I]_{\mathcal{S}}^{\beta}$ is given by

$$\begin{pmatrix} a_{11} & a_{12} & a_{13} \\ a_{21} & a_{22} & a_{23} \\ a_{31} & a_{32} & a_{33} \end{pmatrix},$$

and thus $\det\left([I]_{\mathcal{S}}^{\beta}\right) = v_3 \cdot (v_1 \wedge v_2)$. Therefore, a positively oriented basis of $\mathbf{R}^3$ is such that $\det\left([I]_{\mathcal{S}}^{\beta}\right) > 0$.

Generalizing this condition and using the fact that the determinant of a transition matrix is nonzero, we define the orientation of a basis of $\mathbf{R}^n$.

Definition 5.6.2 *We say that a basis β of $\mathbf{R}^n$ has the* same orientation *as the standard basis $\mathcal{S}$, or is* positively oriented, *if*

$$\det\left([I]_{\mathcal{S}}^{\beta}\right) > 0.$$

If $\det\left([I]_{\mathcal{S}}^{\beta}\right) < 0$, then we say that $\mathcal{S}$ and β have opposite orientation *or that β is* negatively oriented.

The separation of all bases of $\mathbf{R}^n$ into two classes, according to their sign, allows us to define the sign of a rigid motion. Observe first that an orthogonal transformation maps the standard basis onto another orthonormal basis; if this other orthonormal basis is positively oriented, then we say that the orthogonal transformation preserved orientation, or has sign 1. Recalling that a rigid motion has a linear part that is an orthogonal transformation, we make the following definition.

Definition 5.6.3 *A rigid motion T of $\mathbf{R}^n$ is said to be* orientation-preserving *if its linear part is an orientation-preserving orthogonal transformation, that is, it has determinant 1. If the determinant is -1, then we say that the rigid motion is* orientation-reversing.

Motivated by the results of Section 5.5, we define reflections and rotations of $\mathbf{R}^n$.

Definition 5.6.4 *(a) A* rotation *of* $\mathbf{R}^n$ *is an orientation-preserving rigid motion.*
(b) A reflection *of* $\mathbf{R}^n$ *is a rigid motion whose linear part is of the form* S_v*, where* $v \neq 0$ *and*

$$S_v(w) = w - \frac{2(w \cdot v)}{||v||^2} v.$$

Proposition 5.6.5 *A reflection of* $\mathbf{R}^n$ *is an orientation-reversing rigid motion.*

Proof We have to show that $\det(S_v) = -1$. For that we consider v_1 orthogonal to v. If $n > 2$, we then find v_2 such that v_2 is orthogonal to v and v_1. We continue this procedure and obtain vectors $v_1, \ldots, v_{n-1}$ that are orthogonal to v and to each other. Then the set $\beta = \{v_1, \ldots, v_{n-1}, v\}$ is a basis of $\mathbf{R}^n$, by Proposition 5.2.9. Observe that $S_v(v_i) = v_i$ for $i = 1, \ldots, n-1$ and $S_v(v) = -v$. Therefore,

$$[S_v]_\beta = \begin{pmatrix} 1 & \ldots & 0 & 0 \\ \vdots & \vdots & \vdots & \vdots \\ 0 & \ldots & 1 & 0 \\ 0 & \ldots & 0 & -1 \end{pmatrix},$$

which shows that the determinant is -1. □

Exercises

1. Let $T : \mathbf{R}^n \to \mathbf{R}^n$ be defined as $T(v) = -v$. Show that T is a composition of rotations if and only if n is even.

2. Determine if the ordered bases below are positively oriented:
   (a) $\{(1,1,0),\ (0,1,0),\ (0,1,1)\}$.
   (b) $\{(1,1,1),\ (0,-1,0),\ (1,0,-1)\}$.

3. Show that the set of rigid motions of $\mathbf{R}^n$ is a group under the operation of composition of transformations. This group is denoted by $\mathrm{Iso}(\mathbf{R}^n)$.

4. (a) Show that the set of orthogonal transformations of $\mathbf{R}^n$ forms a subgroup of Iso($\mathbf{R}^n$). This is called the *orthogonal group* and is denoted by $O(n)$.
   (b) Show that the set of translations of $\mathbf{R}^n$ forms a subgroup of Iso($\mathbf{R}^n$).
   (c) Show that the set of orientation-preserving transformations of $\mathbf{R}^n$ forms a subgroup of Iso($\mathbf{R}^n$).

5. Let $\circ$ and $\star$ denote the group operations of G_1 and G_2, respectively. A map $\Phi : G_1 \to G_2$ is said to be a *homomorphism* if $\phi(x \circ y) = \phi(x) \star \phi(y)$ for all $x, y \in G_1$. In addition, if Φ is one-to-one and onto, then Φ is called an isomorphism and we say that G_1 and G_2 are *isomorphic*.
   (a) Show that the operation of vector addition defines a group structure in $\mathbf{R}^n$.
   (b) Show that $\mathbf{R}^n$, with structure defined in (a), is isomorphic to the group of translations of $\mathbf{R}^n$.

Chapter 6

PERIMETER, AREA, AND VOLUME

In this chapter we want to assign to subsets of the 3-space positive real numbers that "measure" the set in some sense. Although it is possible to give a precise definition of the k-dimensional measure of "measurable subsets" of $\mathbf{R}^n$, we do not want to discuss such a theory in this text. We will restrict ourselves to finding the length of some special types of curves, areas of polygonal regions and disks of the plane, and volumes of spheres and special solids in space.

6.1 Perimeter and Circumference

Recall that a *polygonal curve* is the union of line segments that are called the *sides* of the polygonal curve. In Chapter 1 we assigned to each line segment, its length, and this was postulated by Axiom S_1. It is natural then to define the *length or perimeter* of any polygonal curve by adding the length of its sides.

We have not yet addressed the question of measuring lengths of curves, and in fact the only types of curves we have studied so far are polygons and circles. We start this section by defining the length of an arc of a circle. We need the lemmas below in order to justify the definition that will be given.

Lemma 6.1.1 *(a) Let $\widehat{AB}$ be an arc of circle γ centered at point O. Let C be on ray $\overrightarrow{OB}$ such that $OB < OC$. Then $AB < AC$.*

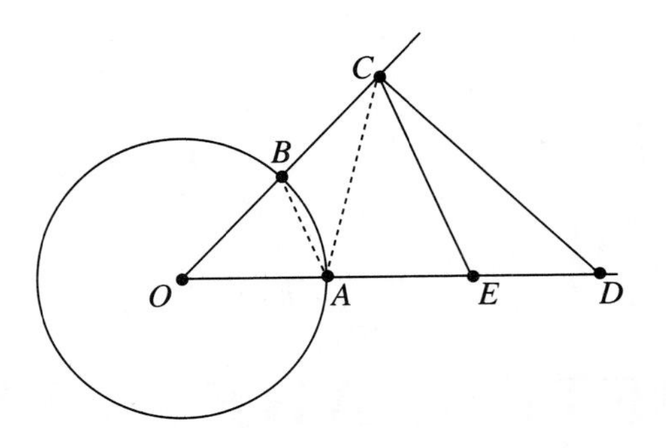

Figure 6.1

(b) Let $D \in \overrightarrow{OA}$ such that $OD > OC$. Then $DC > AB$.

Proof (a) Since $OA = OB$, we have that $\angle OAB \simeq \angle OBA$, and hence each of them is an acute angle; it follows that $\angle ABC$ is obtuse. Consider now triangle $\triangle ABC$. We have $m(\angle ABC) > m(\angle ACB)$, and then $AC > AB$.
(b) Consider point $E \in \overrightarrow{OA}$ such that $OE = OC$. Since $\triangle OEC$ is isosceles, using part (a) we conclude that $EC < DC$. Now observe that $\triangle OAB$ is similar to $\triangle OEC$, and then

$$\frac{EC}{AB} = \frac{OC}{OB} > 1.$$

It follows that $EC > AB$ and hence $DC > AB$. □

Lemma 6.1.2 *Let $\overset{\frown}{AB}$ be an arc of circle γ centered at point O. Let $A_1, A_2, \ldots, A_n$ be a sequence of points on γ such that $A_1 = A$ and $A_n = B$. Let P_n denote the perimeter of the polygonal curve*

$$\overline{A_1A_2} \cup \cdots \cup \overline{A_iA_{i+1}} \cup \cdots \cup \overline{A_{n-1}A_n}$$

and $\mathcal{S}$ the perimeter of a square that contains the circle in its interior. Then $P_n \leq \mathcal{S}$.

Proof Let B_i be the point where ray $\overrightarrow{OA_i}$ intersects the square. From Lemma 6.1.1(b) we get that

$$A_1A_2 + \cdots + A_{n-1}A_n \leq B_1B_2 + \cdots + B_{n-1}B_n.$$

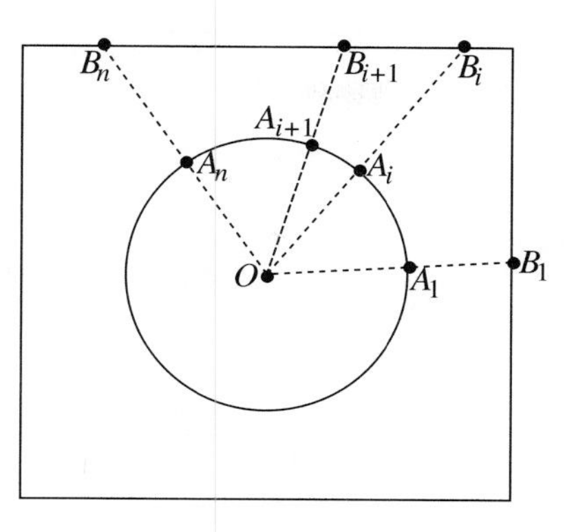

Figure 6.2

Since $B_1B_2 + \cdots + B_{n-1}B_n \leq \mathcal{S}$ (*why?*), we conclude the lemma. $\square$

It follows from Lemma 6.1.2 that the set $\mathcal{L}$ of real numbers which are perimeters of polygonal curves inscribed in arc $\overset{\frown}{AB}$ is bounded above. Therefore, using the completeness axiom of real numbers, we can make the following definition.

Definition 6.1.3 *Let $\overset{\frown}{AB}$ be an arc of circle γ. Then the length of $\overset{\frown}{AB}$, denoted by $L(\overset{\frown}{AB})$, is defined by*

$$L(\overset{\frown}{AB}) = \sup \mathcal{L}.$$

The length of γ, also called the *circumference*, is then defined as the supremum of the set of real numbers that are perimeters of polygons inscribed in the circle. The *real number that is the length of a semicircle of radius* 1 is denoted by π.

The lemma below gives us a formula to estimate the number π.

Lemma 6.1.4 *Let $\mathcal{P}_n$ be a regular n-gon of side s_n inscribed in a circle γ of radius 1 and center O. Consider the regular $2n$-gon $\mathcal{P}_{2n}$ obtained as follows: for each side, say $\overline{AB}$, we consider point C given by the intersection of γ with $\overleftrightarrow{OM}$, M being the midpoint of $\overline{AB}$. Then the side*

s_{2n} *of* $\mathcal{P}_{2n}$ *is given by*

$$s_{2n} = \sqrt{2 - \sqrt{4 - s_n^2}}$$

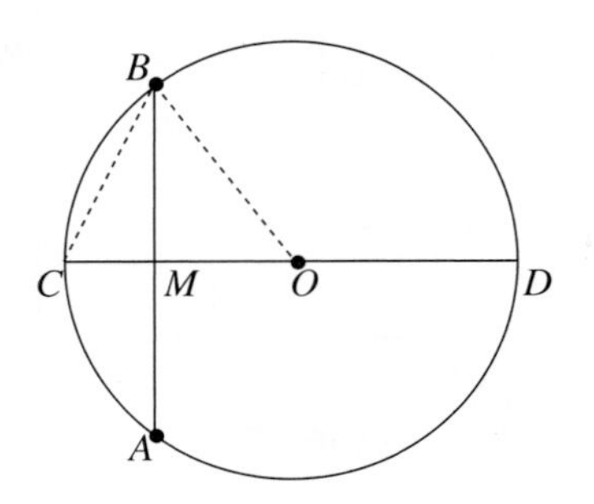

Figure 6.3

Proof Let D be the antipodal point of C on γ. Since $\angle DBC$ is inscribed in a semicircle, we have that $\triangle DBC$ is a right triangle whose altitude to the hypotenuse $\overline{CD}$ is $\overline{BM}$. Then we have $BM^2 = CM \cdot MD = BC^2 - CM^2$ and therefore

$$BC^2 = CM^2 + CM \cdot MD = CM\,(CM + MD) = 2CM.$$

But triangle $\triangle BMO$ is also a right triangle, and hence $MO^2 = BO^2 - BM^2$. Let x denote the length CM; recall that $BM = s_n/2$ and $BC = s_{2n}$. Then the equations above yield:

$$s_{2n}^2 = 2x \qquad \text{and} \qquad (1-x)^2 = 1 - (s_n/2)^2.$$

Eliminating x, we get

$$s_{2n} = \sqrt{2 - \sqrt{4 - s_n^2}}.$$

□

We start our estimate with a square inscribed in a circle of radius 1. Then, using the notation of Lemma 6.1.4, we have $s_4 = \sqrt{2}$. Applying the lemma, we obtain:

$$s_8 = \sqrt{2 - \sqrt{2}},$$

$$s_{16} = \sqrt{2 - \sqrt{2 + \sqrt{2}}}.$$

The following table has correct values up to the fourth decimal place.

n	s_n	$P_n = ns_n$
4	1.41421	5.6568
8	0.76537	6.1229
16	0.39018	6.2428
32	0.19603	6.2730
64	0.09814	6.2806
128	0.04908	6.2825
256	0.02454	6.2830
512	0.01227	6.2831

An approximation of π is then 3.1416; a more precise approximation is 3.14159265. The value of π has been calculated to more than a billion places with the aid of a computer and using power series (studied in calculus).

We now want to derive a formula for the circumference. For that, first we compare the perimeter of two circles.

Proposition 6.1.5 *Let γ and γ' be two circles of radius r and r', respectively. Let L_γ and $L_{\gamma'}$ denote their respective perimeters. Then*

$$\frac{L_\gamma}{r} = \frac{L_{\gamma'}}{r'}.$$

Proof It follows from Exercise 1 that we do not lose generality if we suppose that the two circles have same center O. Now we consider points A and B on γ and points A' and B' on γ obtained by the intersection of rays $\overrightarrow{OA}$ and $\overrightarrow{OB}$ with γ' (see Figure 6.4). Since triangles $\triangle OAB$ and $\triangle OA'B'$ are similar, we get

$$\frac{AB}{A'B'} = \frac{OA}{OA'},$$

and thus

$$AB = \frac{r}{r'} A'B'.$$

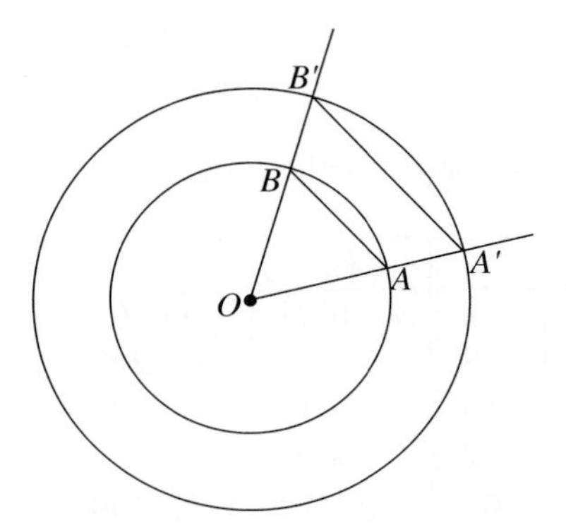

Figure 6.4

Therefore, if $A_1, \ldots, A_n$ are the vertices of a polygon $\mathcal{P}_n$ inscribed in γ, we consider points $A'_1, \ldots, A'_n$ on γ', obtaining the polygon $\mathcal{P}'_n$ which is inscribed in γ'. For these two polygons we have

$$\text{Perimeter}(\mathcal{P}_n) = \frac{r}{r'} \text{ Perimeter}(\mathcal{P}'_n).$$

Using Exercise 2, we conclude that

$$L_\gamma = \sup\mathcal{L} = \frac{r}{r'}\sup\mathcal{L}' = \frac{r}{r'}L_{\gamma'}.$$

□

Notice that the above proposition is saying that the ratio of the circumference to the radius of the circle does not depend on the circle. Then from the definition of π and Proposition 6.1.5 we obtain:

Proposition 6.1.6 *The length of a circle of radius r is $2\pi r$.*

Another consequence of Proposition 6.1.5 is that if $\overset{\frown}{AB}$ and $\overset{\frown}{A'B'}$ are arcs of same degree measure in circles of radius r and r', respectively, then

$$\frac{L(\overset{\frown}{AB})}{r} = \frac{L(\overset{\frown}{A'B'})}{r'}.$$

The ratio

$$\frac{L(\overset{\frown}{AB})}{r}$$

is called *radian measure* of $\widehat{AB}$ and, as in the case of the degree measure, it depends only on the angle and not on the radius. It follows that the radian measure of a semicircle is π, and then we have that 180° corresponds to π radians. Having in mind such a correspondence, we define the *radian measure of the central angle* $\angle AOB$ as $L(\widehat{AB})/r$. It follows that the radian measure of $\measuredangle(AOB)$ is simply $m(\angle AOB)\,\pi/180$.

Exercises

1. Let T be a rigid motion of the plane and γ a circle. Show that $\gamma' = T(\gamma)$ is a circle and $L_\gamma = L_{\gamma'}$.

2. Let S be a bounded set of real numbers and $cS = \{cs \mid s \in S\}$, where $c \in \mathbf{R}$. Show that
   (a) If $c \geq 0$, then $\sup cS = c \sup S$.
   (b) If $c < 0$, then $\sup cS = c \inf S$

6.2 Area of Polygonal Regions

Definition 6.2.1 *The union of a triangle with its interior is called a* triangular region. *The sides of the triangle are called* edges *of the region and the vertices of the triangle are called* vertices *of the region.*

Definition 6.2.2 *A* polygonal region *in the plane is the union of a finite number of triangular regions such that if two of the triangular regions intersect, their intersection is an edge or a vertex of each of them.*

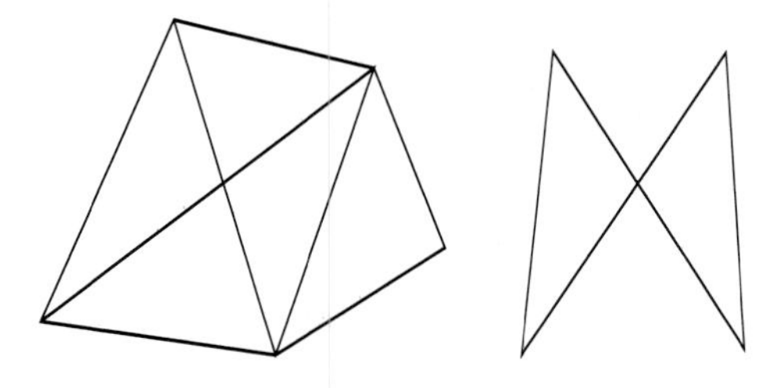

Figure 6.5 Polygonal regions

In Exercise 6(c) the reader is asked to show that a polygon together with its interior form a polygonal region. However, notice that a polygonal region is not necessarily the interior of a polygon (see Figure 6.5).

A point is said to be an *interior point* of a polygonal region if it is an interior point of some triangular region contained in the polygonal region. The set of all interior points is called the *interior* of the polygonal region and the *boundary* consists of the points of the polygonal region that are not interior points.

Given a polygonal region $\mathcal{R}$, we will denote by $\mathcal{T}$ the set whose elements are the triangular regions. The set $\mathcal{T}$ is called a *complex*, and the procedure of cutting the region $\mathcal{R}$ into triangular regions is called a *triangulation* of $\mathcal{R}$.

In elementary geometry, to each polygonal region one assigns a positive number, called its *area*, using the following axioms:

> **$\mathbf{Ar_1}$** If two triangular regions are congruent, then they have the same area.
>
> **$\mathbf{Ar_2}$** If the intersection of two polygonal regions does not contain interior points (only edges or vertices), then the area of their union is the sum of their areas.
>
> **$\mathbf{Ar_3}$** If a square region has edges of length s, then its area is s^2.

Observe that there are different triangulations for the same polygonal region $\mathcal{R}$ (see Figure 6.6). Therefore the axioms above are assuming that the numerical area that is assigned to any polygonal region is independent of the triangulation.

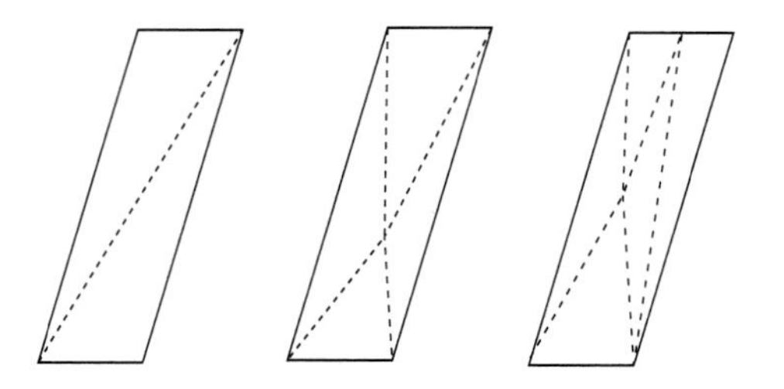

Figure 6.6

In the rest of the section we shall present a geometric definition of area; that is, we define the area of a polygonal region $\mathcal{R}$ instead of assuming the axioms above. It will be defined as the sum of the areas of the triangular regions of a triangulation, and we will prove that this sum is the same for all triangulations of $\mathcal{R}$. To define the area of a polygonal region we shall start with the area of a triangle.

Lemma 6.2.3 *Given triangle $\triangle ABC$, let $\overline{AD}$ be the altitude dropped from A and $\overline{CE}$ be the altitude from C. Then*

$$AD \cdot BC \quad = \quad CE \cdot AB,$$

that is, in a triangle the product of any base by its corresponding altitude is independent of the choice of the base.

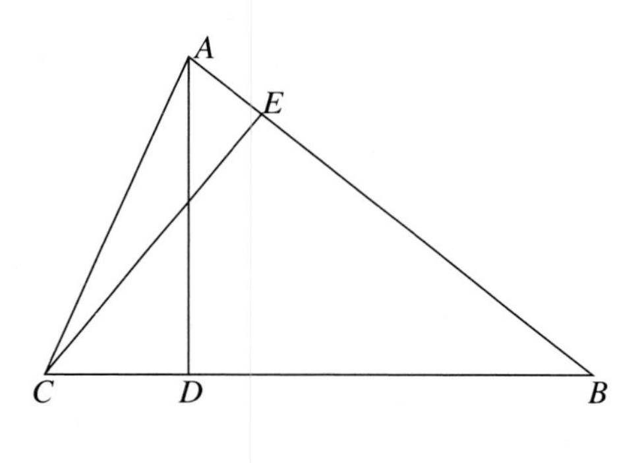

Figure 6.7

Proof If $D = B$, then $\triangle ABC$ is a right triangle, and then $E = B$. In this case the lemma says $AB \cdot BC = BC \cdot AC$, which is obviously true.

Let us suppose $D \neq B$ (see Figure 6.7), which in turn implies $E \neq B$. Then $\triangle ADB$ is similar to $\triangle CEB$, since they have a common angle $\angle B$ and $\angle CEB \simeq \angle ADB$. This implies

$$\frac{CE}{AD} = \frac{CB}{AB},$$

which is the result of the lemma. □

Definition 6.2.4 *The area of a triangle is half of the product of any base and the corresponding altitude.*

Definition 6.2.5 *The area of a complex $\mathcal{T} = \{T_1, T_2, \ldots, T_n\}$ is given by*

$$\text{Area}\,(T_1) + \text{Area}\,(T_2) + \cdots + \text{Area}\,(T_n),$$

where Area(T_i) *denotes the area of the triangle T_i.*

Lemma 6.2.6 *Let $\square ABCD$ be a trapezoid such that $\overline{AB} \parallel \overline{CD}$. Let $b_1 = AB$, $b_2 = CD$, and h denote the altitude of the trapezoid* (see Figure 6.8). *Consider a triangulation of the trapezoid so that all triangles have their vertices on the upper or lower base and $\mathcal{T}$ is its corresponding complex. Then*

$$\text{Area}\,(\mathcal{T}) = \frac{(b_1 + b_2)\,h}{2}.$$

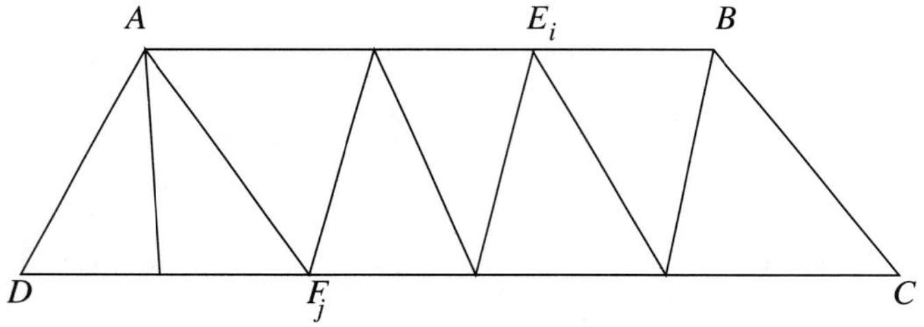

Figure 6.8

Proof Let $A = E_1, \ldots, E_n = B$ denote the vertices of the triangles of $\mathcal{T}$ that lie on $\overline{AB}$. Then the sum of the lengths of their opposite sides is $CD = b_2$. Choosing their opposite sides for bases, their corresponding altitude is the trapezoid's altitude. We obtain a similar conclusion if we consider the vertices $F_1, \ldots, F_m$ lying on $\overline{CD}$. Therefore,

$$\begin{aligned}\text{Area}\,(\mathcal{T}) &= \frac{1}{2}[F_1F_2 + \cdots + F_{m-1}F_m \\ &\quad + E_1E_2 + \cdots + E_{n-1}E_m]\,h \\ &= \frac{(b_1 + b_2)\,h}{2}.\end{aligned}$$

□

It follows from Lemma 6.2.6 that for a given trapezoid, all complexes whose triangles have vertices on its bases have the same area.

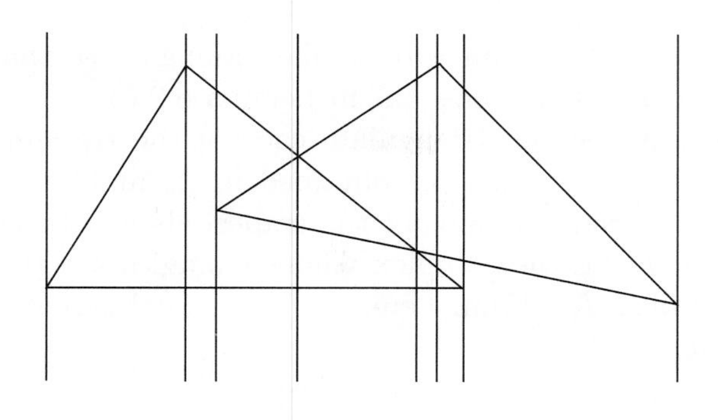

Figure 6.9

Theorem 6.2.7 *If $\mathcal{T}_1$ and $\mathcal{T}_2$ are two complexes given by two triangulations of the same polygonal region $\mathcal{R}$, then* Area$(\mathcal{T}_1) =$ Area$(\mathcal{T}_2)$.

Proof Given two triangulations, let $\mathcal{T}_1$ and $\mathcal{T}_2$ denote their corresponding complexes. We consider parallel lines through all vertices of $\mathcal{T}_1$, all vertices of $\mathcal{T}_2$, and points where edges of $\mathcal{T}_1$ intersect edges of $\mathcal{T}_2$ (see Figure 6.9). Observe that now each triangle of $\mathcal{T}_1$ is covered by triangles or trapezoids. Moreover, each triangle is decomposed in one of the following ways (see Figure 6.10):
(i) the triangle is the union of smaller triangles, so that all vertices are on the base or the opposite vertex;
(ii) the triangle is the union of one or two smaller triangles and a finite number of trapezoids.

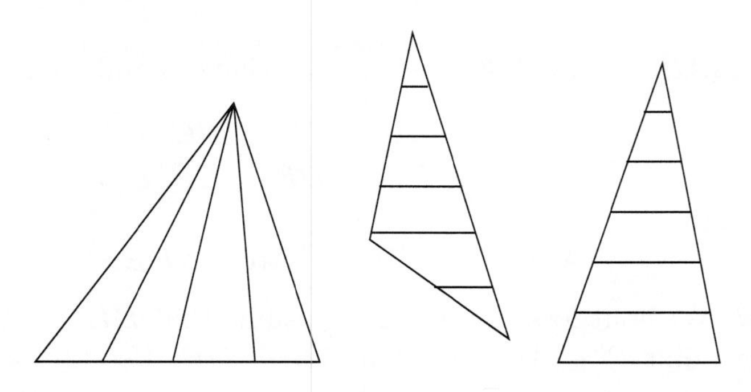

Figure 6.10

The trapezoid is the union of smaller triangles, so that all vertices are on the upper or lower base (as in Lemma 6.2.6).

Observe that one can triangulate each of the trapezoids in (ii) as in Lemma 6.2.6. The triangles obtained in (i) and (ii) and the ones obtained by the triangulations of the trapezoids will be called *smaller triangles*. Let $\mathcal{T}'$ be the complex whose triangles are the smaller triangles of complex $\mathcal{T}_1$. From Lemma 6.2.6, and Exercises 4 and 5 we conclude that

$$\text{Area}\,(\mathcal{T}_1) = \text{Area}\,(\mathcal{T}'),$$

since the area of a complex is the sum of areas of its triangles.

To finish this proof, just observe that the smaller triangles of $\mathcal{T}_2$ are the same as the smaller triangles of $\mathcal{T}_1$ and hence

$$\text{Area}\,(\mathcal{T}_2) = \text{Area}\,(\mathcal{T}') = \text{Area}\,(\mathcal{T}_1),$$

which concludes the theorem. □

Using Theorem 6.2.7, we can finally define the area of a polygonal region.

Definition 6.2.8 *The area of polygonal region $\mathcal{R}$ is the sum of the areas of the triangles of a triangulation of $\mathcal{R}$.*

Exercises

1. Use Definition 6.2.4 to show that if $\triangle ABC \simeq \triangle DEF$, then

$$\text{Area}\,(\triangle ABC) = \text{Area}\,(\triangle DEF).$$

2. Let $\triangle ABC$ and $\triangle DEF$ be similar triangles and let ρ denote the ratio

$$\rho = \frac{AB}{DE} = \frac{AC}{DF} = \frac{BC}{EF}.$$

Show that

$$\text{Area}\,(\triangle ABC) = \rho^2\,\text{Area}\,(\triangle DEF).$$

Hint: Without loss of generality assume that $DE < AB$ and consider point $B' \in \overline{AB}$ such that $AB' = DE$. Use the parallel projection theorem to conclude that the ratio between the altitudes of $\triangle ABC$ and $\triangle DEF$ is also ρ. Then use Exercise 1 above.

3. Given a right triangle $\triangle ABC$ such that $\overline{AB}$ is the hypotenuse, construct squares $\square ABFG$, $\square ACHE$, and $\square BCIJ$. Let $\overline{CD}$ be the altitude corresponding to the hypotenuse and K the point where ray $\overrightarrow{CD}$ intersects $\overline{FG}$ (see the next figure). Show the following:
(a) Area $(\square ADKF) = 2\,$Area $(\triangle CAF)$.
(b) Area $(\square ACHE) = 2\,$Area $(\triangle EAB)$.
(c) Area $(\square ACHE) =$ Area $(\square ADKF)$.
(d) Use part (c) to conclude the Pythagorean theorem.
This was Euclid's proof of the Pythagorean theorem in the *The Elements.*

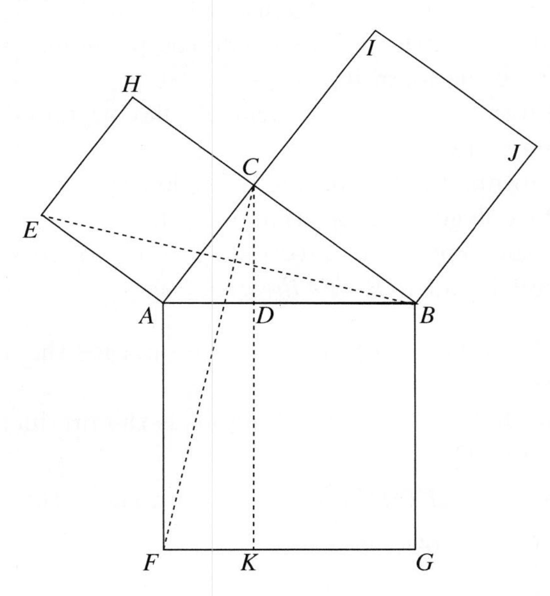

4. Consider $\triangle ABC$ and D a point between B and C. Use Definition 6.2.4 to show that

$$\text{Area}\,(\triangle ABC) = \text{Area}\,(\triangle ABD) + \text{Area}\,(\triangle ADC).$$

5. Consider $\triangle ABC$, D a point between A and B, and E a point between A and C such that $\overline{BC} \parallel \overline{DE}$. Let h denote the altitude of the trapezoid $\square DECB$. Use Definition 6.2.4 and Lemma 6.2.6

to show that

$$\text{Area}\,(\triangle ABC) = \text{Area}\,(\triangle ADE) + \frac{(DE + CB)h}{2}.$$

6. Consider a set of points in the Euclidean plane with integer coordinates. Such a set is called a *lattice* and each point is called a *lattice point*. Show that if a polygon P has vertices that are lattice points, then

$$\text{Area}\,(P) = \frac{1}{2}p + q - 1,$$

where p is the number of lattice points on the boundary and q is the number of lattice points inside the polygon.
Hint: Prove the following steps:
(a) The formula holds for a triangle having no lattice points between the vertices.
(b) The formula holds for any triangle (use (a)).
(c) Every polygon can be triangulated.
(d) Add the expressions involving p and q for these triangles.
This result is called *Pick's theorem.*

7. (a) Use Definition 6.2.8 to find a formula for the area of the parallelogram $\square ABCD$.
(b) Show that the area of a rhombus is the product of the lengths of its diagonals.
(c) Consider $\square ABCD$ in the Euclidean space $\mathbf{R}^3$ and let $u = \overset{\longmapsto}{AB}$ and $v = \overset{\longmapsto}{CD}$. Show that

$$|u \wedge v| = \text{Area}\,(\square ABCD).$$

8. Given a right triangle $\triangle ABC$, with legs of length b and c and hypotenuse a, consider a square $\square DEFG$ of side $b + c$ and consider the square $\square HIJK$ as in the figure below.
(a) Show that

$$\text{Area}\,(\square DEFG) = \text{Area}\,(\square HIJK) + 4\,\text{Area}\,(\triangle ABC).$$

(b) Use part (a) to conclude the Pythagorean theorem.

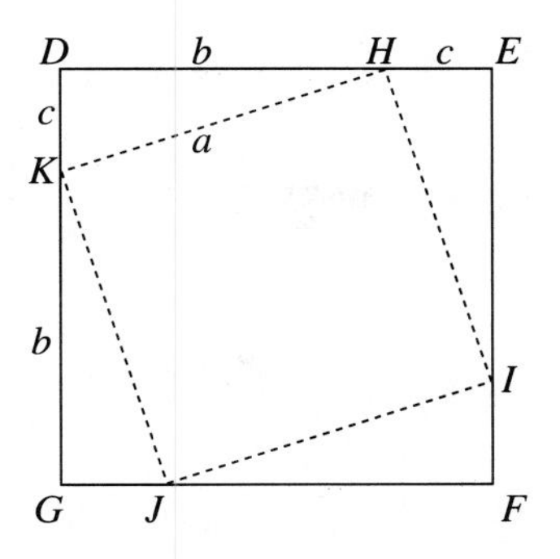

9. The following problem is a generalization of the Pythagorean theorem proved by Pappus of Alexandria in the fourth century.

   Let $\triangle ABC$ be any triangle. Construct rectangles $\square ACHE$ and $\square BCIJ$ (see the next figure). Let P be the point where $\overleftrightarrow{EH}$ intersects $\overleftrightarrow{JI}$. Construct a parallelogram such that one side is $\overline{AB}$ and the other $\overline{AF}$ so that $\overrightarrow{AF} \parallel \overrightarrow{PC}$ and $PC = AF$. Show that

   $$\text{Area}\,(\square ACHE) + \text{Area}\,(\square BCIJ) = \text{Area}\,(\square ABGF).$$

   Hint: Consider point Q where ray $\overrightarrow{FA}$ intersects $\overline{EH}$ and point R the intersection of $\overrightarrow{BG}$ and $\overline{IJ}$.

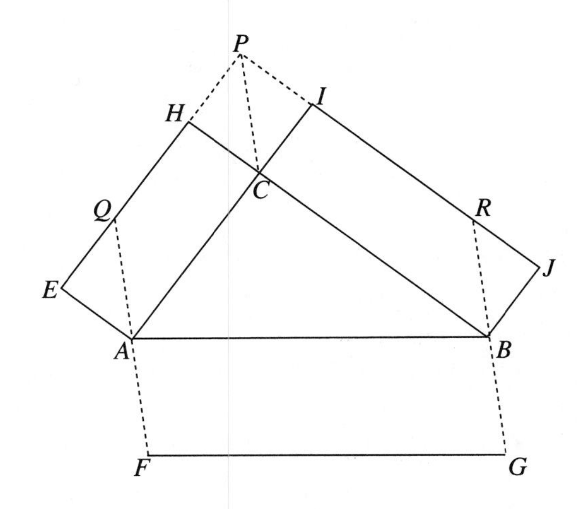

6.3 Area of Circles

Let γ be a circle of radius r and $\mathcal{P}_n$ a polygon of n sides inscribed in γ. Let $A_1, \ldots, A_n$ be its vertices (see Figure 6.11). The polygon $\mathcal{P}_n$ is the boundary of a polygonal region whose area is

$$\text{Area}\,(\mathcal{P}_n) = \frac{1}{2}\,[A_1A_2 \,+\, A_2A_3 \,+\cdots+\, A_nA_1]\,r = \frac{1}{2}P_n r,$$

where P_n denotes the perimeter of the polygon. Now we consider the set $\mathcal{A}$ of the real numbers that are areas of polygons inscribed in circle γ. Notice that $\mathcal{A}$ is bounded above, because for any polygon $\mathcal{P}_n$ inscribed in γ we have $P_n < 2\pi r$. Then, using the completeness axiom of the real numbers, we define the *area of the region bounded by a circle* as $A_\gamma = \sup \mathcal{A}$.

In the next theorem we derive a formula for the area of the region bounded by a circle of radius r, called the *disk* of radius r.

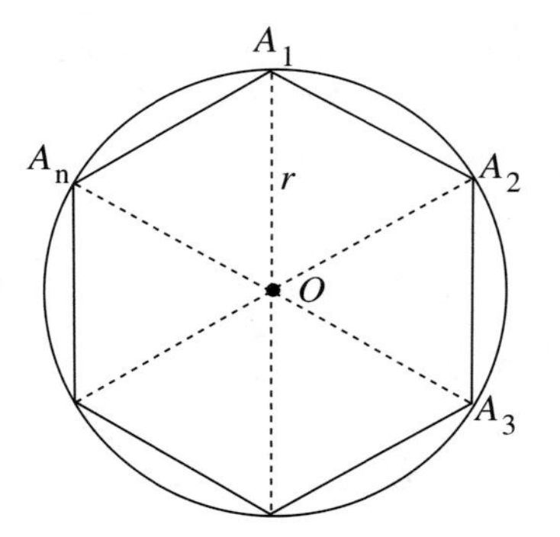

Figure 6.11

Theorem 6.3.1 *The area of the disk of radius r is πr^2, that is, 1/2 of the product of its circumference and its radius.*

Proof Given a positive number ϵ, we consider a polygon $\mathcal{P}^1$ of side $s < \epsilon$. The definition of L_γ implies that there exists a polygon $\mathcal{P}^2$ such that $L_\gamma - L(\mathcal{P}^2) < \epsilon$. Again, from the definition of A_γ we obtain a polygon $\mathcal{P}^3$ such that $A_\gamma - A(\mathcal{P}^3) < r\epsilon$. We then construct another polygon $\mathcal{P}$ whose vertices are the vertices of all three polygons above.

For polygon $\mathcal{P}$ we have:

(i) $s(\mathcal{P}) < \epsilon$.

(ii) $L_\gamma - L(\mathcal{P}) < \epsilon$.

(iii) $A_\gamma - A(\mathcal{P}) < r\epsilon$.

Let O denote the center of the circle. By connecting all vertices of $\mathcal{P}$ to O we obtain a triangulation of the region bounded by $\mathcal{P}$. Let A and B denote two consecutive vertices of the polygon $\mathcal{P}$ and let C be the midpoint of $\overline{AB}$ (see Figure 6.12). We then have

$$\text{Area}\,(\triangle OAB) = \frac{1}{2} OC \cdot AB.$$

But we have $OA - AC < OC < OA$, and then

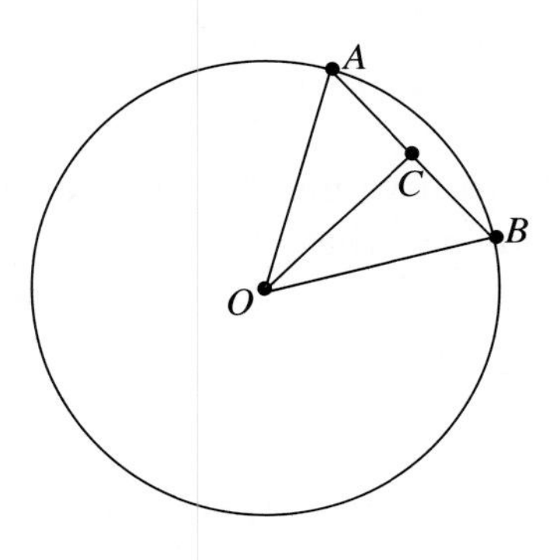

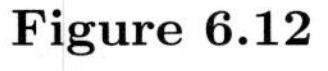

Figure 6.12

$$\frac{1}{2} AB \cdot (OA - AC) < \frac{1}{2} AB \cdot OC < \frac{1}{2} AB \cdot OA.$$

Since $OA = r$ and $AC < \epsilon$, we obtain

$$\frac{1}{2} AB \cdot (r - \epsilon) < \text{Area}\,(\triangle OAB) < \frac{1}{2} AB \cdot r.$$

Adding for all triangles, we get

$$\frac{1}{2} L(\mathcal{P}) \cdot (r - \epsilon) < A(\mathcal{P}) < \frac{1}{2} L(\mathcal{P}) \cdot r.$$

We know that $L(\mathcal{P}) < L_\gamma$ (by definition) and we are assuming $L_\gamma - \epsilon < L(\mathcal{P})$ (condition (ii)). Substituting above, we have

$$\frac{1}{2}(L_\gamma - \epsilon) \cdot (r - \epsilon) < A(\mathcal{P}) < \frac{1}{2} L_\gamma \cdot r$$

which implies

$$|A(\mathcal{P}) - \frac{1}{2} L_\gamma \cdot r| < \frac{1}{2}(r\epsilon + \epsilon L_\gamma - \epsilon^2).$$

To conclude the proof we observe that

$$\begin{aligned} |A_\gamma - \frac{1}{2} L_\gamma \cdot r| &= |A_\gamma - A(\mathcal{P}) + A(\mathcal{P}) - \frac{1}{2} L_\gamma \cdot r| \\ &\leq |A_\gamma - A(\mathcal{P})| + |\frac{1}{2} L_\gamma \cdot r| \\ &< r\epsilon + \frac{1}{2}(r\epsilon + \epsilon L_\gamma - \epsilon^2). \end{aligned}$$

Since ϵ is an arbitrary positive number, it can be taken as small as we please, and then we conclude that

$$A_\gamma = \frac{1}{2} L_\gamma \cdot r = \pi r^2.$$

□

The formula πr^2 is quite often referred to as the *area of a circle.*

6.4 Volumes

Definition 6.4.1 *A (convex)* polyhedron P *is a 3-dimensional figure composed of polygonal regions such that:*
(i) if two of the regions intersect, the intersection is an edge or a vertex of each.
(ii) if $\mathcal{R}$ *is any of the polygonal regions of* P *and* $\mathcal{P}$ *is the plane containing region* $\mathcal{R}$*, then all other points of the polyhedron* P *are on the same side of plane* $\mathcal{P}$*.*

The polygonal regions are called *faces* of the polyhedron, and the edges of the polygonal regions are also called the *edges* of the polyhedron.

Example

The polyhedron of six faces such that opposite faces are parallel and congruent parallelograms is called a *parallelepiped.* Given a face, its corresponding altitude is the distance between the the two parallel planes that contain the opposite faces.

In this section we will assign real numbers to regions of 3-space, also called *solids*, that express their 3-dimensional measure. Such a measure is called *volume.* We are interested in the volumes of regions bounded by polyhedra, cylinders, cones, and spheres. For simplicity we say, for example, "the volume of a sphere," instead of "the volume of the region bounded by the sphere."

We will assume that there exists a collection $\mathcal{V}$ of subsets of 3-space such that:
(a) The regions bounded by polyhedra, cylinders, cones, and spheres are in $\mathcal{V}$.
(b) If $A, B \in \mathcal{V}$, then $A \cup B$, $A \cap B$, and $A - B$ are also in $\mathcal{V}$.
(c) For each $A \in \mathcal{V}$, we assign a positive number, called the *volume* of A, denoted by $V(A)$, which satisfies the following axioms:

> $\mathbf{V_1}$ If $A, B \in \mathcal{V}$ and $A \subset B$, then $V(A) \leq V(B)$.
>
> $\mathbf{V_2}$ If $A, B \in \mathcal{V}$ and $V(A \cap B) = 0$, then $V(A \cup B) = V(A) + V(B)$.
>
> $\mathbf{V_3}$ The volume of a parallelepiped is the product of the area of a face and its altitude.
>
> $\mathbf{V_4}$: *Cavalieri's principle*: Given two solids $\mathcal{S}_1$, $\mathcal{S}_2$, and a plane $\mathcal{P}$. If for every plane that intersects $\mathcal{S}_1$ and $\mathcal{S}_2$ and is parallel to $\mathcal{P}$, the two intersections determine regions that have the same area, then $V(\mathcal{S}_1) = V(\mathcal{S}_2)$.

The axioms above immediately imply the following:

Proposition 6.4.2 *(a) Let T denote a triangular region of the plane. Then $V(T) = 0$.*
(b) Let $\mathcal{R}$ be a polygonal region of the plane. Then $V(\mathcal{R}) = 0$.

Proof (a) Let $\triangle ABC$ denote the boundary of T. Consider line l through A and parallel to $\overleftrightarrow{BC}$. Let $D \in l$ such that $AD = BC$. Then

$\square ABCD$ is a parallelogram that we denote by P. Let us suppose that Area$(P) = a$. Given any positive number ϵ, let us consider $\epsilon' = \epsilon/a$ and a parallelepiped $\tilde{P}$ of base P and altitude ϵ'. Then from Axioms $\mathbf{V_1}$ and $\mathbf{V_3}$ we have

$$V(T) < V(P) < V(\tilde{P}) < a \cdot \epsilon' = \epsilon.$$

Therefore, $V(T)$ is less than any positive number and hence $V(T) = 0$.
(b) Triangulate the region $\mathcal{R}$ and use Axiom $\mathbf{V_2}$. □

Volumes of prisms and cylinders

Definition 6.4.3 *Let $\mathcal{P}_1$ and $\mathcal{P}_2$ be two parallel planes and $\mathcal{R}_1$ a polygonal region contained in $\mathcal{P}_1$. Let l be a line intersecting $\mathcal{P}_1$ and $\mathcal{P}_2$. Through each point $P \in \mathcal{R}_1$ we consider a line segment $\overline{PP'}$, parallel to l with $P' \in \mathcal{P}_2$. The union of all segments $\overline{PP'}$ is called the* prism *of* base *$\mathcal{R}_1$* and directrix *l* (see Figure 6.13). *The distance between planes $\mathcal{P}_1$ and $\mathcal{P}_2$ is called the* altitude *of the prism.*

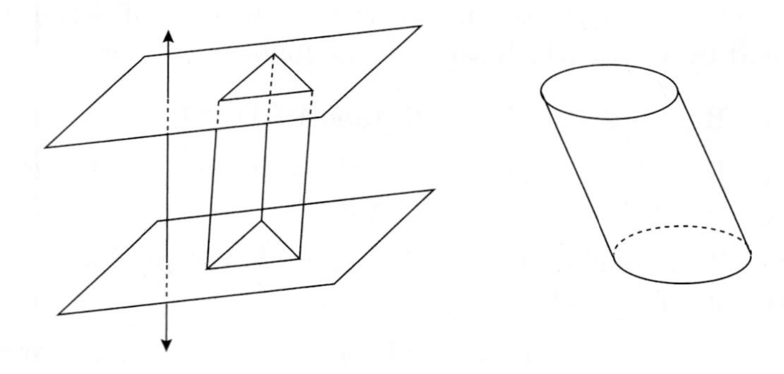

Figure 6.13

If the plane region is a disk (a circle together with its interior), the 3-dimensional figure obtained is then called a *cylinder*. The radius of the circle is said to be the *radius of the cylinder*.

If line l is perpendicular to plane $\mathcal{P}_1$ (and hence to plane $\mathcal{P}_2$), then the prism or cylinder is said to be *right*.

Proposition 6.4.4 *(a) Let P be a right prism of altitude h and triangular base T. Then*

$$V(P) = \text{Area}(T) \cdot h.$$

(b) Let P be a right prism of altitude h and base B. Then

$$V(P) = \text{Area}\,(B) \cdot h.$$

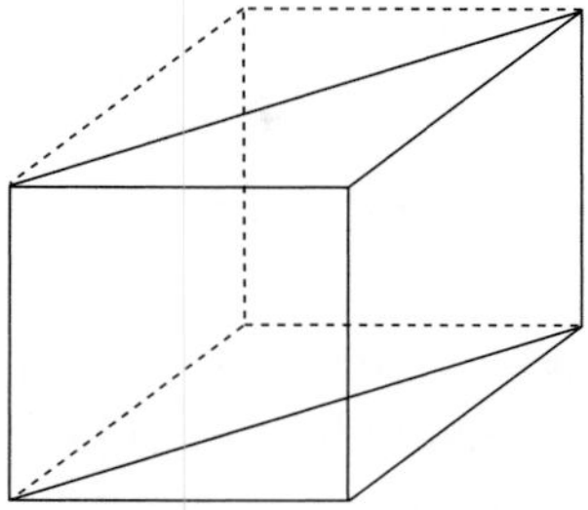

Figure 6.14

Proof *(a)* Let P be another prism of base T' such that $T \cup T'$ is a parallelogram B. Recall that T and T' are congruent triangles and $\text{Area}\,(B) = 2\text{Area}(T)$. Notice that we have constructed a parallelepiped $P \cup P'$, and then from Axiom $\mathbf{V_3}$ we get

$$V(P \cup P') = \text{Area}\,(B) \cdot h = 2\,\text{Area}\,(T) \cdot h.$$

By Cavalieri's principle we have $V(P) = V(P')$; moreover $P \cap P'$ is a plane region and then using Lemma 6.4.2 and Axiom $\mathbf{V_2}$ we obtain

$$V(P \cup P') = 2\,V(P),$$

yielding $V(P) = \text{Area}\,(T) \cdot h$.
(b) We consider a triangulation of B; that is, we cover B with a finite collection of triangles. It is not difficult now to see that the result of part (a) generalizes to B, and we leave the details to the reader. □

Theorem 6.4.5 *Let P be any prism (not necessarily right) of base B and altitude h. Then*

$$V(P) = \text{Area}\,(B) \cdot h.$$

Proof Let $\mathcal{P}_0$ be the plane containing the base B. Let P' be a right prism with same base and same altitude (see Figure 6.15). We consider $\mathcal{P}$ a plane parallel to $\mathcal{P}_0$. It follows from Exercise 1 of this section that

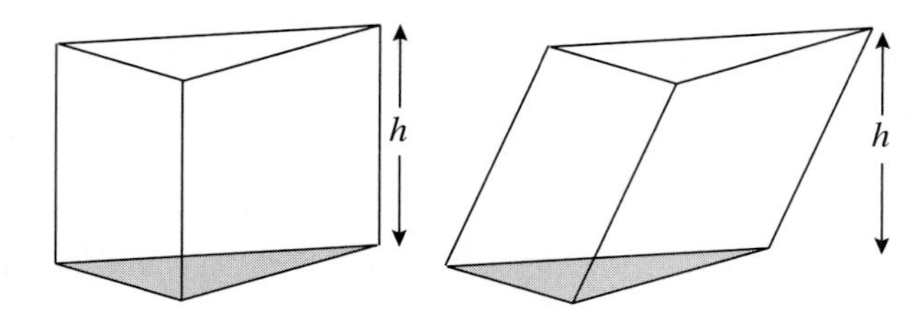

Figure 6.15

the area of all plane regions given by the intersections of $\mathcal{P}$ and P is Area (B). Likewise, the area of all intersections of $\mathcal{P}$ and P' is Area (B). Therefore, by Cavalieri's principle we have

$$V(P) = V(P')$$

and the result follows from Proposition 6.4.4. □

Theorem 6.4.6 *Let C be any cylinder (not necessarily right) of altitude h and radius r. Then*

$$V(C) = \pi r^2 \cdot h.$$

Proof Let $\mathcal{P}_0$ be the plane containing the base of the cylinder; on this plane consider a parallelogram of area πr^2 and let P be a right prism whose base is B and altitude is h. Then Proposition 6.4.4 implies that $V(P) = \pi r^2 \cdot h$. Let $\mathcal{P}$ be a plane parallel to $\mathcal{P}_0$. Using Exercise 1, we conclude that all intersections of $\mathcal{P}$ and C and all intersections of $\mathcal{P}$ and P have the same area πr^2. Then Cavalieri's principle implies $V(C) = V(P)$. □

Volumes of pyramids and cones

Definition 6.4.7 *Let $\mathcal{R}$ be a polygonal region contained in a plane $\mathcal{P}$ and A a point not in $\mathcal{P}$; through each point $P \in \mathcal{R}$ we consider a line segment $\overline{PA}$. The union of all segments $\overline{PA}$ such that $P \in \mathcal{R}$ is called* pyramid of base $\mathcal{R}$ and apex A. *The distance between planes A and $\mathcal{P}$ is called the* altitude *of the pyramid.*

If the plane region is a disk, then we call the figure a *cone.* The radius of the cone is the radius of the circle. If line $\overleftrightarrow{AO}$ is perpendicular to $\mathcal{P}$, where O is the center of the circle, then we say that the cone is a right cone.

Lemma 6.4.8 *Let P_1 and P_2 be two pyramids whose bases are in the same plane and whose apexes are on the same side of this plane. If they have same altitude and their bases have same area, then the two pyramids have same volume* (see Figure 6.16).

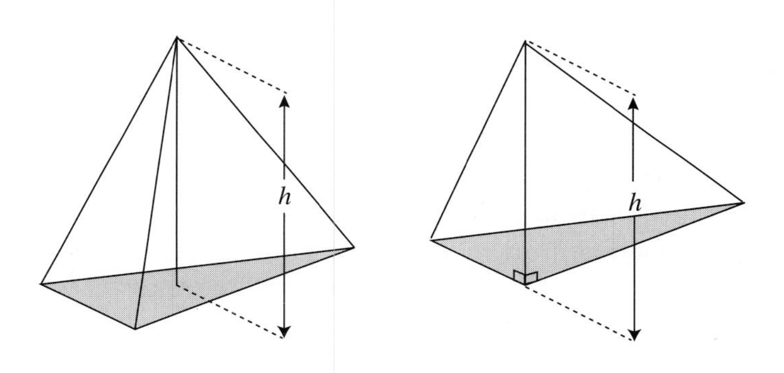

Figure 6.16

Proof The result follows from Cavalieri's principle and Exercise 3. We leave the details to the reader. □

Now we want to derive a formula for the volume of a pyramid. For that we will study first a right triangular prism (see Figure 6.17). The bases are the triangles $\triangle ABC$ and $\triangle DEF$, and the lateral faces are the rectangles $\square CBFE$, $\square CFDA$, and $\square ABDE$. Then we have

$$\triangle BCF \simeq \triangle BEF \quad \text{and} \quad \triangle ABE \simeq \triangle ADE.$$

Let us consider now the planes $\mathcal{P}_1$ containing points A, B, F, and plane $\mathcal{P}_2$ containing points A, E, F. They cut the prism in three triangular pyramids as shown in Figures 6.17 and 6.18.

We look at pyramids P_1 and P_3 as having the same apex A and bases $\triangle BCF$ and $\triangle BEF$, respectively. Their altitude, the distance from point A to the plane containing the rectangle $\square CBFE$, is also the same. Since they have congruent bases, from Lemma 6.4.8 we conclude that $V(P_1) = V(P_3)$.

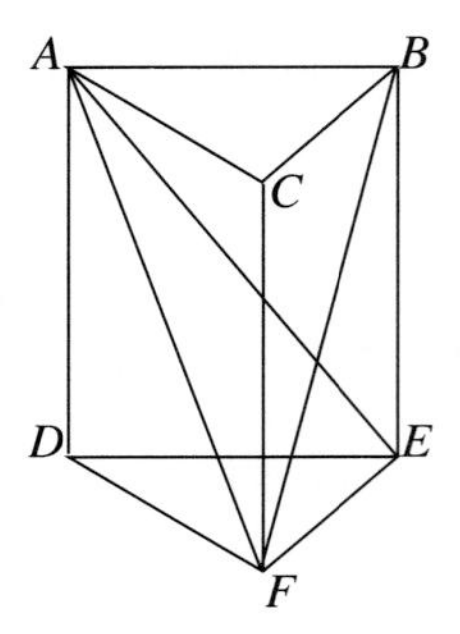

Figure 6.17

Now we look at pyramids P_3 and P_2. We consider point F as their apex, and their bases are $\triangle ABE$ and $\triangle ADE$, respectively. Their altitude is the distance from F to the plane that contains the rectangle $\square ABED$. Using again Lemma 6.4.8, we conclude that $V(P_2) = V(P_3)$, and the three pyramids have the same volume.

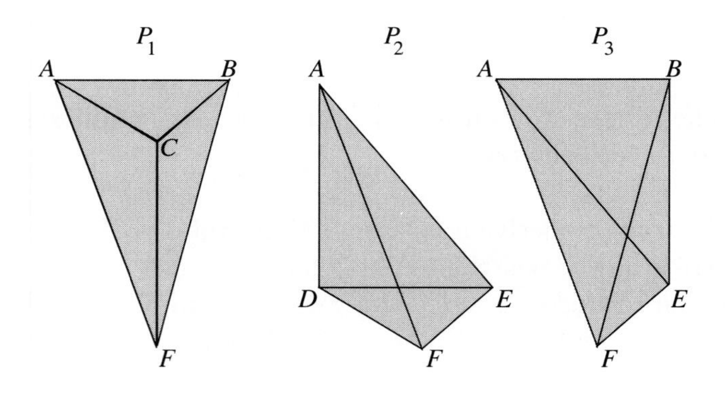

Figure 6.18

We then prove the following result.

Proposition 6.4.9 *Let P be a triangular pyramid of altitude h. Let A denote the area of its base. Then*

$$V(P) = \frac{1}{3} A \cdot h.$$

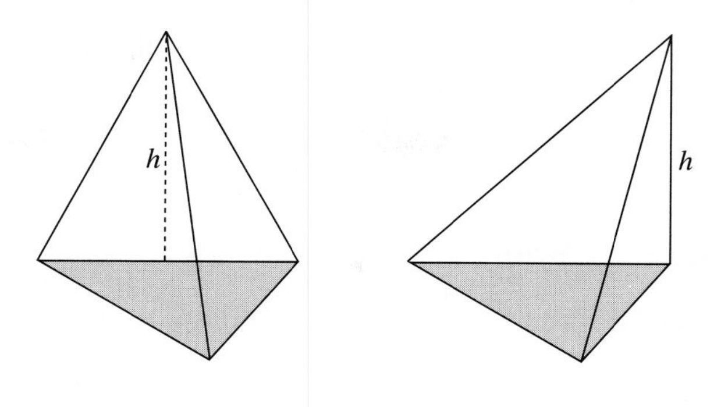

Figure 6.19

Proof Let P' be a pyramid of the same base and vertical edge of length h. Then, consider a prism $\overline{P}$ with the same base and the same altitude as P'. Observe that if we cut prism $\overline{P}$ as above, pyramid P' is pyramid P_3 and hence $V(P_3) = V(P')$. Further, pyramids P_1, P_2, P_3 intersect each other in planar regions, and using Lemma 6.4.2 and Axiom $\mathbf{V_2}$ we conclude that

$$V(\overline{P}) = V(P_1) + V(P_2) + V(P_3) = 3V(P_3) = 3V(P')$$

which yields

$$V(P') = \frac{1}{3}V(\overline{P}) = \frac{1}{3}A \cdot h.$$

Now, Lemma 6.4.8 implies the result that $V(P') = V(P)$. □

Theorem 6.4.10 *The volume of a pyramid of base area A and altitude h is $1/3$ of the product of A and h.*

Proof If the base is any polygonal region, we triangulate the region. Then the theorem follows from Lemma 6.4.2, Axiom $\mathbf{V_2}$, and Proposition 6.4.9. □

Theorem 6.4.11 *The volume of a cone of radius r and altitude h is*

$$V(C) = \frac{1}{3}\pi r^2 h.$$

Proof Let $\mathcal{P}$ be the plane that contains the base of the cone. Consider pyramid P with base contained in $\mathcal{P}$, of area πr^2 and altitude h. Now from Exercises 2 and 3 we obtain that planes parallel to $\mathcal{P}$ intersect the cone and P in regions whose areas are the same. Therefore the volume of the cone is equal to the volume of the P, which is the result of the theorem. $\square$

The Volume of a Sphere

Definition 6.4.12 *Let O be a point in 3-space and r a positive real number. The set $S = \{P \mid OP = r\}$ is called a* sphere *of center O and radius r.*

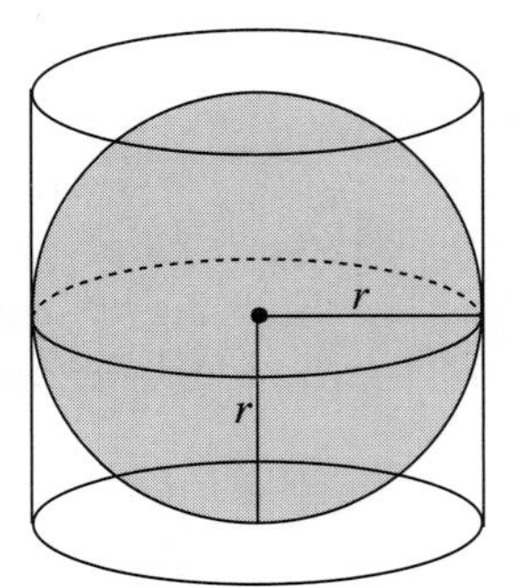

Figure 6.20

To find the volume of a sphere, we will use again Cavalieri's principle, but this time, we will apply it to the solid D given by

$$D = C - S,$$

where C is the cylinder that circumscribes the sphere S. Observe that the equality above makes sense only if we think of C and S as solids, that is, the union of the surfaces with their interiors. The cylinder C has radius r and altitude $2r$. Then, from our assumptions for the collection of measurable sets, $D \in \mathcal{V}$ and we get

$$V(S) = V(C) - V(D).$$

Lemma 6.4.13 *Let $\mathcal{P}_0$ be a plane through the center O of the sphere S. Consider a plane $\mathcal{P}$ parallel to $\mathcal{P}_0$ such that the distance between them is $s < r$. Then the intersection of $\mathcal{P}$ and D has area πs^2.*

Proof The intersection of $\mathcal{P}$ and C is a circle of radius r, and the intersection of $\mathcal{P}$ and S is also a circle but of radius $r' = \sqrt{r^2 - s^2}$ by the Pythagorean theorem (see Figure 6.21). Since the intersection of $\mathcal{P}$ and D is the plane region $\mathcal{R}$ bounded by these two concentric circles, we have that

$$\text{Area}\,(\mathcal{R}) = \pi(r^2 - [r^2 - s^2)] = \pi s^2.$$

□

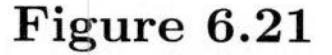

Figure 6.21

Now we consider another solid D' given by the union of two cones with common apex O and bases coinciding with the bases of the cylinder. The region $\mathcal{R}'$ given by the intersection of $\mathcal{P}$ and D' is a disk of radius s, and hence Area $(\mathcal{R}') = \pi s^2$. Using Cavalieri's principle, we show the following theorem.

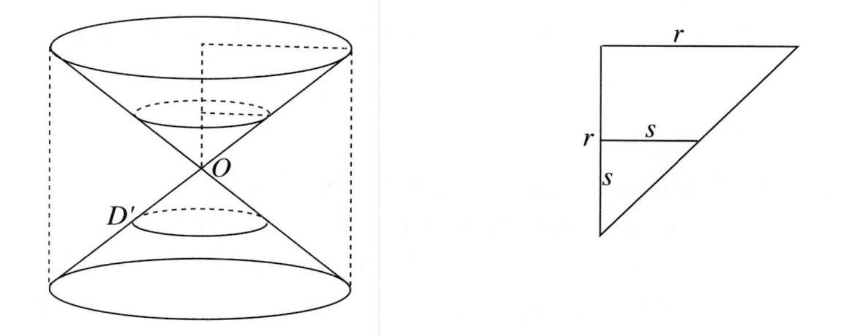

Figure 6.22

Theorem 6.4.14 *The volume of a sphere of radius r is*

$$V(S) = \frac{4}{3}\pi r^3.$$

Proof We use the formula for the volume of a cone (see Figure 6.22) and obtain

$$V(D) = V(D') = 2\frac{1}{3}\pi r^2 \cdot r = \frac{2}{3}\pi r^3.$$

Using the equality above and the formula for the volume of a cylinder, we get

$$\begin{aligned} V(S) &= V(C) - V(D) \\ &= \pi r^2 \cdot 2r - \frac{2}{3}\pi r^3 \\ &= \frac{4}{3}\pi r^3. \end{aligned}$$

□

Exercises

1. With the notation of Definition 6.4.3, let $\mathcal{R}_2$ be the plane region which is the intersection of plane $\mathcal{P}_2$ with the interior of a prism (or of a cylinder). Show that

   $$\text{Area}\,(\mathcal{R}_1) = \text{Area}\,(\mathcal{R}_2).$$

2. Let C be a cone of radius r and altitude h. Let $\mathcal{P}$ denote the plane containing the base B of the cone and $\mathcal{P}'$ a plane parallel to $\mathcal{P}$. Let $B' = \mathcal{P}' \cap C$. Show that B' is a circle of radius

   $$r' = r\frac{h - h'}{h},$$

   where h' denotes the distance between $\mathcal{P}$ and $\mathcal{P}'$.
   Hint: Let A denote the apex of the cone, O the center of the base, and X any point of the base B. Consider plane $\mathcal{P}_1$ determined by $\overleftrightarrow{AX}$ and $\overleftrightarrow{AO}$, and plane $\mathcal{P}_2$ determined by $\overleftrightarrow{AX}$ and $\overleftrightarrow{AF}$, where F denotes the foot of the altitude of the cone. Use Exercise 5 of Section 4.1 to conclude that $\mathcal{P}_1$ and $\mathcal{P}_2$ intersect $\mathcal{P}$ and $\mathcal{P}'$ in a pair of parallel lines. To finish the exercise, find similar triangles in this configuration and write the proportion between their sides.

3. Let P be a pyramid of base B contained in plane $\mathcal{P}$ and altitude h; let $\mathcal{P}'$ be a plane parallel to $\mathcal{P}$ and h' the distance between $\mathcal{P}$ and $\mathcal{P}'$. Let $B' = \mathcal{P}' \cap C$. Show that

$$\text{Area}\,(B') = [\frac{h - h'}{h}]^2 \cdot \text{Area}\,(B).$$

Hint: Consider first the case that the base is a triangle $\triangle XYZ$. Let A denote the apex of the pyramid. Repeat the procedure of Exercise 2 for each vertex X, Y, and Z; that is, consider the planes determined by $\overleftrightarrow{AX}$ and $\overleftrightarrow{AF}$, $\overleftrightarrow{AY}$ and $\overleftrightarrow{AF}$, and $\overleftrightarrow{AZ}$ and $\overleftrightarrow{AF}$. Conclude then that $\triangle XYZ$ is similar to $\triangle X'Y'Z'$, where X', Y', and Z' are the points where $\mathcal{P}'$ intersects the pyramid P. Use Exercise 2 of Section 6.2 to finish this case. Then triangulate base B to completely finish the exercise.

Chapter 7

SPHERICAL GEOMETRY

7.1 Arc Length

In this section we generalize the procedure used in Chapter 6 when we derived the formula for the circumference. We will assume that the reader is familiar with some concepts from calculus of a single variable. Recall that an *open interval* is defined as

$$I = (a, b) = \{x \in \mathbf{R} \mid a < x < b\}.$$

A real function is said to be *smooth* on I if it has derivatives of all orders at all points of I. It then follows that all derivatives are continuous.

Definition 7.1.1 *Let I be an open interval of the real line $\mathbf{R}$. A* parametrized smooth *curve is a function*

$$\gamma : I \to \mathbf{R}^n, \quad \gamma(t) = (\gamma_1(t), \ldots, \gamma_n(t))$$

such that each $\gamma_i : I \to \mathbf{R}$ is a smooth function.

Example 1

A line is the simplest example of a smooth curve in $\mathbf{R}^2$. Let us consider a line given by $y = mx + b$. It can be parametrized as

$$\gamma(t) = (t, mt + b), \quad t \in \mathbf{R}.$$

Here $\gamma_1(t) = t$ and $\gamma_2(t) = mt + b$. It is clear that a line is a smooth curve.

Example 2

The circle of center $(0,0)$ and radius 1 in the plane $\mathbf{R}^2$ can be parametrized as

$$\gamma(t) = (\cos(t), \sin(t)), \quad 0 - \epsilon < t < 2\pi + \epsilon.$$

Since $\cos(t)$ and $\sin(t)$ are smooth functions, we have then that $\gamma(t)$ is a smooth curve.

Example 3

The parametrized curve in $\mathbf{R}^2$ given by $\gamma(t) = (t, |t|), \ \ t \in \mathbf{R}$ is not smooth, since $|t|$ is not differentiable at $t = 0$.

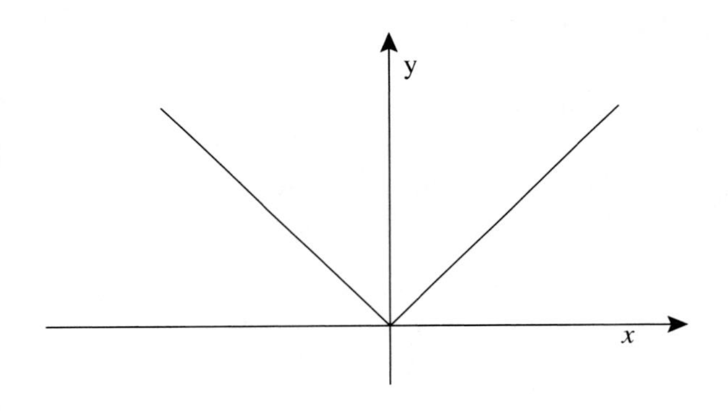

Figure 7.1

Definition 7.1.2 *Let* $\gamma : I \to \mathbf{R}^n$ *be a smooth curve. For each* $t \in I$, *the* velocity vector of γ *at* t *is the vector given by*

$$\gamma'(t) = (\gamma_1'(t), \dots, \gamma_n'(t)).$$

The geometric interpretation of the velocity vector is in the next figure. Since

$$\begin{aligned}
\gamma'(t) &= (\gamma_1'(t), \dots, \gamma_n'(t)) \\
&= \left(\lim_{h\to 0} \frac{\gamma_1(t+h) - \gamma_1(t)}{h}, \dots, \lim_{h\to 0} \frac{\gamma_n(t+h) - \gamma_n(t)}{h}\right) \\
&= \lim_{h\to 0} \frac{\gamma(t+h) - \gamma(t)}{h},
\end{aligned}$$

the line in $\mathbf{R}^n$ tangent to the curve γ at the point $\gamma(t)$ is in the direction of the vector $\gamma'(t)$. The velocity vector is also called the *tangent vector* of γ at $\gamma(t)$.

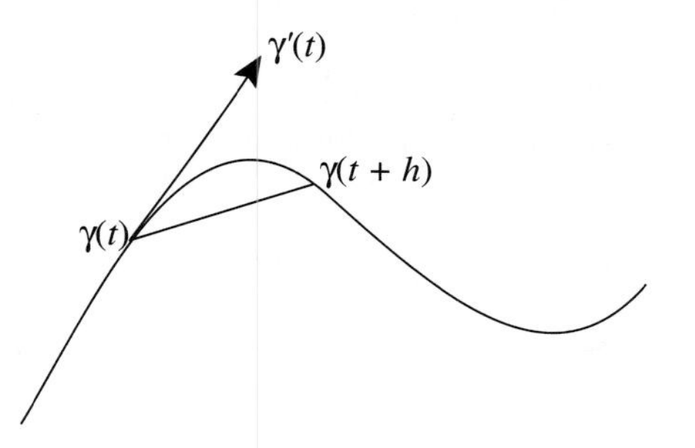

Figure 7.2

Definition 7.1.3 *Let $\gamma : I \to \mathbf{R}^n$ be a smooth curve. Then γ is said to be* regular *if*

$$\gamma'(t) \neq 0, \quad \forall t \in I.$$

Lines and circles are examples of regular curves. Another interesting example is described below.

Example 4

A *helix* is a curve obtained by letting a circle in the xy-plane rise or fall at a constant rate (it resembles a spring). A helix is given by the formula

$$\gamma(t) = (a\cos t, a\sin t, b\,t), \quad t \in \mathbf{R}, \quad a, b \neq 0.$$

Therefore its velocity vector is given by

$$\gamma'(t) = (-a\sin t, a\cos t, b).$$

Recall that $\gamma'(t) \neq 0$ if and only if $||\gamma'(t)|| \neq 0$, and thus

$$||\gamma'(t)|| = \sqrt{a^2\cos^2 t + a^2\sin^2 t + b^2} = \sqrt{a^2 + b^2} \neq 0.$$

Let $[a, b] = \{x \in \mathbf{R} |\ a \leq x \leq b\}$ be a closed interval contained in an open interval I. The length of a *regular* curve $\gamma : I \to \mathbf{R}^n$ from $\gamma(a)$ to $\gamma(b)$ is given by:

$$L(\gamma) = \int_a^b ||\gamma'(t)||\,dt.$$

We explain now the geometric meaning of the formula above. Suppose we choose points on the interval $[a, b]$ such that

$$a = t_0 < t_1 < \cdots < t_{n-1} < t_n = b,$$

and corresponding points on the curve γ, $\gamma(t_i), 0 \leq i \leq n$. The polygonal curve connecting these points has length given by

$$L(\mathcal{P}_n) = \sum_{i=1}^{n} ||\gamma(t_i) - \gamma(t_{i-1})||.$$

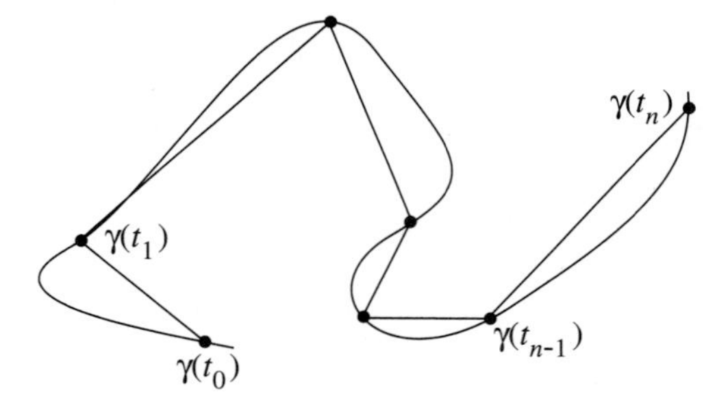

Figure 7.3

Theorem 7.1.4 *For every $\epsilon > 0$ there exists a polygonal $\mathcal{P}_n$ curve such that*

$$|\int_a^b ||\gamma'(t)||\, dt \quad - \quad L(\mathcal{P}_n)| < \epsilon.$$

As in the case of the circumference, the length of a regular curve is the "limit" of the lengths of inscribed polygonal curves. The proof of Theorem 7.1.4 is very technical and elaborate; it will be relegated to the appendix of this chapter.

Using the notion of arc length, we prove an important fact that we will use later to compare Euclidean and some of the non-Euclidean geometries.

Theorem 7.1.5 *Let $\gamma : (a, b) \to \mathbf{R}^n$ be a regular curve. Then*

$$||\gamma(b) - \gamma(a)|| \leq L(\gamma),$$

i.e., the line segment $\overline{\gamma(a)\gamma(b)}$ is the curve of shortest length from $\gamma(a)$ to $\gamma(b)$

Proof Let $u = (u_1, \ldots, u_n)$ denote a unit vector in $\mathbf{R}^n$. Consider the function

$$f(t) = \gamma(t) \cdot u = \gamma_1(t)\, u_1 \;+\; \cdots \;+\; \gamma_n(t)\, u_n.$$

Then

$$f'(t) = \gamma_1'(t)\, u_1 \;+\; \cdots \;+\; \gamma_n'(t)\, u_n = \gamma'(t) \cdot u.$$

From the fundamental theorem of calculus we obtain

$$f(b) - f(a) = \int_a^b f'(t)\, dt = \int_a^b \gamma'(t) \cdot u \;\, dt.$$

Let $\theta(t)$ denote the angle between $\gamma'(t)$ and u. Since u is a unit vector, we have

$$\gamma'(t) \cdot u = ||\gamma'(t)|| \;\; ||u|| \;\cos\theta(t) \leq ||\gamma'(t)||.$$

Substituting above, we obtain

$$f(b) - f(a) \leq \int_a^b ||\gamma'(t)||\, dt,$$

that is,

$$\gamma(b) \cdot u - \gamma(a) \cdot u = \Big(\gamma(b) - \gamma(a)\Big) \cdot u \leq L(\gamma).$$

Since the inequality above holds for all unit vectors u, we set

$$u = \frac{\gamma(b) - \gamma(a)}{||\gamma(b) - \gamma(a)||},$$

which gives

$$(\gamma(b) - \gamma(a)) \cdot \frac{\gamma(b) - \gamma(a)}{||\gamma(b) - \gamma(a)||} = ||\gamma(b) - \gamma(a)||,$$

implying that

$$||\gamma(b) - \gamma(a)|| \leq L(\gamma).$$

□

Now we fix $n = 3$ and consider a special subset of $\mathbf{R}^3$, the sphere

$$S^2 = \{(x, y, z) \in \mathbf{R}^3 \mid x^2 + y^2 + z^2 = 1\}.$$

We choose the sphere of radius 1 centered at the origin to simplify the notation. All concepts that we will define and the results that will be proved in the rest of the chapter can be extended to a sphere of any center and radius.

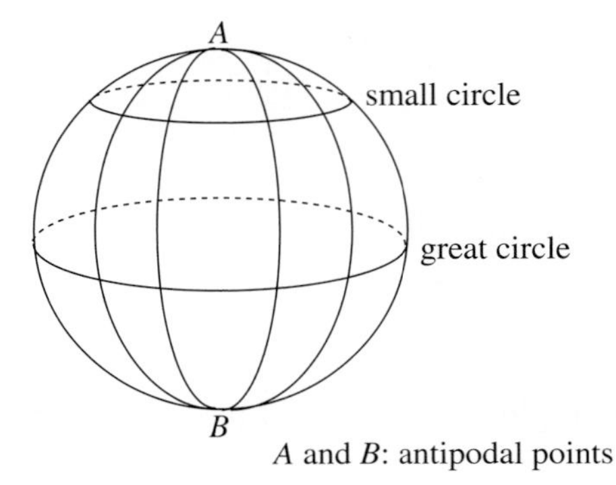

Figure 7.4

Definition 7.1.6 *A* great circle *is the curve obtained by intersecting the sphere with a plane that passes through the center of the sphere.*

Given two points X and Y on S^2, there is always a great circle through X and Y. It is a circle obtained by intersecting a plane $\mathcal{P}$ of $\mathbf{R}^3$ spanned by vectors $\vec{OX}$ and $\vec{OY}$ with the sphere. The points X and Y are called *antipodal* points if $Y = -X$. Notice that two points determine a unique great circle if and only if they are not antipodal. If X and Y are antipodal points, then all arcs of circles connecting these points are semicircles and therefore they all have same length, π.

If X and Y are not antipodal points, they determine two arcs in the great circle, called minor and major arcs. The minor arc, which is the shorter one, will be denoted by $\overset{\frown}{XY}$. Given two nonantipodal points, we want to consider all curves between them that lie entirely on the sphere and find the one of shortest length. It is convenient to use the *geographical coordinates* or *spherical coordinates* (θ, φ).

Definition 7.1.7 *Let P be a point on the unit sphere S^2 and (x, y, z) its rectangular coordinates. We say that P has* spherical coordinates *(θ, φ) if*

$$\begin{aligned} x &= \cos\theta \sin\varphi, \\ y &= \sin\theta \sin\varphi, \\ z &= \cos\varphi, \end{aligned}$$

$$0 \leq \theta < 2\pi, \qquad 0 \leq \varphi \leq \pi,$$

where φ, usually called the colatitude, *measures the angle between $\overrightarrow{OP}$ and the positive part of the z-axis and θ, called the* longitude, *measures the angle $\overrightarrow{OP'}$ and the positive part of the x-axis, with P' denoting the orthogonal projection of $\overrightarrow{OP}$ onto the xy-plane.*

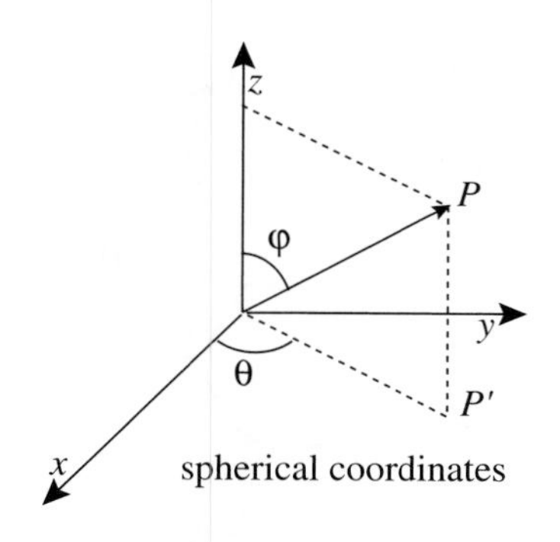

Figure 7.5

Observe that the equator is given by $\varphi(t) = \pi/2$ and the curve $\theta =$ *constant* is a great semicircle.

Theorem 7.1.8 *The shortest curve between two points on a sphere is an arc of a great circle.*

Proof Let A and B be two points on S^2. Without loss of generality we suppose that A and B have the same longitude (why? Use Exercises 8 and 9 of this section). Let $\gamma : (a, b) \to S^2$ be a regular curve, $\gamma(a) = A$ and $\gamma(b) = B$. We write

$$\gamma(t) = (x(t), y(t), z(t)),$$

and use the spherical coordinates, that is,

$$\begin{aligned} x(t) &= \cos\theta(t)\ \sin\varphi(t), \\ y(t) &= \sin\theta(t)\ \sin\varphi(t), \\ z(t) &= \cos\varphi(t). \end{aligned}$$

From the chain rule we obtain

$$\begin{aligned} x'(t) &= -\sin\theta\sin\varphi\cdot\theta'(t)+\cos\theta\cos\varphi\cdot\varphi'(t), \\ y'(t) &= \cos\theta\sin\varphi\cdot\theta'(t)+\sin\theta\cos\varphi\cdot\varphi'(t), \\ z'(t) &= -\sin\varphi\cdot\varphi'(t). \end{aligned}$$

Therefore

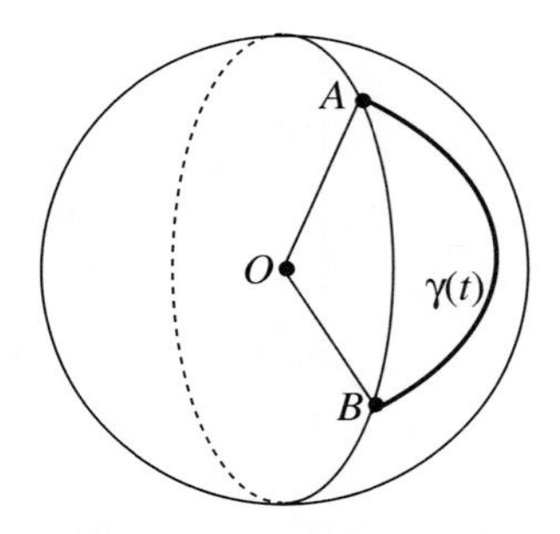

Figure 7.6

$$||\gamma'(t)|| = \sqrt{x'(t)^2+y'(t)^2+z'(t)^2} = \sqrt{\varphi'(t)^2+\sin^2\varphi(t)\theta'(t)^2},$$

and

$$\begin{aligned} L(\gamma) &= \int_a^b \sqrt{\varphi'(t)^2+\sin^2\varphi(t)\theta'(t)^2}\,dt \\ &\geq \int_a^b \sqrt{\varphi'(t)^2}dt \\ &= \int_a^b |\varphi'(t)|dt \geq \left|\int_a^b \varphi'(t)dt\right| \\ &= |\varphi(b)-\varphi(a)| = m(\angle BOA). \end{aligned}$$

Consider now the great circle passing through A and B (see Figure 7.6). Since the sphere has radius 1, we have

$$m(\angle BOA) = L(\widehat{AB}),$$

and then from the above we get

$$L(\gamma) \geq L(\widehat{AB}).$$

□

Exercises

1. Find a parametrization for each of the following curves. In addition, determine if they are smooth.
(a) $4x^2 + y^2 = 1$.
(b) $2x + 3y = 1$.
(c) $(x-1)^2 + y^2 = 1$.
(d) $y^2 = |x|$.

2. Determine if the following curves are regular.
(a) $\gamma(t) = (e^t, e^{-t}, \sqrt{2}t), \quad t \in \mathbf{R}$.
(b) $\gamma(t) = a(t - \sin t, 1 - \cos t, 4\cos(t/2)), \quad -2\pi \leq t \leq 2\pi$.

3. For a fixed t, the *tangent line* to a regular curve γ at $\gamma(t)$ is the line with equation $s \mapsto \gamma(t) + s\gamma'(t)$. Find the tangent lines to the following curves:
(a) the line $\gamma(t) = (t, 2t)$ at $\gamma(1)$.
(b) the circle $\gamma(t) = (2\cos t, 2\sin t)$ at $\gamma(\pi/3)$.
(c) the helix $\gamma(t) = (2\cos t, 2\sin t, t)$ at the points $\gamma(0)$ and $\gamma(\pi/4)$.

4. Write a circle γ as a parametrized curve. Then use the dot product to show that the tangent line at point A is perpendicular to the radius $\overline{OA}$.

5. Find the unique curve γ such that:
(a) $\gamma(1) = (1, 0, -1)$ and $\gamma'(t) = (2t, t^{-1}, -1)$.
(b) $\gamma(0) = (1, 0, -5)$ and $\gamma'(t) = (t^2, t, e^t)$.

6. Use the arc-length formula to find the length of circle of radius r.

7. Let $T : \mathbf{R}^n \to \mathbf{R}^n$ be a one-to-one linear map and $\gamma : I \to \mathbf{R}^n$ a regular curve. Show that $T(\gamma(t))$ is a regular curve.
   Hint: Use Lemma 7.6.6 of the appendix of this chapter.

8. Let $\gamma : I \to \mathbf{R}^n$ be a regular curve and F a rigid motion of $\mathbf{R}^n$. Let $\sigma(t) = (F \circ \gamma)(t) = F(\gamma(t))$. Show that
$$L(\sigma) = L(\gamma).$$
   Hint: Use the result that a rigid motion is the composition of an orthogonal transformation and a translation to find $\sigma'(t)$.

9. Given points A and B on the sphere S^2, show that there exists a rigid motion T of $\mathbf{R}^3$ such that $T(A)$ and $T(B)$ have the same longitude.
   Hint: Let $\mathcal{P}$ be the plane that contains the origin O, A, and B. Show that there exists rigid motion T that fixes O and maps $\mathcal{P}$ to a plane perpendicular to the xy-plane.

7.2 Metric Spaces

In Chapter 1 we assigned a real number to each pair of points A and B of the plane, namely, the length of line segment $\overline{AB}$, which we denoted by AB. Using the axioms of neutral geometry, we proved that the function $(A, B) \mapsto AB$ satisfies the following properties:

(i) $AB \geq 0$ and $AB = 0$ if and only if $A = B$.
(ii) $AB = BA$.
(iii) $AC \leq AB + BC$.

Property (iii) is implied by the *triangle inequality* and the result that if B is between A and C, then $AC = AB + BC$.

In Chapter 3, using the axioms of Euclidean geometry, we coordinatized the plane and as a consequence of the *Pythagorean theorem* we obtained that if $A = (x_1, x_2)$ and $B = (y_1, y_2)$, then
$$AB = \sqrt{(x_1 - y_1)^2 + (x_2 - y_2)^2},$$
which we called *Euclidean length.* In Chapter 5 we generalized this definition for points of the set all n-tuples of real numbers $\mathbf{R}^n$; that is, if $X = (x_1, \ldots, x_n)$ and $Y = (y_1, \ldots, y_n)$, we defined
$$d(X, Y) = ||X - Y|| = \sqrt{(x_1 - y_1)^2 + \cdots + (x_n - y_n)^2}$$

and called it the *Euclidean distance.*

Lemma 7.2.1 *The map $d : \mathbf{R}^n \times \mathbf{R}^n \to \mathbf{R}$ that to (X, Y) assigns the real number $d(X, Y)$ satisfies the following:*
(i) $d(X, Y) \geq 0$ and $d(X, Y) = 0$ if and only if $X = Y$.
(ii) $d(X, Y) = d(Y, X)$, for all $X, Y \in \mathbf{R}^n$
(iii) $d(X, Y) \leq d(X, Z) + d(Z, Y)$, for all $X, Y, Z \in \mathbf{R}^n$.

Proof It is easy to see that (i) and (ii) are satisfied. To verify (iii), we first observe that for any vectors A and B we have

$$\begin{aligned} ||A + B||^2 &= (A + B) \cdot (A + B) \\ &= ||A||^2 + ||B||^2 + 2A \cdot B \\ &\leq ||A||^2 + ||B||^2 + 2||A|| \, ||B|| \\ &= \Big(||A|| + ||B||\Big)^2 \end{aligned}$$

yielding

$$||A + B|| \leq ||A|| + ||B||.$$

Applying this inequality below, we obtain

$$||X - Y|| = ||X - Z + Z - Y|| \leq ||X - Z|| + ||Z - Y||.$$

□

In this chapter we start studying some types of non-Euclidean geometry by giving a more general definition of *distance* between elements of a set, which we shall call *points.*

Definition 7.2.2 *Let M be a nonempty set. A map $d : M \times M \to \mathbf{R}$ is called a* metric *or a* distance *on M if for all $X, Y, Z \in M$ we have*
(i) $d(X, Y) \geq 0$ and $d(X, Y) = 0$ if and only if $X = Y$.
(ii) $d(X, Y) = d(Y, X)$.
(iii) $d(X, Y) \leq d(X, Z) + d(Z, Y)$.
A set M on which a metric has been defined is called a metric space.

Example 1

Let $M = \mathbf{R}^n$ and $d(X, Y) = |x_1 - y_1| + \cdots + |x_n - y_n|$. The map d satisfies (i), (ii), and (iii) of Definition 7.2.2. This is often called the *taxi cab metric.* The verification that d defines a metric on $\mathbf{R}^n$ is an exercise (Exercise 1).

Example 2

Let M be any set. For $X, Y \in M$, we define

$$d(X,Y) = 0, \text{ if } X = Y, \quad \text{and} \quad d(X,Y) = 1, \text{ if } X \neq Y.$$

This is called the *discrete metric.*

Example 3

Let M be a subset of $\mathbf{R}^n$ such that for any two points X and Y in M there exists a *piecewise* regular curve connecting X to Y. By a piecewise regular curve we mean that the domain (t_0, t_n) of γ is divided in subintervals $(t_0, t_1], (t_1, t_2], \ldots, (t_{n-2}, t_{n-1}], (t_{n-1}, t_n)$, and γ restricted to each open subinterval (t_{i-1}, t_i) is a regular curve. The length of such a curve is defined by

$$L(\gamma) = \sum_{i=1}^{n} \int_{i-1}^{i} ||\gamma'(t)||\, dt.$$

Let $\mathcal{C}_{(X,Y)}$ denote the set of all piecewise regular curves from X to Y contained in M. We define

$$d(X,Y) = \inf\{\mathcal{L}(\gamma) \mid \gamma \in \mathcal{C}_{(X,Y)}\}.$$

We will show that d satifies $(i), (ii)$, and (iii) and hence d is metric on M.
(i) Since $L(\gamma) \geq XY$, where XY denotes the length of the line segment from X to Y, we have $L(\gamma) \geq 0$, for all $\gamma \in \mathcal{C}_{(X,Y)}$. Then 0 is a lower bound for the set $\{L(\gamma) \mid \gamma \in \mathcal{C}_{(X,Y)}\}$, and then its infimum must be nonnegative. Now suppose that $X \neq Y$; then if γ is a curve that connects X to Y, since $M \subset \mathbf{R}^n$, $XY \leq L(\gamma)$. Since $XY > 0$, XY is a positive lower bound for the set $\{L(\gamma) \mid \gamma \in \mathcal{C}_{(X,Y)}\}$, which implies that its infimum is positive. Therefore $d(X,Y) = 0$ implies $X = Y$.
(ii) This condition is obviously verified.

(iii) Given X, Y, Z let γ^{XZ} and γ^{ZY} denote piecewise regular curves from X to Z and from Z to Y, respectively. Then

$$\gamma = \gamma^{XZ} \cup \gamma^{ZY}$$

is a piecewise regular curve from X to Y. It follows from the definition of d that

$$d(X,Y) \leq L(\gamma) = L(\gamma^{XZ}) + L(\gamma^{ZY}).$$

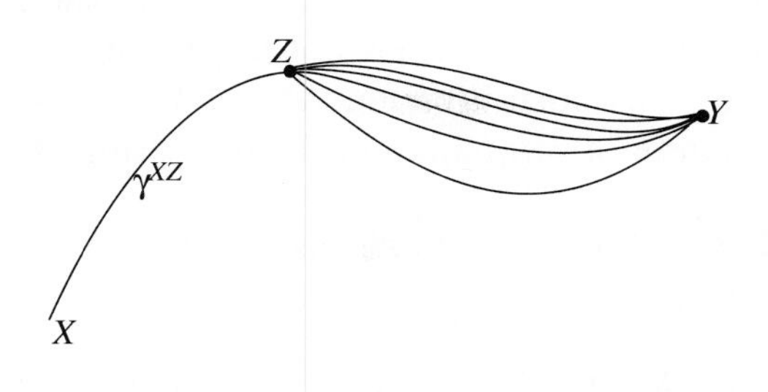

Figure 7.7

Now we fix γ^{XZ} and consider all curves in $\mathcal{C}_{(Z,Y)}$ (see Figure 7.7). We have

$$d(X,Y) - L(\gamma^{XZ}) \leq L(\gamma^{ZY}), \quad \forall \gamma^{ZY} \in \mathcal{C}_{(Z,Y)}$$

which implies that $d(X,Y)-L(\gamma^{XZ})$ is a lower bound of the set $\{L(\gamma) \mid \gamma \in \mathcal{C}_{(Y,Z)}\}$, and then

$$d(X,Y) - L(\gamma^{XZ}) \leq d(Z,Y),$$

because $d(Z,Y)$ is the greatest lower bound. Since the fixed curve γ^{XZ} is an arbitrary curve of $\mathcal{C}_{(X,Z)}$, we have

$$d(X,Y) - d(Z,Y) \leq L(\gamma^{XZ}), \quad \forall \gamma^{XZ} \in \mathcal{C}_{(X,Z)}.$$

Therefore $d(X,Y) - d(Z,Y)$ is a lower bound of the set $\{L(\gamma) \mid \gamma \in \mathcal{C}_{(X,Z)}\}$ and hence

$$d(X,Y) - d(Z,Y) \leq d(X,Z),$$

which implies

$$d(X,Y) \leq d(X,Z) + d(Z,Y).$$

Example 4

Let $M = S^2 - \{(0,0,1)\}$; that is, the unit sphere where we removed the north pole. Let X, Y be two points on S^2 such that the minor arc of great circle $\widehat{XY}$ goes through $(0,0,1)$. Notice that it is impossible to find a curve of shortest length connecting X to Y in M, since we no longer can take $\widehat{XY}$, because $\widehat{XY}$ is not in M, even though we can still define the distance between X and Y in M using the metric described in Example 3. It follows that there is not a curve in the metric space M whose length is the distance between X and Y. In other words, there does not exist a curve of *minimal* length connecting these points.

Definition 7.2.3 *Let M be a metric space. A map $f : M \to M$ is said to be an* isometry *if*

$$d(f(X), f(Y)) = d(X, Y), \quad \forall\, X, Y \in M.$$

If M is $\mathbf{R}^n$ endowed with the Euclidean metric, then the isometries of M are the rigid motions studied in Chapter 5.

Exercises

1. Show that d defined in Example 1 of Section 7.2 is a metric. Explain why it is called the taxi cab metric.

2. Let $S^n = \{X = (x_1, \ldots, x_n) \in \mathbf{R}^n \mid x_1^2 + \cdots + x_n^2 = 1\}$ be the unit n-dimensional sphere and $\mathbf{RP}^n$ the set whose elements are $[X] = \{X, -X\}$; that is, the sphere where antipodal points are identified. $\mathbf{RP}^n$ is called the *real projective space.* Define

$$d([X], [Y]) \leq \min\{||X - Y||, ||X + Y||\}.$$

(a) Show that d is a metric in $\mathbf{RP}^n$.
(b) Let d' be the Euclidean metric in R^n. Show that $d([X], [Y]) \leq d'(X, Y)$.

7.3 Spherical Distance and Its Isometries

We start this section applying Example 3 to the following sets:

Example 3(a): $M = \mathbf{R}^3$

From Theorem 7.1.5 we conclude that the length of line segment $\overline{XY}$ satisfies

$$XY \leq \mathcal{L}(\gamma), \quad \forall \gamma \in \mathcal{C}_{(X,Y)}$$

and hence

$$d(X, Y) = \inf\{\mathcal{L}(\gamma) \,|\, \gamma \in \mathcal{C}_{(X,Y)}\} = XY,$$

that is, d is the Euclidean distance.

Example 3(b): $M = S^2 = \{(x,\ y,\ z) \in \mathbf{R}^3 \mid x^2 + y^2 + z^2 = 1\}$
It follows from Theorem 7.1.8 that given $X, Y \in S^2$,

$$\mathcal{L}(\overset{\frown}{XY}) \leq \mathcal{L}(\gamma), \quad \forall \gamma \in \mathcal{C}_{(X,Y)}.$$

Therefore we make the following definition.

Definition 7.3.1 *Given $X, Y \in S^2$, $d(X, Y) = \mathcal{L}(\overset{\frown}{XY})$ defines a metric on S^2, called the* spherical distance *between X and Y.*

The reader should think of this distance as if the sphere were the surface of our planet Earth and one wanted to find the shortest path to travel from a point of the planet to another. We can choose only curves that lie entirely on the planet, that is, on the sphere. Among them, the shortest is arc $\overset{\frown}{XY}$, and then its length is the spherical distance.

We study now the isometries of the spherical metric.

Lemma 7.3.2 *Let $f : S^2 \to S^2$ be an isometry. Then*

$$\angle(X, Y) = \angle(f(X), f(Y)),$$

where $\angle(X, Y)$ is the angle between the corresponding vectors $\overrightarrow{0X}, \overrightarrow{0Y}$, with 0 denoting the origin.

Proof Let α denote $\angle(X, Y)$ and $\alpha' = \angle(f(X), f(Y))$. We have

$$d(X, Y) = \alpha \quad \text{and} \quad d(f(X), f(Y)) = \alpha'.$$

Since f is an isometry, we conclude that $\alpha = \alpha'$. $\square$

Theorem 7.3.3 *Let us consider on S^2 and $\mathbf{R}^3$ the spherical and the Euclidean metrics, respectively.*
(a) Let $T : \mathbf{R}^3 \to \mathbf{R}^3$ be an orthogonal transformation and define $f : S^2 \to S^2$ by $f(X) = T(X)$. Then f is an isometry of S^2.
(b) Given $P, Q \in S^2$, there exists an isometry f of S^2 such that $f(P) = Q$.
(c) Let f be an isometry of the sphere S^2. Then f can be extended to an orthogonal transformation of $\mathbf{R}^3$.

Proof *(a)* Let $X \in S^2$. Since T is an orthogonal transformation, $||X|| = ||T(X)||$, and this implies that $f(X) \in S^2$. We also have that

$$d(X,Y) = \mathcal{L}(\widehat{XY}) = \angle(X,Y),$$

$$d\,(f(X), f(Y)) = \mathcal{L}(\widehat{f(X)f(Y)}) = \angle(f(X), f(Y)) = \angle(T(X), T(Y)).$$

Since T also preserves angles, we obtain $d(f(X), f(Y)) = d(X,Y)$.
(b) Let T be a rotation of $\mathbf{R}^3$ centered at 0 mapping P to Q. Let $f : S^2 \to S^2$ be given by $f(X) = T(X)$. It follows from part (a) that f is an isometry. Then we have $f(P) = Q$.
(c) Let $f : S^2 \to S^2$ be an isometry. Consider points $X, Y, Z \in S^2$ such that

$$\angle(X,Y) = \angle(X,Z) = \angle(Y,Z) = \pi/2.$$

Let $T : \mathbf{R}^3 \to \mathbf{R}^3$ such that

$$T(X) = f(X), \quad T(Y) = f(Y), \quad \text{and} \quad T(Z) = f(Z).$$

Notice that $\{X, Y, Z\}$ is an orthonormal basis and so is $\{f(X), f(Y), f(Z)\}$ by Lemma 7.3.2. Thus the map defined at a point $P = aX + bY + cZ$ of $\mathbf{R}^3$ by

$$T(P) = aT(X) + bT(Y) + cT(Z)$$

is an orthogonal transformation, since it is linear and it maps orthonormal basis to orthonormal basis.

It remains to prove that $T(P) = f(P)$ for all $P \in S^2$. Let us suppose that

$$f(P) = rf(X) + sf(Y) + tf(Z).$$

Since $\{f(X), f(Y), f(Z)\}$ is an orthonormal basis of $\mathbf{R}^3$ and $f(P)$ is a unit vector,

$$r = \cos\angle(f(P), f(X)), \quad s = \cos\angle(f(P), f(Y)), \quad t = \cos\angle(f(P), f(Z)),$$

and Lemma 7.3.2 implies

$$r = \cos\angle(P,X), \quad s = \cos\angle(P,Y), \quad t = \cos\angle(P,Z).$$

On the other hand,

$$a = \cos\angle(P,X), \quad b = \cos\angle(P,Y), \quad c = \cos\angle(P,Z),$$

and therefore $f(P) = T(P)$. □

Exercises

1. Show that any two great circles of the sphere intersect each other.

2. Find an isometry of the unit sphere S^2 that maps the circle $x^2 + y^2 = 1$ onto $x^2 + z^2 = 1$.

3. Let $\gamma(t)$ be a curve on the unit sphere S^2 and let us denote $\gamma(t_0) = (x_0, y_0, z_0)$. Show that the tangent vector $\gamma'(t_0)$ is orthogonal to vector $v = (x_0, y_0, z_0)$.
   Hint: Consider the function $f(x, y, z) = x^2 + y^2 + z^2$ and the composition $(f \circ \gamma)(t) = x^2(t) + y^2(t) + z^2(t) = 1$. Use the chain rule to find $(f \circ \gamma)'(t)$.

4. Let $P = (x_0, y_0, z_0)$ be a point on the unit sphere S^2 and let $T_P S^2$ denote the plane passing through P and orthogonal to vector $v = (x_0, y_0, z_0)$. Show that for each vector $v \in T_P S^2$ there exists a unique geodesic of S^2 that goes through P, and its tangent vector at P is v.

7.4 Geodesics and Triangles on Spheres

In Section 7.3 we saw that given two points $X, Y \in \mathbf{R}^3$ there is a curve, namely the line segment from X to Y, whose length is the Euclidean distance $d(X, Y)$. The spherical metric also has the same property; the length of arc $\overset{\frown}{XY}$ on S^2 is the spherical distance between X and Y.

In geometry curves of minimal length are called *geodesics*. Therefore, the minor arc of a great circle is a geodesic of a sphere in $\mathbf{R}^3$. Similarly, straight lines are geodesics in Euclidean spaces.

The fact that line segments and minor arcs of great circles realize the Euclidean and the spherical distances, respectively, suggests a kind of "plane geometry," in which the plane is a sphere. In this "plane," *lines will be the geodesics of the sphere*, i.e., the *great circles*. Points will be called *collinear* if they lie on the same great circle. There are some striking differences between the geometry of geodesics on the sphere and the geometry of straight lines in the plane. In Exercise 1 the reader is asked to verify the fact that parallel lines do not exist. In this section we show a result on the angle sum of triangles quite different from the

Saccheri-Legendre theorem proved in Chapter 1 using the axioms of neutral geometry.

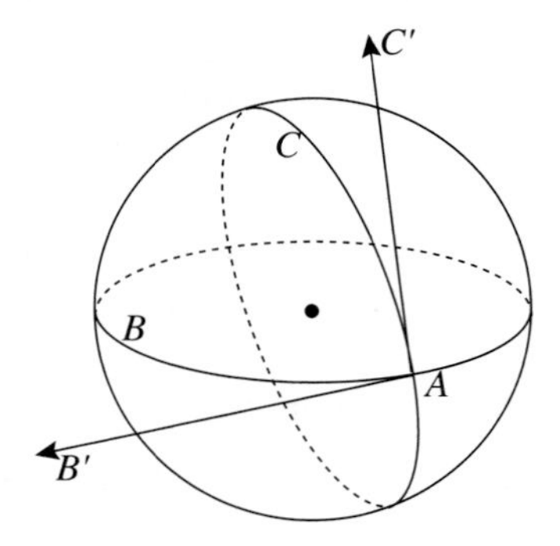

Figure 7.8

Definition 7.4.1 *A* spherical angle $\angle_S BAC$ *is the union of the arcs* $\overset{\frown}{AB}$ *and* $\overset{\frown}{AC}$*, which are called the sides of the angle. Point A is called the vertex of the angle . Its measure is given by*

$$m(\angle_S BAC) = m(\angle B'AC'),$$

where $\overrightarrow{AB'}$ *and* $\overrightarrow{AC'}$ *are vectors tangent to* $\overset{\frown}{AB}$ *and* $\overset{\frown}{AC}$*, respectively, at point* A (see Figure 7.8).

Observe that the definition above makes sense only if points A, B and A, C are nonantipodal, otherwise the sides of the angle would not be well determined.

Recall that, given three noncollinear points in neutral geometry, there exists only one triangle determined by these points. Notice below that another condition must be assumed to uniquely determine a triangle in spherical geometry.

Definition 7.4.2 *Let* $A, B,$ *and* C *be three noncollinear and pairwise nonantipodal points. A* spherical triangle *of vertices* $A, B,$ *and* C *is the geometric figure containing these points, three sides, which are arcs of geodesics joining the pairs of vertices, and three angles* (see Figure 7.9).

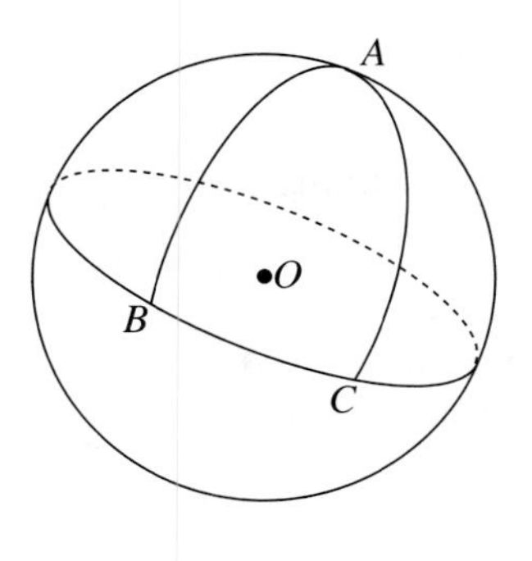

Figure 7.9

Before we state the result on the angle sum of triangles we need to make some observations. In Chapter 6 we constructed a geometric definition of the area of a region of the Euclidean plane by triangulating the region. Another geometric definition of the area of a plane region comes from calculus. If $\mathcal{R} \subset \mathbf{R}^2$ is a region, we consider n small rectangles R_i of area $\Delta_i x \; \Delta_i y$ and such that $\mathcal{R} \subset \cup R_i$ and the intersection of the interior of two such rectangles is empty. The union $\mathcal{P}$ of these rectangles is called a *partition* of $\mathcal{R}$, and the norm δ of $\mathcal{P}$ is the length of the largest diagonal of the R_i's. Then we consider the sum

$$A_\delta = \sum_{i=1}^{n} \Delta_i x \; \Delta_i y.$$

If, as $\delta \to 0$ (partitioning each rectangle into smaller rectangles), the limit of A_δ exists in "some sense," we say that $\mathcal{R}$ is *measurable* and define the limit, the double integral, to be the area of the region, that is,

$$\text{Area}(\mathcal{R}) = \int \int_{\mathcal{R}} dx \, dy.$$

One can prove that if $\mathcal{R}$ is measurable, then the value of the integral does not depend on the chosen partition of the region. We do not intend to rigorously discuss these facts, since they go beyond the scope of this text. Such a discussion can be found in textbooks of real analysis. Likewise, we will not give a geometric definition of the surface area of bounded regions of regular surfaces. Such a theory is studied in differential geometry courses. We shall compute a formula for the surface area of bounded regions of the sphere and give an intuitive explanation for it.

Let $f : (0, 2\pi) \times (0, \pi) \to \mathbf{R}^3$ be given by

$$f(\theta, \varphi) = (\cos\theta \sin\varphi, \sin\theta \sin\varphi, \cos\varphi).$$

The image of f covers the unit sphere S^2, except for the north and south poles and points on a semicircle in the xz-plane. We have

$$\begin{aligned} \frac{\partial f}{\partial \theta} &= (\frac{\partial f_1}{\partial \theta}, \frac{\partial f_2}{\partial \theta}, \frac{\partial f_3}{\partial \theta}) \\ &= (-\sin\theta \sin\varphi, \cos\theta \sin\varphi, 0), \\ \frac{\partial f}{\partial \varphi} &= (\frac{\partial f_1}{\partial \varphi}, \frac{\partial f_2}{\partial \varphi}, \frac{\partial f_3}{\partial \varphi}) \\ &= (\cos\theta \cos\varphi, \sin\theta \cos\varphi, -\sin\varphi). \end{aligned}$$

Their vector product is

$$\frac{\partial f}{\partial \theta} \wedge \frac{\partial f}{\partial \varphi} =$$
$$(-\cos\theta \sin^2\varphi, \sin\theta \sin^2\varphi, -\sin^2\theta \sin\varphi \cos\varphi - \cos^2\theta \sin\varphi \cos\varphi),$$

and then

$$\begin{aligned} \left\| \frac{\partial f}{\partial \theta} \wedge \frac{\partial f}{\partial \varphi} \right\| &= (\sin^4\varphi + \sin^2\varphi \cos^2\varphi)^{1/2} \\ &= (\sin^2\varphi)^{1/2} = \sin\varphi, \quad 0 < \varphi < \pi. \end{aligned}$$

Definition 7.4.3 *The area of region $\mathcal{R}(\theta, \varphi)$ of the unit sphere is defined as*

$$\int\int_{\mathcal{R}} \sin\varphi \, d\theta \, d\varphi.$$

Therefore the surface area of the unit sphere is

$$\int_0^{\pi} \int_0^{2\pi} \sin\varphi \, d\theta \, d\varphi = 4\pi.$$

We now justify, intuitively, the formula of Definition 7.4.3. We consider a partition $\mathcal{P}$ of $\mathcal{R}$ into n smaller regions R_i such that the intersection of two of them is either empty or contains only points on their boundary. The norm δ of $\mathcal{P}$ is the largest distance between any two points that are in the same R_i. For each R_i we choose a point P_i and project R_i onto the plane that is tangent to the sphere at P_i. Let A_i denote this projection. The region A_i is a plane region and its area is given by a double integral. The sum of the area of the plane regions A_i's approximates the surface area of $\mathcal{R}$. If we refine the partition, that is, we partition each region R_i, the norm δ of the new partition is smaller. Then the limit of $\sum \text{Area}(A_i)$, as $\delta \to 0$, is the area of $\mathcal{R}$.

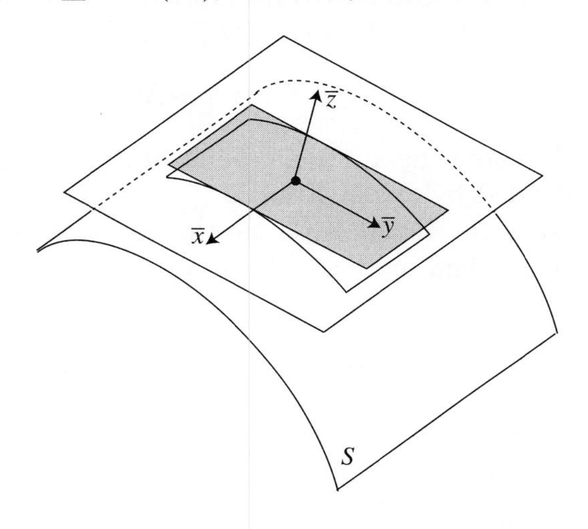

Figure 7.10

In order to compute Area(A_i) we consider a new rectangular system of coordinates $\bar{x}, \bar{y}, \bar{z}$, placing the origin at P_i, translating, and then rotating the axes x, y, and z such that the $\bar{z}$-axis is orthogonal to the tangent plane. The area of A_i is computed by the integral

$$\text{Area}(A_i) = \int\int_{R_i} d\bar{x}\, d\bar{y}.$$

If $P_i = f(\theta_0, \varphi_0)$, $f(\theta, \varphi_0)$ depends only on parameter θ, and then it is a parametrized smooth curve passing through P_i; likewise $f(\theta_0, \varphi)$ is a smooth curve through P_i. Their tangent vectors at P_i are given, respectively, by

$$\frac{\partial f}{\partial \theta}(\theta_0, \varphi_0) \qquad \text{and} \qquad \frac{\partial f}{\partial \varphi}(\theta_0, \varphi_0).$$

Since they are on the tangent plane and $||\frac{\partial f}{\partial \theta} \wedge \frac{\partial f}{\partial \varphi}||$ gives the area of the parallelogram of sides $\partial f/\partial\theta$ and $\partial f/\partial\varphi$, one can prove that the "element of area" is given by

$$d\bar{x}\ d\bar{y} = \left\|\frac{\partial f}{\partial \theta} \wedge \frac{\partial f}{\partial \varphi}\right\| d\theta\, d\varphi.$$

Theorem 7.4.4 (The Gauss-Bonnet theorem in spherical geometry) *Let $\triangle ABC$ be a spherical triangle. Then*

$$m(\angle A) + m(\angle B) + m(\angle C) = \pi + \text{Area}(\triangle ABC).$$

Proof Let $-A$ denote the antipodal point of A and S_A the sector of the sphere subtended by angles $\angle A$ and $\angle(-A)$. In order to find the area of S_A, we suppose that A is the north pole of the sphere (surface area is preserved by rigid motions). Such a region is then described in spherical coordinates by

$$0 < \varphi < \pi \quad \text{and} \quad \theta_0 < \theta < \theta_0 + m(\angle A),$$

where θ_0 corresponds to point B. Using Formula 7.4.3, we obtain

$$\text{Area}(S_A) = \int\int_{\mathcal{R}} \sin\varphi\ d\theta\, d\varphi = m(\angle A)\int_0^{\pi} \sin\varphi\ d\varphi = 2m(\angle A).$$

On the other hand,

$$\text{Area}(S_A) = \text{Area}(\triangle ABC) + \text{Area}(\triangle(-A)BC),$$

$$\text{Area}(S_B) = \text{Area}(\triangle ABC) + \text{Area}(\triangle A(-B)C),$$

$$\text{Area}(S_C) = \text{Area}(\triangle ABC) + \text{Area}(\triangle AB(-C)).$$

The great circle through A and B divides the sphere in two hemispheres. Let H_C denote the one that contains point C. Then

$$\begin{aligned}\text{Area}(H_C) \quad = \quad & \text{Area}(\triangle ABC) + \text{Area}(\triangle(-A)BC) + \\ & \text{Area}(\triangle A(-B)C) + \text{Area}(\triangle(-A)(-B)C).\end{aligned}$$

Notice that $\triangle(-A)BC$ is mapped onto $\triangle ABC$ by a reflection through the plane that contains $\overset{\frown}{BC}$, $\triangle A(-B)C$ onto $\triangle ABC$ by a reflection

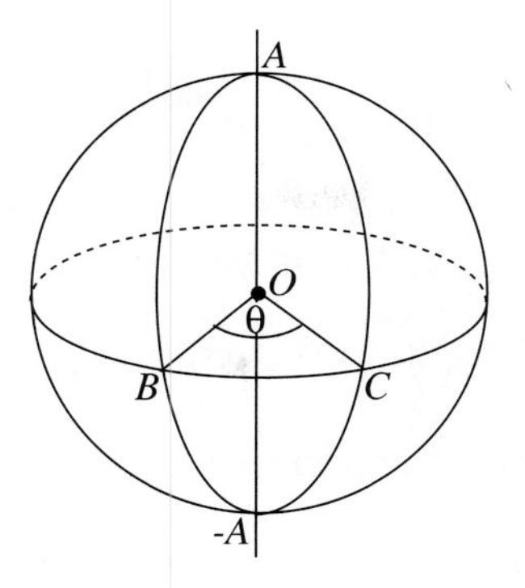

Figure 7.11

through the plane that contains $\widehat{AC}$, and $\triangle AB(-C)$ onto $\triangle ABC$ by a reflection through the plane that contains $\widehat{AB}$. Further a rotation of $180°$ maps $\triangle AB(-C)$ onto $\triangle(-A)(-B)C$. Since area is preserved by isometries (Exercise 2), we conclude that

$$\begin{aligned}\text{Area}(H_C) &= 6\,\text{Area}(\triangle ABC) \\ &= \text{Area}(S_A) + \text{Area}(S_B) + \text{Area}(S_C) + 2\text{Area}(\triangle ABC),\end{aligned}$$

implying

$$2\pi = 2m(\angle A) + 2m(\angle B) + 2m(\angle C) + 2\text{Area}(\triangle ABC),$$

which gives the theorem. □

Proposition 7.4.5 (The law of cosines) *Let $\triangle ABC$ be a spherical triangle with side a opposite vertex A, side b opposite vertex B, and side c opposite vertex C. Then*

$$\cos a = \cos b\ \cos c + \sin b\ \sin c\ \cos(\angle A).$$

Proof Identify each point X with vector $\overrightarrow{OX}$. From Exercise 1 we get that $\angle A = \angle(A \wedge B, A \wedge C)$ and hence

$$(A \wedge B) \cdot (A \wedge C) = ||(A \wedge B)||\ \ ||(A \wedge C)||\ \cos \angle A.$$

Recall from Chapter 4 that

$$||(A\ \wedge\ B)|| = ||A||\ ||B||\ \sin(\angle(A, B))$$

and

$$||(A \wedge C)|| = ||A||\ ||C||\ \sin(\angle(A, C)).$$

But $\angle(A, C) = L(\overset{\frown}{AC}) = c$ and $\angle(A, B) = L(\overset{\frown}{AB}) = b$. Since $A, B, C \in S^2$, $||A|| = ||B|| = ||C|| = 1$. Then we have

$$(A \wedge B) \cdot (A \wedge C) = \sin b\ \sin c\ \cos(\angle A).$$

On the other hand, using the identity

$$(A \wedge B) \cdot (C \wedge D) = (A \cdot C)(B \cdot D) - (A \cdot D)(B \cdot C)$$

(see Exercise 10 of Section 4.4), we obtain

$$\begin{aligned}(A \wedge B) \cdot (A \wedge C) &= ||A||^2\ ||B||\ ||C||\ \cos a - \\ &\quad ||A||\ ||C||\ \cos b\ ||A||\ ||B||\ \cos c \\ &= \cos a - \cos b\ \cos c.\end{aligned}$$

Therefore

$$\cos a = \cos b\ \cos c + \sin b\ \sin c\ \cos\,(\angle A).$$

□

Exercises

1. Let $\triangle ABC$ be a spherical triangle with side a opposite to vertex A, side b opposite to vertex B, and side c opposite to vertex C. Again we identify each point X with vector $\overrightarrow{OX}$. Show the following:
(a) $a = m(\angle BOC)$, $b = m(\angle AOC)$, and $c = m(\angle AOB)$.
(b) $\angle A \simeq \angle(A \wedge B, A \wedge C)$, $\angle B \simeq \angle(B \wedge C, B \wedge A)$, and $\angle C \simeq \angle(C \wedge A, C \wedge B)$.
Hint: Recall that the angle between two planes is defined as the angle between their normal vectors.

2. Define congruence of triangles in the usual way. Then
(a) State and prove an SSS theorem.
(b) State and prove an SAS theorem.
Hint: Use the law of cosines.

3. Review the proof of the Saccheri-Legendre theorem and explain why it does not hold in spherical geometry.

4. Find the angle sum of the spherical triangles below:
   (a) With vertices at $(1, 0, 0)$, $(\sqrt{2}/2, \sqrt{2}/2, 0)$, $(0, 0, 1)$.
   (b) With vertices at $(0, 1, 0)$, $(0, \sqrt{2}/2, \sqrt{2}/2)$, $(1, 0, 0)$.

5. Let f be an isometry of S^2. Show that for all measurable regions $\mathcal{R} \subset S^2$
$$\text{Area}(f(\mathcal{R})) = \text{Area}(\mathcal{R}).$$

7.5 The Stereographic Projection and Conformal Maps

We start this section by pointing out that we cannot find maps between S^2 and $\mathbf{R}^2$ that preserve length and angle measure, for if such a map existed, it would carry a spherical triangle onto a congruent triangle that is Euclidean. But Theorem 7.4.4 implies that they have different angle sums.

However, there exists a map $\pi : S^2 - \{P\} \to \mathbf{R}^2$, where $P \in S^2$, that preserves angle measure in the sense described in Lemma 7.5.1 below. The map π is called the *stereographic projection.*

Stereographic projection Let $P = (0, 0, 1)$. Geometrically, $\pi(X)$ is the point where ray $\overrightarrow{PX}$ intersects the xy-plane, which is identified with $\mathbf{R}^2$. We can find a formula for π as follows:

The equation of the line through points P and X is $P + t(X - P)$. The third coordinate of the points on this line is given by $1 + t(z - 1)$. But the point of this line that belongs to the xy-plane has the third coordinate equal to zero, and hence $t = 1/(1 - z)$. Substituting in the equation of the line, we obtain

$$\pi(X) = (0, 0, 1) - \frac{1}{1 - z}\Big((x, y, z) - (0, 0, 1)\Big) = \frac{1}{1 - z}(x, y, 0).$$

We then write
$$\pi(X) = \frac{1}{1 - z}(x, y).$$

Lemma 7.5.1 *Let $\gamma_1(t)$ and $\gamma_2(t)$ be two regular curves on S^2 meeting at $Q = \gamma_1(t_0) = \gamma_2(t_1) \neq P$. Then the angle betwen $\gamma_1'(t_0)$ and $\gamma_2'(t_1)$ is*

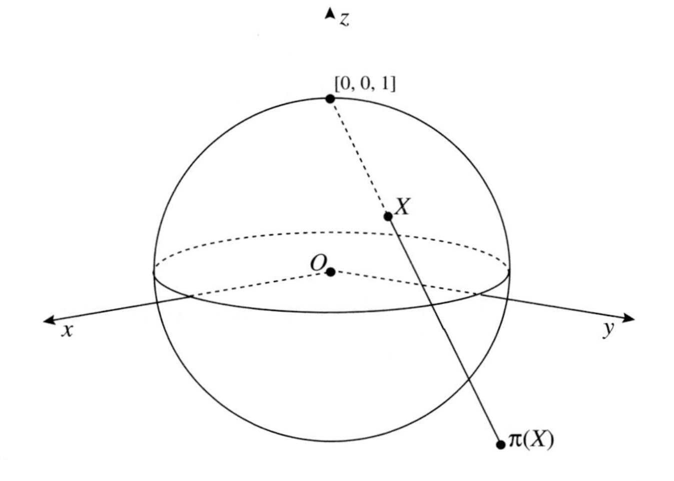

Figure 7.12 Stereographic projection

congruent to the angle between $\sigma_1'(t_0)$ *and* $\sigma_2'(t_1)$*, where* $\sigma_i(t) = \pi(\gamma_i(t))$.

Proof Let $\gamma_i(t) = (x_i(t), y_i(t), z_i(t))$.

$$\gamma_1'(t_0) = (x_i'(t_0), y_i'(t_0), z_i'(t_0)) \quad \text{and} \quad \gamma_2'(t_1) = (x_i'(t_1), y_i'(t_1), z_i'(t_1)).$$

Using the chain rule, we get

$$\sigma_1'(t_0) = \Big(\frac{x_1'(t_0)}{1 - z_1(t_0)} + \frac{x_1(t_0) \cdot z_1'(t_0)}{(1 - z_1(t_0))^2}, \frac{y_1'(t_0)}{1 - z_1(t_0)} + \frac{y_1(t_0) \cdot z'(t_0)}{(1 - z_1(t_0))^2}\Big)$$

and

$$\sigma_2'(t_1) = \Big(\frac{x_2'(t_1)}{1 - z_2(t_1)} + \frac{x_2(t_1) \cdot z_2'(t_1)}{(1 - z_2(t_0))^2}, \frac{y_2'(t_1)}{1 - z_2(t_1)} + \frac{y_2(t_1) \cdot z'(t_1)}{(1 - z_2(t_1))^2}\Big).$$

We will show that

$$\sigma_1'(t_0) \cdot \sigma_2'(t_1) = \frac{1}{(1 - z)^2} \ \gamma_1'(t_0) \cdot \gamma_2'(t_1).$$

Since $\gamma_1(t_0) = \gamma_2(t_1)$, let

$$x = x_1(t_0) = x_2(t_1), \quad y = y_1(t_0) = y_2(t_1), \quad z = z_1(t_0) = z_2(t_1).$$

$$\begin{aligned}\sigma_1'(t_0)\cdot\sigma_2'(t_1) &= \frac{1}{(1-z)^2}\Big(x_1'x_2' + y_1'y_2' + \frac{(x^2+y^2)z_1'z_2'}{(1-z)^2}\\ &+ \frac{(x_1'x + y_1'y)z_2' + (x_2'x + y_2'y)z_1'}{1-z}\Big).\end{aligned}$$

But γ_1 and γ_2 are curves on S^2, and then $x_i^2(t) + y_i^2(t) + z_i^2(t) = 1$, which in turn implies

$$x_i(t)x_i'(t) + y_i(t)y_i'(t) + z_i(t)z_i'(t) = 0.$$

Therefore

$$x_1'x + y_1'y = -z_1'z \quad \text{and} \quad x_2'x + y_2'y = -z_2'z.$$

Substituting above, we have

$$\begin{aligned}\sigma_1'(t_0)\cdot\sigma_2'(t_1) &= \frac{1}{(1-z)^2}\Big(x_1'x_2' + y_1'y_2' + \frac{(1-z^2)z_1'z_2'}{(1-z)^2} + \frac{-zz_1'z_2' + -zz_2'z_1'}{1-z}\Big)\\ &= \frac{1}{(1-z)^2}\Big(x_1'x_2' + y_1'y_2' + \frac{(1+z)z_1'z_2'}{1-z} + \frac{-2zz_1'z_2'}{1-z}\Big)\\ &= \frac{1}{(1-z)^2}\Big(x_1'x_2' + y_1'y_2' + \frac{(1-z)z_1'z_2'}{1-z}\Big)\\ &= \frac{1}{(1-z)^2}\Big(x_1'x_2' + y_1'y_2' + z_1'z_2'\Big)\\ &= \frac{1}{(1-z)^2}\gamma_1'(t_0)\cdot\gamma_2'(t_1).\end{aligned}$$

Now, using the notation $\lambda^2 = (1-z)^2$, we have

$$\begin{aligned}\cos\angle(\gamma_1'(t_0), \gamma_2'(t_1)) &= \frac{\gamma_1'(t_0)\cdot\gamma_2'(t_1)}{||\gamma_1'(t_0)||\;||\gamma_2'(t_1)||}\\ &= \frac{\lambda^2\;\sigma_1'(t_0)\cdot\sigma_2'(t_1)}{|\lambda|\;||\sigma_1'(t_0)||\;|\lambda|\;||\sigma_2'(t_1)||}\\ &= \frac{\sigma_1'(t_0)\cdot\sigma_2'(t_1)}{||\sigma_1'(t_0)||\;||\sigma_2'(t_1)||}\\ &= \cos\angle(\sigma_1'(t_0), \sigma_2'(t_1)).\end{aligned}$$

□

Some geometric transformations of Euclidean spaces have the same angle-preserving property. We then make the definition below.

Definition 7.5.2 *A smooth map $f : \mathbf{R}^n \to \mathbf{R}^n$ is called a* conformal map *if it preserves angle measure; that is, whenever $\gamma_1(t)$ and $\gamma_2(t)$ are two regular curves meeting at $P = \gamma_1(t_0) = \gamma_2(t_1)$, the angle between $\gamma_1'(t_0)$ and $\gamma_2'(t_1)$ is congruent to the angle between $\sigma_1'(t_0)$ and $\sigma_2'(t_1)$, where $\sigma_i(t) = f(\gamma_i(t))$, $i = 1, 2$.*

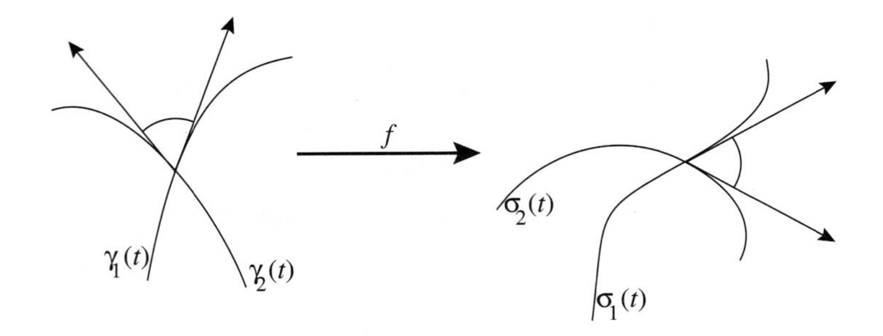

Figure 7.13

Lemma 7.5.3 *With the notation of Definition 7.5.2, if*

$$\gamma_1'(t_0) \cdot \gamma_2'(t_1) = \lambda^2 \sigma_1'(t_0) \cdot \sigma_2'(t_1),$$

then $f : \mathbf{R}^n \to \mathbf{R}^n$ is conformal.

This proof is essentially the last paragraph in the proof of Lemma 7.5.1.

Theorem 7.5.4 *A rigid motion of $\mathbf{R}^n$ is a conformal map.*

Proof Let F be a rigid motion. Recall that we can write $F(X) = T(X) + P$, where T is an orthogonal transformation and P is a point of $\mathbf{R}^n$. Let $\gamma(t)$ be a regular curve and $\sigma(t) = F(\gamma(t))$. Then

$$\begin{aligned} \sigma'(t) &= \lim_{h \to 0} \frac{F(\gamma(t+h) - F(\gamma(t))}{h} \\ &= \lim_{h \to 0} \frac{T(\gamma(t+h)) - T(\gamma(t))}{h} = \left[T(\gamma(t))\right]'. \end{aligned}$$

Since T is a linear map, Lemma 7.6.6 of the appendix implies $\left[T(\gamma(t))\right]' = T(\gamma'(t))$. Now, if we have two curves, γ_1 and γ_2, we obtain

$$\sigma_1'(t_0) \cdot \sigma_2'(t_1) = T(\gamma_1'(t_0)) \cdot T(\gamma_2'(t_1)).$$

But T is an orthogonal transformation, and then

$$T(\gamma_1'(t_0)) \cdot T(\gamma_2'(t_1)) = \gamma_1'(t_0) \cdot \gamma_2'(t_1).$$

Now Lemma 7.5.3 concludes the proof. □

Exercises

1. Let $\pi : S^2 - \{(0,0,1)\} \to \mathbf{R}^2$ denote the stereographic projection.
   (a) Show that π is onto $\mathbf{R}^2$.
   (b) Show that π is injective.
   (c) Show that $\pi^{-1} : \mathbf{R}^2 \to S^2 - \{(0,0,1)\}$ is given by

$$\pi^{-1}(u,v) = \Big(\frac{2u}{u^2+v^2+1}, \frac{2v}{u^2+v^2+1}, \frac{u^2+v^2-1}{u^2+v^2+1}\Big).$$

   Hint: Find the equation of the line through points $(u,v,0)$ and $(0,0,1)$ and the points where such a line intersects the sphere.

2. Let π denote the stereographic projection and $P = (0,0,1)$; let γ be a curve on the sphere S^2. Show the following:
   (a) $\pi(\gamma)$ is a line if and only if γ is a circle and $P \in \gamma$.
   (b) $\pi(\gamma)$ is a circle if and only if γ is a circle and $P \notin \gamma$.
   Hint: A circle is a plane curve; that is, there exists a plane $ax + by + cz + d = 0$ containing γ. Write

$$(x,y,z) = \Big(\frac{2u}{u^2+v^2+1}, \frac{2v}{u^2+v^2+1}, \frac{u^2+v^2-1}{u^2+v^2+1}\Big),$$

   and the equation of the plane in terms of u and v; then get the equation for the corresponding curve in $\mathbf{R}^2$.

3. (a) Show that if f is a conformal map of $\mathbf{R}^n$, then so is f^{-1}.
   (b) Show that if f and g are conformal maps of $\mathbf{R}^n$, then so is $f \circ g$.

4. (a) Show that the homothety $\mathcal{H}_0^r : \mathbf{R}^n \to \mathbf{R}^n$ is a conformal map.
   Hint: Use Lemma 7.6.6
   (b) Show that a similarity of $\mathbf{R}^n$ is a conformal map.

7.6 Appendix

In this section we prove some of the results of advanced calculus used in this chapter. We start by recalling the mean value theorem, which is a fundamental tool in the proofs of this section.

Theorem 7.6.1 (The mean value theorem) *Let $a < b$ and f a real function that is continuous on the closed interval $a \leq x \leq b$ and differentiable on the open interval (a, b). Then, there exists a point $c \in (a, b)$ such that*

$$f(b) - f(a) = f'(c)(b - a).$$

Since the study of curves requires vector functions, we give now some definitions of multivariable calculus.

Definition 7.6.2 *An open ball in $\mathbf{R}^n$ of center $X_0 = (x_1^0, \ldots, x_n^0)$ and radius r is the set*

$$B_r(X_0) = \{X = (x_1, \ldots, x_n) \in \mathbf{R}^n \mid ||X - X_0|| < r\}.$$

Definition 7.6.3 *Let $f : B_r(X_0) \to \mathbf{R}$. The* partial derivative of f with respect to x_i *at X_0, denoted by $(\partial f/\partial x_i)(X_0)$, is (when it exists) the derivative of the function*

$$(x_1^0, \ldots, x_i, \ldots, x_n^0) \mapsto f(x_1^0, \ldots, x_i, \ldots, x_n^0)),$$

which is a function of only one variable, namely, x_i.

Example 1 Consider the function $f : \mathbf{R}^4 \to \mathbf{R}$ given by

$$f(x_1, x_2, x_3, x_4) = 3x_1x_2^2 - x_1x_2x_3 + 2(x_3 - x_4)^2.$$

To find the partial derivative of f with respect to x_3 at the point $(1, -1, 1, 0)$ is to find the derivative of the function

$$g(x_3) = f(1, -1, x_3, 0) = 3 + x_3 + 4x_3^2$$

at point $x_3 = 1$. From the definition of the derivative of a single-variable function we have

$$\begin{aligned} g'(1) &= \lim_{h \mapsto 0} \frac{g(1+h) - g(1)}{h} \\ &= \lim_{h \mapsto 0} \frac{f(1, -1, 1+h, 0) - f(1, -1, 1, 0)}{h} = 9. \end{aligned}$$

For any point of the type $(1,-1,x,0)$ we have

$$\begin{aligned} g'(x) &= \lim_{h\mapsto 0} \frac{g(x+h)-g(x)}{h} \\ &= \lim_{h\mapsto 0} \frac{f(1,-1,x+h,0)-f(1,-1,x,0)}{h} \\ &= 1+8x. \end{aligned}$$

It follows that the *partial derivative of* $f : B_r(X_0) \subset \mathbf{R}^n \to \mathbf{R}$ *with respect to* x_i, denoted by $\frac{\partial f}{\partial x_i}$, at the point $(x_1^0, \ldots, x_n^0)$ is given by

$$\frac{\partial f}{\partial x_i}((x_1^0, \ldots, x_n^0)) = \lim_{h\to 0} \frac{f(x_1^0, \ldots, x_i^0 + h, \ldots, x_n^0) - f(x_1^0, \ldots, x_i^0, \ldots, x_n^0)}{h}.$$

When f has partial derivatives in all points of an open ball B, we can consider the *second* partial derivatives at all points of B. If they exist, we can consider partial derivatives of order 3, and so on.

Definition 7.6.4 *Let B be an open ball and $f : B \to \mathbf{R}$. We say that f is* smooth *on B, if f has all partial derivatives of all orders. It follows that they are all continuous functions.*

Proposition 7.6.5 (The chain rule) *Let I be an open interval of $\mathbf{R}$ and $\gamma : I \to \mathbf{R}^n$,*

$$\gamma(t) = (x_1(t), \ldots, x_n(t)).$$

Let B be an open ball containing point $\gamma(t_0) = (x_1(t_0), \ldots, x_n(t_0))$ and $f : B \to \mathbf{R}$ a smooth function defined on B. Then

$$\frac{df(\gamma(t))}{dt}(t_0) = \frac{\partial f}{\partial x_1} x_1'(t_0) + \cdots + \frac{\partial f}{\partial x_n} x_n'(t_0).$$

Proof For simplicity, we prove it for $n = 2$.

$$\frac{df(\gamma(t))}{dt}(t_0) = \lim_{h\to 0} \frac{f(x_1(t_0+h), x_2(t_0+h)) - f(x_1(t_0), x_2(t_0))}{h}$$

$$\begin{aligned} &= \lim_{h\to 0}\Big[\frac{f(x_1(t_0+h),x_2(t_0+h)) - f(x_1(t_0),x_2(t_0+h))}{h} + \\ &\quad \frac{f(x_1(t_0),x_2(t_0+h)) - f(x_1(t_0),x_2(t_0))}{h}\Big] \\ &= \lim_{h\to 0}\frac{f(x_1(t_0+h),x_2(t_0+h)) - f(x_1(t_0),x_2(t_0+h))}{h} + \\ &\quad \lim_{h\to 0}\frac{f(x_1(t_0),x_2(t_0+h)) - f(x_1(t_0),x_2(t_0))}{h}. \end{aligned}$$

Now let $a = x_1(t_0)$, $b = x_1(t_0+h)$, and g the one-variable function

$$g(x_1) = f(x_1, x_2(t_0+h))$$

(observe that the second variable is fixed). Applying the mean value theorem to the function g defined on the interval $[a, b]$, we obtain that

$$g(b) - g(a) = g'(c)(b-a) \quad \text{for} \;\; c \in (a, b),$$

and since there exists $0 < t_1 < h$ such that $c = x_1(t_1)$, we have

$$\begin{aligned} g(b) - g(a) &= f(x_1(t_0+h),x_2(t_0+h)) - f(x_1(t_0),x_2(t_0+h)) \\ &= g'(x_1(t_1))(x_1(t_0+h) - x_1(t_0)) \\ &= \frac{\partial f}{\partial x_1}(x_1(t_1),x_2(t_0+h))(x_1(t_0+h) - x_1(t_0)). \end{aligned}$$

Therefore

$$\frac{f(x_1(t_0+h),x_2(t_0+h)) - f(x_1(t_0),x_2(t_0+h))}{h}$$

$$= \frac{\partial f}{\partial x_1}\,(x_1(t_1),x_2(t_0+h))\,\frac{x_1(t_0+h) - x_1(t_0)}{h}\,.$$

Similarly,

$$\frac{f(x_1(t_0),x_2(t_0+h)) - f(x_1(t_0),x_2(t_0))}{h}$$

$$= \frac{\partial f}{\partial x_2}\,(x_1(t_0),x_2(t_2))\,\frac{x_2(t_0+h) - x_2(t_0)}{h}\,,$$

for $0 < t_2 < h$. Observe that $t_1, t_2 \to t_0$, as $h \to 0$. Since f is smooth, its partial derivatives are continuous functions, and then

$$\begin{aligned} \lim_{h\to 0} \frac{\partial f}{\partial x_1}(x_1(t_1), x_2(t_0+h)) &= \frac{\partial f}{\partial x_1}(x_1(t_0), x_2(t_0)), \\ \lim_{h\to 0} \frac{\partial f}{\partial x_2}(x_1(t_0), x_2(t_2)) &= \frac{\partial f}{\partial x_2}(x_1(t_0), x_2(t_0)). \end{aligned}$$

Therefore

$$\frac{df(\gamma(t))}{dt}(t_0) = \frac{\partial f}{\partial x_1}(x_1(t_0), x_2(t_0))x_1'(t_0) + \frac{\partial f}{\partial x_2}(x_1(t_0), x_2(t_0))x_2'(t_0).$$

□

The chain rule has the following geometric application.

Lemma 7.6.6 *Let $\gamma(t)$ be a smooth curve in $\mathbf{R}^n$, $T : \mathbf{R}^n \to \mathbf{R}^n$ a linear map and $\sigma(t) = T(\gamma(t))$. Then $\sigma'(t) = T(\gamma'(t))$.*

Proof We write $\gamma(t) = (x_1(t), \ldots, x_n(t))$ and $T(X) = (T_1(X), \ldots, T_n(X))$. The tangent vector of σ is expressed as

$$\sigma'(t) = \left(\left[T_1(\gamma(t))\right]', \ldots, \left[T_n(\gamma(t))\right]'\right).$$

The chain rule implies

$$\left[T_i(\gamma(t))\right]' = \frac{\partial T_i}{\partial x_1}x_1'(t) + \cdots + \frac{\partial T_i}{\partial x_n}x_n'(t).$$

Therefore, in matrix notation, $\sigma'(t)$ is expressed as

$$\sigma'(t) = \begin{pmatrix} \frac{\partial T_1}{\partial x_1} & \cdots & \frac{\partial T_1}{\partial x_n} \\ \vdots & & \vdots \\ \frac{\partial T_n}{\partial x_1} & \cdots & \frac{\partial T_n}{\partial x_n} \end{pmatrix} \begin{pmatrix} x_1'(t) \\ \vdots \\ x_n'(t) \end{pmatrix}.$$

Now we compute $\partial T_i/\partial x_j$. Since T is a linear transformation, so is each $T_i : \mathbf{R}^n \to \mathbf{R}$ (the reader should prove this). Then

$$\frac{\partial T_i}{\partial x_j} = \lim_{h\to 0} \frac{T_i(x_1, \ldots, x_j + h, \ldots, x_n) - T_i(x_1, \ldots, x_j, \ldots, x_n)}{h}.$$

Using the fact that T_i is a linear map, we obtain that

$$T_i(x_1, \ldots, x_j + h, \ldots, x_n = T_i(x_1, \ldots, x_j, \ldots, x_n) \; + \; T_i(0, \ldots, h, \ldots, 0),$$

and then

$$\begin{aligned} T_i(x_1, \ldots, x_j + h, \ldots, x_n) - T_i(x_1, \ldots, x_j, \ldots, x_n) &= T_i(0, \ldots, h, \ldots, 0) \\ &= hT_i(0, \ldots, 1, \ldots, 0), \end{aligned}$$

where the last equality was implied by the linearity of T. Therefore

$$\begin{aligned} \frac{\partial T_i}{\partial x_j} &= \lim_{h \to 0} \frac{h(T_i(0, \ldots, 1, \ldots, 0))}{h} \\ &= \lim_{h \to 0} T_i(0, \ldots, 1, \ldots, 0) = T_i(e_j), \end{aligned}$$

where e_j denotes the vector $(0, \ldots, 1, \ldots, 0)$, with 1 as the j-th coordinate. Thus

$$\sigma'(t) = \begin{pmatrix} T_1(e_1) & \ldots & T_1(e_n) \\ \vdots & & \vdots \\ T_n(e_1) & \ldots & T_n(e_n) \end{pmatrix} \begin{pmatrix} x'_1(t) \\ \vdots \\ x'_n(t) \end{pmatrix}.$$

Notice that $(a_{ij}) = (T_i(e_j))$ is the matrix that represents the transformation T in the standard basis of $\mathbf{R}^n$. Since $(x'_1(t), \ldots, x'_n(t))$ are coordinates of $\gamma'(t)$ relative to the standard basis, we conclude that $\sigma'(t) = T(\gamma'(t))$. □

We finish this section by proving Theorem 7.1.4. First we recall the definition of the Riemann integral.

A *partition* of an interval $[a, b]$ is a collection of points $\{t_0 < t_1 < \cdots < t_{n-1} < t_n\}$ in $[a, b]$ such that $t_0 = a$ and $t_n = b$. The *norm* μ of the partition is defined as

$$\mu = \max\{|t_i - t_{i-1}|, i = 1, \ldots n\}.$$

Then, given a regular γ defined on (a, b), we consider a partition of $[a, b]$; choosing a point $t_i^j \in [t_i - t_{i-1}]$, the sum

$$\sum_{i=1}^{n} ||\gamma'(t_i^j)|| \;\; |t_i - t_{i-1}|$$

is a Riemann sum of the function $||\gamma'(t)||$. The integral is then defined as the limit of a Riemann sum, as $\mu \to 0$. More precisely, for every $\epsilon > 0$, there exists $\delta > 0$, such that if a partition has norm $\mu < \delta$, then

$$\left| \int_a^b ||\gamma'(t)||\, dt - \sum_{i=1}^{n} ||\gamma'(t_i^\star)|| \ \ ||t_i - t_{i-1}|| \right| < \epsilon.$$

Proof of Theorem 7.1.4 For simplicity, we prove the theorem for curves in $\mathbf{R}^3$; we write $\gamma(t) = (x(t), y(t), z(t))$. Since x, y, z are smooth functions, we can apply the mean value theorem, and then

$$x(t_i) - x(t_{i-1}) = x'(t_i^\star)(t_i - t_{i-1}),$$

$$y(t_i) - y(t_{i-1}) = y'(t_i^\star)(t_i - t_{i-1}),$$

$$z(t_i) - z(t_{i-1}) = z'(t_i^\star)(t_i - t_{i-1}).$$

Here $t_i^\star$ is denoting a real number between t_{i-1} and t_i; its value may be different for the different functions x, y, z. We compute the terms of the Riemann sum.

$$\begin{aligned} ||\gamma(t_i) - \gamma(t_{i-1})||^2 &= (x(t_i) - x(t_{i-1}))^2 + (y(t_i) - y(t_{i-1}))^2 \\ &\quad + (z(t_i) - z(t_{i-1}))^2 \\ &= x'(t_i^\star)^2(t_i - t_{i-1})^2 + y'(t_i^\star)^2(t_i - t_{i-1})^2 \\ &\quad + z'(t_i^\star)^2(t_i - t_{i-1})^2 \\ &= (t_i - t_{i-1})^2[x'(t_i^\star))^2 + y'(t_i^\star))^2 + z'(t_i^\star))^2], \end{aligned}$$

which implies

$$||\gamma(t_i) - \gamma(t_{i-1})|| = |t_i - t_{i-1}|\sqrt{x'(t_i^\star)^2 + y'(t_i^\star)^2 + z'(t_i^\star)^2}.$$

Now let A_i and B_i denote

$$A_i = ||\gamma'(t_i)|| = \sqrt{x'(t_i)^2 + y'(t_i)^2 + z'(t_i)^2},$$

$$B_i = \sqrt{x'(t_i^\star)^2 + y'(t_i^\star)^2 + z'(t_i^\star)^2} - \sqrt{x'(t_i)^2 + y'(t_i)^2 + z'(t_i)^2}.$$

Therefore

$$||\gamma(t_i) - \gamma(t_{i-1})|| = |t_i - t_{i-1}|(A_i + B_i).$$

Using the inequality $\Big| ||a|| - ||b|| \Big| \leq ||a - b||$ for all vectors of $\mathbf{R}^3$, we obtain that

$$B_i \leq |x'(t_i^\star) - x'(t_i)| + |y'(t_i^\star) - y'(t_i)| + |z'(t_i^\star) - z'(t_i)|.$$

Moreover, since x', y', z' are continous on $[a, b]$, they are uniformly continuous, and hence, given $\epsilon > 0$, there exists $\delta_1 > 0$ such that

$$\begin{aligned} |x'(u) - x'(v)| &< \frac{\epsilon}{6(b-a)}, \\ |y'(u) - y'(v)| &< \frac{\epsilon}{6(b-a)}, \\ |z'(u) - z'(v)| &< \frac{\epsilon}{6(b-a)}, \end{aligned}$$

whenever $|u - v| < \delta_1$. Then, if a partition has norm $\mu < \delta_1$,

$$B_i \leq |x'(t_i^\star) - x'(t_i)| + |y'(t_i^\star) - y'(t_i)| + |z'(t_i^\star) - z'(t_i)| < \frac{\epsilon}{2(b-a)}$$

and

$$\begin{aligned} \Big| \sum_{i=1}^{n} B_i \ (t_i - t_{i-1}) \Big| &\leq \sum_{i=1}^{n} |B_i| \ |t_i - t_{i-1}| \\ &< \sum_{i=1}^{n} \frac{\epsilon}{2(b-a)} \ |t_i - t_{i-1}| \\ &= \frac{\epsilon}{2(b-a)} \sum_{i=1}^{n} \ |t_i - t_{i-1}| = \frac{\epsilon}{2}. \end{aligned}$$

Now we compute

$$\begin{aligned} \Big| L(\gamma) - L(\mathcal{P}_n) \Big| &= \Big| L(\gamma) - \sum_{i=1}^{n} ||\gamma(t_i) - \gamma(t_{i-1})|| \Big| \\ &= \Big| L(\gamma) - \sum_{i=1}^{n} |t_i - t_{i-1}| (A_i + B_i) \Big| \\ &\leq \Big| L(\gamma) - \sum_{i=1}^{n} |t_i - t_{i-1}| |A_i| \Big| \\ &\quad + \Big| L(\gamma) - \sum_{i=1}^{n} |t_i - t_{i-1}| |B_i| \Big| \\ &< \Big| L(\gamma) - \sum_{i=1}^{n} |t_i - t_{i-1}| \ ||\gamma'(t_i)|| \Big| + \frac{\epsilon}{2}. \end{aligned}$$

But

$$\sum_{i=1}^{n} |t_i - t_{i-1}| \ ||\gamma'(t_i)||$$

is a Riemann sum for $||\gamma'(t)||$, and then, given $\epsilon > 0$, there exists δ_2 such that if a partition has norm $\mu < \delta_2$,

$$\left| L(\gamma) - \sum_{i=1}^{n} |t_i - t_{i-1}| \ ||\gamma'(t_i)|| \right| < \frac{\epsilon}{2}.$$

Let $\delta = \min\{\delta_1, \delta_2\}$. If a partition has norm $\mu < \delta$, the polygonal curve obtained using such a partition satisfies

$$\left| L(\gamma) - L(\mathcal{P}_n) \right| < \epsilon.$$

□

Chapter 8

HYPERBOLIC GEOMETRY

8.1 Introduction

When in the beginning of the nineteenth century it was noted that Euclid's fifth postulate was independent of the first four, C. F. Gauss called the new geometry a *non-Euclidean geometry.* The first non-Euclidean geometry was studied simultaneously and independently by J. Bolyai (1826–1860) and N. Lobachevsky (1792–1856). Bolyai and Lobachevsky developed their geometry based on a new axiom that will be presented in this section.

Gauss did not focus on the axiomatic approach; he studied surfaces by considering them to be embedded in the usual three-dimensional space, and this led him to introduce the concept of the *curvature* on a surface. This idea was greatly generalized by Gauss's most remarkable student, G. F. B. Riemann (1826–1866). Riemann's ideas, although intuitive, incorporated all the new geometries in a much more general context. He introduced the notion of curved space that in modern language is called a *manifold.* The concept of a differentiable manifold, necessary for the formalization of the work of Riemann, only appeared in 1913 in the work of H. Weyl. An important source of stimulation for the development of Riemannian geometry was the application of its ideas to the theory of relativity by A. Einstein in 1916.

Despite the importance of Riemannian geometry, its applications and its ability to describe our physical space, we will not discuss general results along these lines, since the understanding of Riemannian geometry requires knowledge of more advanced mathematics.

Our focus in this chapter is the geometry that differs from Euclidean geometry in only one axiom, namely the parallel postulate. For this reason, it seems very natural that we first study it following the axiomatic methods of Euclid.

Recall that in Chapter 2 we showed the equivalence between Euclid's fifth postulate and Playfair's (Euclidean) parallel postulate. A simple negation of the Euclidean parallel postulate implies two alternatives:

> (i) There exists a line l such that for some point $P \notin l$ there is no line through P parallel to l.
>
> (ii) There exists a line l such that for some point $P \notin l$ there is more than one line through P parallel to l.

Note that Axiom (i) cannot be incorporated in the axiom system of neutral geometry, since the alternate interior angle theorem of neutral geometry implies that there exists at least one parallel to a given line. Spherical geometry, studied in Chapter 7, is an example of a plane with no parallel "lines." Elliptic plane geometry, which we will briefly discuss in the next chapter, is another example of a geometry satisfying Axiom (i). Such geometries are best studied in the context of differential geometry, and in particular Riemannian geometry.

Axiom (ii) is not inconsistent with the axioms of neutral geometry and hence can be added to them to compose an axiom system for some geometry. Such a geometry is the one studied by Bolyai and Lobachevsky, called hyperbolic geometry. The German mathematician F. Klein (1849–1925) named it *hyperbolic*, from the Greek *hyperballein* which means "to exceed." The reason for the word "exceed" will be clear during this chapter. Hyperbolic geometry will be studied again in a more concrete way in Chapters 9 and 10.

It is not our intention here to do a historial study of the development of hyperbolic geometry. Although we will explore some of the ideas of Clairaut, Saccheri, Lambert, and Legendre in their attempts to prove Euclid's fifth postulate, we will not enter into any discussion about the gaps in their proofs. A detailed discussion of their results as well as the history of the discovery of hyperbolic geometry in the beginning of the nineteenth century can be found in the book by Greenberg.[1] The

---

[1]M. J. Greenberg, *Euclidean and Non-Euclidean Geometries, Development and History*, W. H. Freeman and Co., New York, 1996.

special interest in studying their proofs resides in the fact that the flawed step is usually equivalent in neutral geometry to the Euclidean parallel postulate.

8.2 Neutral Geometry Revisited

In Chapter 2 we proved that the Euclidean parallel postulate implies that any triangle has angle sum equal to $180°$. We start by recalling the *Saccheri-Legendre theorem* proved in Chapter 1, namely, that the axioms of neutral geometry imply that *the angle sum of any triangle is less than or equal to* $180°$.

Lemma 8.2.1 *Let $\triangle ABC$ be any triangle and D a point between B and C. Then the angle sum of $\triangle ABC$ is $180°$ if and only if the angle sums of both $\triangle ABD$ and $\triangle ACD$ are equal to $180°$.*

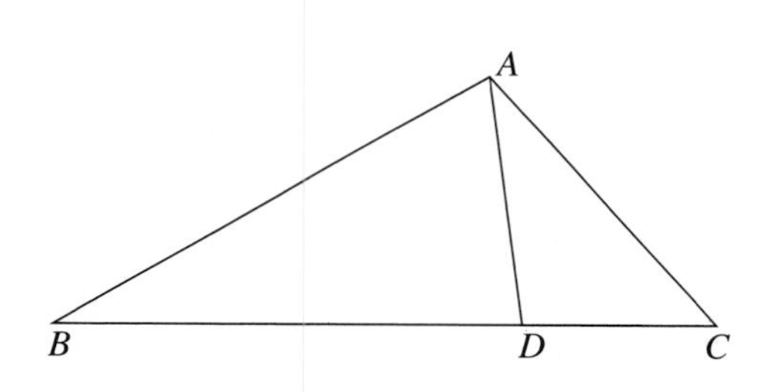

Figure 8.1

Proof Let us assume first that angle sum of $\triangle ABC$ is $180°$. We have

$$\begin{aligned} 180° &= m(\angle B) + m(\angle A) + m(\angle C) \\ &= m(\angle B) + m(\angle BAD) + m(\angle CAD) + m(\angle C) \end{aligned}$$

and

$$180° = m(\angle ADB) + m(\angle ADC)$$

since they are supplementary angles. Therefore

$$\begin{aligned} m(\angle B) + m(\angle BAD) + m(\angle ADB) + \\ m(\angle CAD) + m(\angle ADC) + m(\angle C) &= 360°. \end{aligned}$$

Notice that above we have the sum of the angle sum of $\triangle ABD$ with the angle sum of $\triangle ACD$. Since

$$m(\angle B) + m(\angle BAD) + m(\angle ADB) \leq 180^\circ,$$

$$m(\angle CAD) + m(\angle ADC) + m(\angle C) \leq 180^\circ,$$

by the Saccheri-Legendre theorem, we conclude that

$$m(\angle B) + m(\angle BAD) + m(\angle ADB) = 180^\circ,$$

and

$$m(\angle CAD) + m(\angle ADC) + m(\angle C) = 180^\circ.$$

To prove the converse, just observe that

$$\begin{aligned} m(\angle B) + m(\angle BAD) + m(\angle ADB) + \\ m(\angle CAD) + m(\angle ADC) + m(\angle C) &= 360^\circ \end{aligned}$$

and $m(\angle A) = m(\angle BAD) + m(\angle CAD)$, and since $\angle ADB$ and $\angle ADC$ are supplementary angles, we obtain

$$m(\angle B) + m(\angle A) + m(\angle C) = 360^\circ - 180^\circ = 180^\circ.$$

□

Lemma 8.2.2 *If there exists an isosceles right triangle $\triangle ABC$ with angle sum 180°, then for any positive real number y there exists an isosceles right triangle with angle sum 180° and such that its legs have length greater than y.*

Proof Let us suppose that $\angle A$ denotes the right angle of triangle $\triangle ABC$ and $x = AB = AC$. Then $m(\angle B) = m(\angle C) = 45^\circ$. Consider point D in one of the half-planes determined by line $\overleftrightarrow{BC}$ such that $m(\angle DCB) = 45^\circ$, D and A are on opposite sides of $\overleftrightarrow{BC}$, and $CD = AC = x$. Then line $\overleftrightarrow{CD}$ is parallel to line $\overleftrightarrow{AB}$ and $\triangle DCB \simeq \triangle ABC$ by the alternate interior angle theorem and SAS. It follows that $\angle D$ is a right angle and $\square ABDC$ is a square.

Now, let point C_2 be on ray $\overrightarrow{AC}$ and B_2 on $\overrightarrow{AB}$ such that $AC_2 = AB_2 = 2x$. We then have that $\triangle BB_2D$ and $\triangle CDC_2$ are both congruent

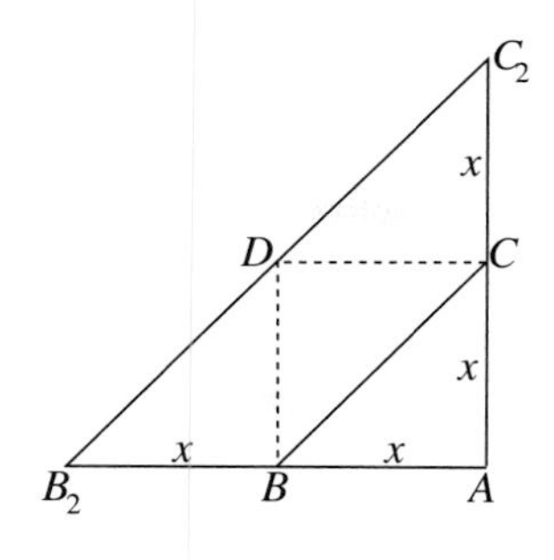

Figure 8.2

to $\triangle ABC$ (by SAS) and therefore their angle sum is equal to $180°$. It follows that

$$m(\angle C_2) = m(\angle C_2DC) = m(\angle B_2DB) = m(\angle B_2) = 45°,$$

implying that

$$m(\angle C_2DC) + m(\angle CDB) + m(\angle B_2DB) = 180°,$$

and then rays $\overrightarrow{DC_2}$ and $\overrightarrow{DB_2}$ are opposite rays, which in turn implies that B_2, D, and C_2 are collinear points. We then have that $\triangle AB_2C_2$ is an isosceles right triangle whose angle sum $180°$ and whose legs have length $2x$.

This procedure can be repeated as many times as we want. Since the Archimedean property of the real numbers says that there exists a positive integer n such that $nx > y$, we repeat the procedure above n times to obtain an isosceles right triangle with legs of length nx and angle sum $180°$. □

Lemma 8.2.3 *If there exists a triangle $\triangle ABC$ with angle sum $180°$, then there exists an isosceles right triangle whose angle sum is equal to $180°$.*

Proof Let $\angle A$ denote the greatest angle and let $\overline{AD}$ be the altitude dropped from A (see Figure 8.3). Then triangles $\triangle ADB$ and $\triangle ADC$ are right triangles, and Lemma 8.2.1 implies that both have angle sum equal to $180°$. If one of them is isosceles, the result is proved. If not, we choose one of them, say $\triangle ADB$. Lets us suppose that $AD > BD$

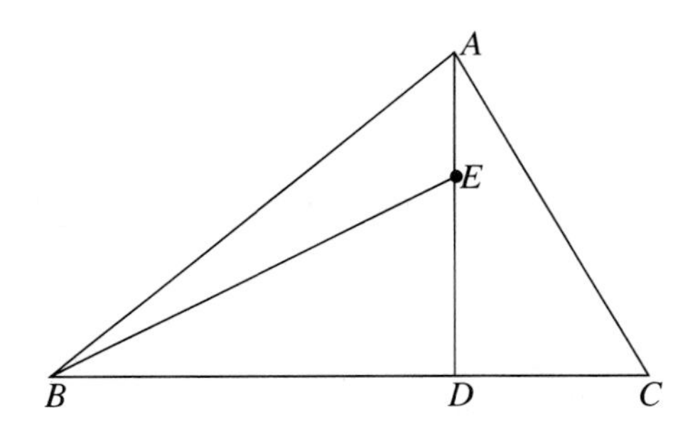

Figure 8.3

and let point E be on $\overline{AD}$ such that $DE = DB$. The triangle $\triangle DBE$ is an isosceles right triangle, and Lemma 8.2.1 implies that it has angle sum equal to $180°$. □

Theorem 8.2.4 *If there exists a triangle whose angle sum is equal to* $180°$, *then every triangle has angle sum* $180°$.

Proof Let $\triangle ABC$ be any triangle and D such that $\overline{AD}$ is an altitude of the triangle. We will show that triangles $\triangle ADB$ and $\triangle ADC$ have angle sum $180°$, and the result will follow from Lemma 8.2.1.

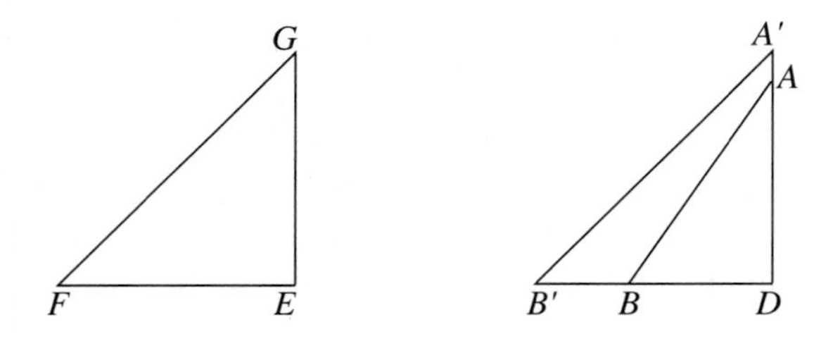

Figure 8.4

For that let $y \in \mathbf{R}$ such that $y > AD$ and $y > DB$. Since we are assuming that there exists a triangle of angle sum $180°$, Lemma 8.2.3 implies that there exists an isosceles right triangle whose angle sum is $180°$. Further, Lemma 8.2.2 implies that an isosceles right triangle $\triangle EFG$ can be constructed such that its angle sum is also $180°$ and its legs $EF = EG > y$ (see Figure 8.4). Now we consider a point A' on ray $\overrightarrow{DA}$ and $B' \in \overrightarrow{DB}$ such that $DA' = DB' = EF = EG$. Then

$\triangle DA'B' \simeq \triangle EFG$, and hence $\triangle DA'B'$ has angle sum $180°$. It follows now from Lemma 8.2.1 that $\triangle A'B'B$ and $\triangle A'B'D$ have angle sum $180°$ and then from the same Lemma that $\triangle ADB$ has angle sum $180°$. Similarly, one shows that the angle sum of $\triangle ADC$ is equal to $180°$. □

Lemma 8.2.5 *Let us suppose that the angle sum of a triangle is equal to* $180°$. *Given a line* l *and a point* P *not lying on* l, *let* Q *be the foot of the perpendicular droped from* P *to* l. *Then for any positive real number* ϵ *there exists a line* n' *through* P *which intersects line* l *at a point* U *and such that* $m(\angle PUQ) < \epsilon$.

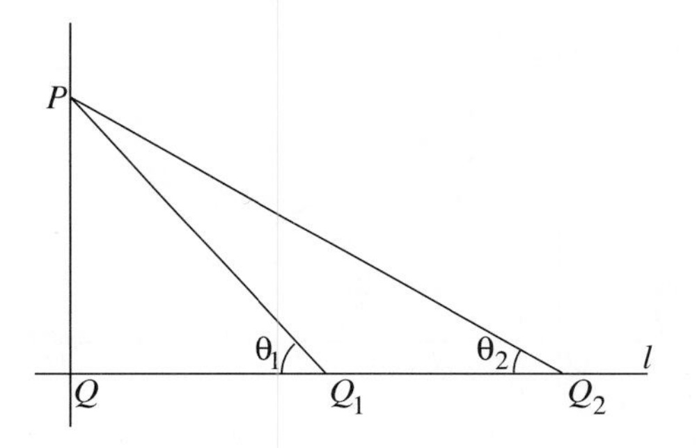

Figure 8.5

Proof Let Q_1 be a point on l such that $PQ = QQ_1$. Then $\triangle PQQ_1$ is isosceles and $m(\angle PQ_1Q) = m(\angle QPQ_1)$. Since we are assuming the existence of a triangle of angle sum $180°$, Theorem 8.2.4 implies that the angle sum of $\triangle PQQ_1$ is $180°$ and hence $m(\angle PQ_1Q) = 45°$, for $\angle PQQ_1$ is a right angle. Then we have

$$\theta_1 = m(\angle PQ_1Q) = \frac{1}{2}\, m(\angle PQQ_1) = 45°.$$

Now consider point Q_2 on l such that $PQ_1 = Q_1Q_2$. Notice that $m(\angle PQ_2Q_1) = m(\angle Q_1PQ_2)$. Let θ_2 denote their measure. Then

$$\theta_1 = 180 - m(\angle PQ_1Q_2) = 2\theta_2,$$

yielding $\theta_2 = (1/2)\theta_1$. This argument can be repeated as many times as we need, and after n times we have

$$\theta_n = \frac{1}{2}\theta_{n-1} = \frac{1}{2^{n-1}}45°.$$

Since, given $\epsilon > 0$, we may find a natural number N such that $(1/2^{N-1})\ 45 < \epsilon$, it follows that there exists a point Q_N on l such that

$$\theta_N = m(\angle PQ_NQ) < \epsilon.$$

Denoting point Q_N by U, the line through P and U is line n', which proves the lemma. □

Theorem 8.2.6 *Euclidean parallel postulate* $\Leftrightarrow$ *"there exists a triangle whose angle sum is equal to* 180°*."*

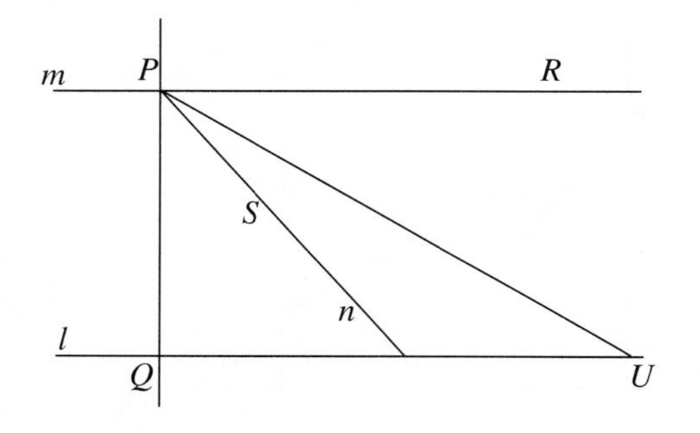

Figure 8.6

Proof ($\Rightarrow$) This part was proved in Chapter 2.
($\Leftarrow$) Consider line l, $P \notin l$, and lines t and m such that $t \perp l$ through P, $m \perp t$ also passing through P, and let $Q = t \cap l$. If n is any other line through P, consider points $R \in m$ and $S \in n$ on the same side of t. Let $\epsilon = m(\angle RPS)$. It follows from Lemma 8.2.5 that there exists a line n' through P that intersects line l at a point, denoted by U, and such that

$$m(\angle PUQ) < \epsilon.$$

Now, since $\angle PQU$ is a right angle,

$$m(\angle UPQ) = 90 - m(\angle PUQ) > 90 - \epsilon = m(\angle SPQ).$$

Therefore ray $\overrightarrow{PS}$ is between rays $\overrightarrow{PU}$ and $\overrightarrow{PQ}$ and hence intersects line segment $\overline{QU}$ by the crossbar theorem. □

Corollary 8.2.7 *If a rectangle exists, then every triangle has angle sum equal to* 180°*. It follows that the Euclidean parallel postulate holds.*

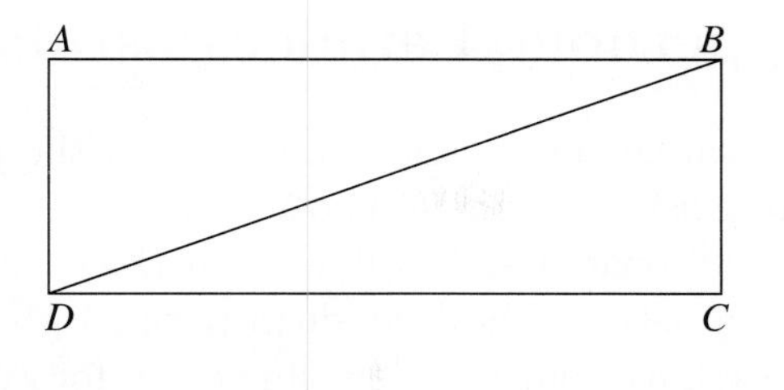

Figure 8.7

Proof Let $\square ABCD$ be a rectangle and consider diagonal $\overline{BD}$. From the Saccheri-Legendre theorem we obtain that

$$m(\angle ABD) + m(\angle ADB) \leq 90^\circ,$$

$$m(\angle CBD) + m(\angle CDB) \leq 90^\circ,$$

for $\angle A$ and $\angle C$ are right angles. Since $\angle B$ and $\angle D$ are also right angles, we have

$$m(\angle ABD) + m(\angle CBD) = 90^\circ,$$

$$m(\angle ADB) + m(\angle CDB) = 90^\circ,$$

and hence

$$m(\angle ABD) + m(\angle ADB) + m(\angle CBD) + m(\angle CDB) = 360^\circ.$$

Note that either

$$m(\angle ABD) + m(\angle ADB) < 90^\circ$$

or

$$m(\angle CBD) + m(\angle CDB) < 90^\circ$$

contradicts that the measures of these four angles sum to 360°.

It follows that $m(\angle ABD) + m(\angle ADB) = 90^\circ$, and then the angle sum of $\triangle ABD$ is 180°. Now Theorem 8.2.4 implies that every triangle has angle sum 180°. $\square$

8.3 The Hyperbolic Parallel Postulate

The axioms of neutral geometry and the results of the previous section imply that if there exists a triangle whose angle sum is *less* than 180°, then this is true for all triangles. It follows that the geometry where this result holds does not assume the Euclidean parallel postulate. Since in neutral geometry we have that for every line l and for every point P not lying on l there exists at least one line through P that is parallel to l, assuming the axioms of neutral geometry, the negation of the Euclidean parallel postulate is:

> **Hyperbolic parallel postulate** There exists a line l such that for some point P not on l at least two lines parallel to l pass through P.

The hyperbolic parallel postulate together with the axioms of neutral geometry form the foundations of *hyperbolic geometry*, which we study in the rest of this chapter. Therefore, from now on, we assume the axioms of Chapter 1 and the hyperbolic parallel postulate.

Our first result is that, although the hyperbolic parallel postulate refers to a particular line and a particular point, in fact this implies that much more is true:

Proposition 8.3.1 *(a) In hyperbolic geometry, given any line l and any point P not lying on l, there exist at least two distinct lines through P parallel to l.*
(b) In hyperbolic geometry, for every line l and for every point P not lying on l, there exist infinitely many parallels to l through P.

Proof (a) As before, let us consider t and m passing through P such that $t \perp l$ and $m \perp t$. Then we have $m \parallel l$. Let $Q = t \cap l$ and $R \in l, R \neq Q$. Let t' be the line that is perpendicular to l, and passes through R. Let S denote the foot of the perpendicular dropped from P to t'. It follows from the alternate interior angle theorem that line $\overleftrightarrow{PS}$ is parallel to l. We claim that $m \neq \overleftrightarrow{PS}$. In fact, suppose that $m = \overleftrightarrow{PS}$; then $S \in m$ and hence $\square PSRQ$ is a rectangle. But the existence of rectangles implies the Euclidean parallel postulate.
(b) Observe that the proof of part (a) implies that to each point $R \in$

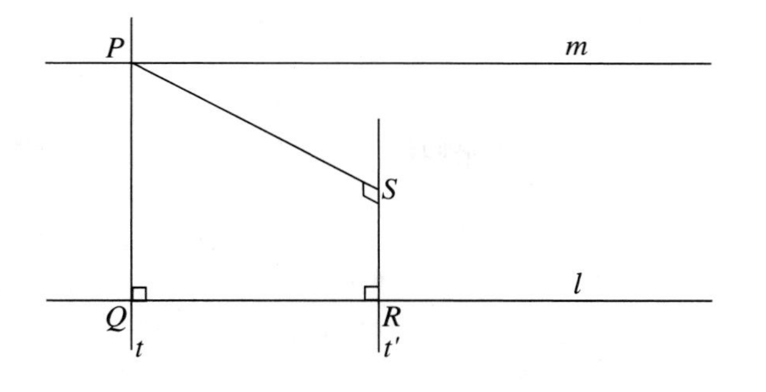

Figure 8.8

$l, R \neq Q$, there corresponds a line through P that is parallel to l; observe also that different points on l give rise to different parallel lines (why?). □

Recall that the results of the previous section imply the first important result of hyperbolic geometry, namely, that the *angle sum of a triangle is less than* $180°$. As a consequence of this result we have:

Proposition 8.3.2 *The angle sum of a convex quadrilateral is less than* $360°$.

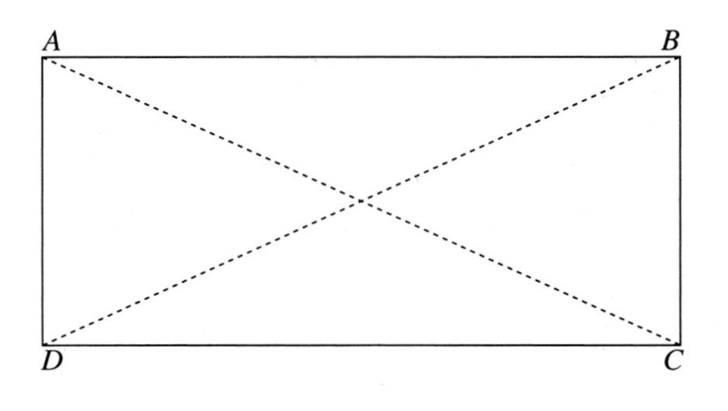

Figure 8.9

Proof Let $\square ABCD$ be a convex quadrilateral with diagonals $\overline{AC}$ and $\overline{DC}$. From the angle measurement axioms of Chapter 1 we have

$$m(\angle D) = m(\angle ADB) + m(\angle ADC),$$

and

$$m(\angle B) = m(\angle ABD) + m(\angle ABC).$$

Therefore

$m(\angle A) + m(\angle B) + m(\angle C) + m(\angle D) =$

angle sum$(\triangle ABD)$ + angle sum$(\triangle BDC) < 180° + 180° = 360°$. □

In Euclidean geometry we can construct two distinct equilateral triangles; for instance, one has sides that are 1 *in* long and the other has sides of length 1 *cm*. We then say that they are similar, but not congruent. One of the most striking results of hyperbolic geometry is the following theorem.

Theorem 8.3.3 *Let $\triangle ABC$ and $\triangle DEF$ be similar triangles. Then (in hyperbolic geometry) they are congruent.*

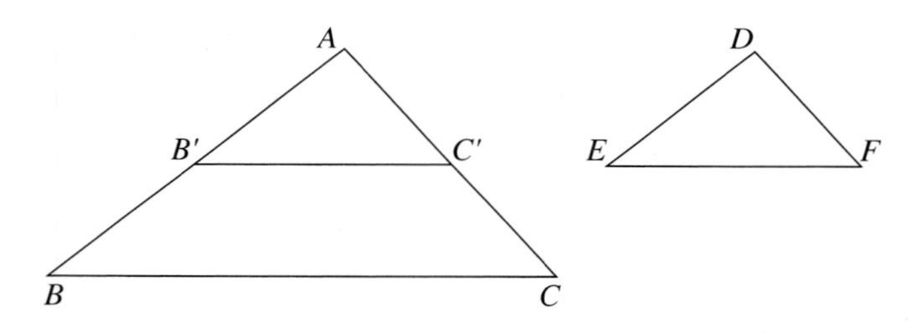

Figure 8.10

Proof Suppose that they are not congruent; then, without loss of generality, we suppose that $AB > DE$ and $AC > DF$. Let $B' \in \overline{AB}$ and $C' \in \overline{AC}$ such that $AB' = DE$ and $AC' = DF$. Then $\triangle AB'C' \simeq \triangle DEF$. It follows that $\angle AB'C' \simeq \angle B$, and hence line $\overleftrightarrow{B'C'}$ is parallel to $\overleftrightarrow{BC}$. Therefore $\square B'C'CB$ is convex. Moreover

$$m(\angle B) + m(\angle BB'C') = m(\angle AB'C) + m(\angle BB'C') = 180°$$

and

$$m(\angle C) + m(\angle CC'B') = m(\angle AC'B') + m(\angle CC'B') = 180°,$$

implying that angle sum of $\square B'C'CB$ is 360°. This contradicts Proposition 8.3.2. Therefore $\triangle ABC \simeq \triangle DEF$. □

8.4 Classification of Parallels

Before we distinguish different types among the infinitely many parallels, we identify two types of quadrilaterals that are rectangles in Euclidean geometry. They are the *Saccheri* and *Lambert* quadrilaterals, so called because Saccheri and Lambert used such quadrilaterals in their attempts to prove Euclid's fifth postulate.

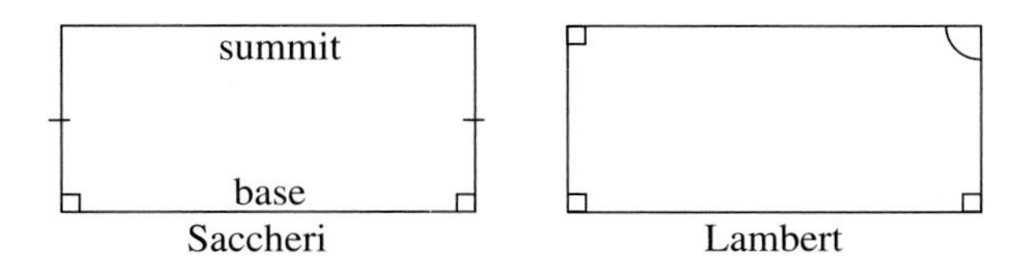

Figure 8.11

Definition 8.4.1 *(i) A* Saccheri *quadrilateral is a 4-gon whose base angles are right angles and sides adjacent to the bases, called arms, are congruent to each other. The side opposite the base is called the* summit *side and its adjacent angles,* summit *angles.*
(ii) A Lambert *quadrilateral is a 4-gon having at least three right angles.*

In Exercises 2 and 6 the reader is asked to prove some results on these quadrilaterals that will be used in the proofs of this section.

Theorem 8.4.2 *Given line l and point $P \notin l$, let Q denote the foot of the perpendicular dropped from P to Q. Then there exist two rays $\overrightarrow{PR}$ and $\overrightarrow{PS}$ on opposite sides of $\overleftrightarrow{PQ}$ such that:*
(i) The rays $\overrightarrow{PR}$ and $\overrightarrow{PS}$ do not intersect l.
(ii) A ray $\overrightarrow{PX}$ intersects line l if and only if $\overrightarrow{PX}$ is between $\overrightarrow{PR}$ and $\overrightarrow{PS}$.
(iii) $\angle QPR \simeq \angle QPS$.

Proof Axiom $\mathbf{A_1}$ of Chapter 1 implies that for every real number $x \in [0, 180]$ there exists a point X on one side of line $\overleftrightarrow{PQ}$ such that $m(\angle XPQ) = x$. Let $\mathcal{S}$ be the subset of $[0, 180]$ given by

$$\mathcal{S} = \{x \in [0, 180] \mid \overrightarrow{PX} \text{ intersects } l\}.$$

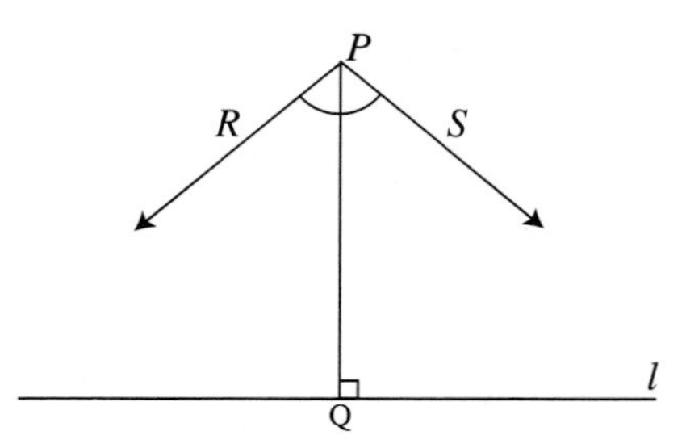

Figure 8.12

Since $\mathcal{S}$ is nonempty ($0 \in \mathcal{S}$) and bounded, we consider

$$s = \sup \mathcal{S}.$$

Let S be a point on one side of $\overleftrightarrow{PQ}$ such that $m(\angle SPQ) = s$.

We claim that ray $\overrightarrow{PS}$ does not intersect l. In fact, suppose it does intersect at a point Y (see Figure 8.13). Let $Z \in l$ such that Y is between Q and Z. Then ray $\overrightarrow{PZ}$ intersects l and $z = m(\angle QPZ) \in \mathcal{S}$. But $z > s$, which contradicts that s is an upper bound of $\mathcal{S}$.

We also claim that if ray $\overrightarrow{PX}$ is such that $m(\angle XPQ) < s$, then $\overrightarrow{PX}$ intersects l. To see this, observe first that the fact that s is the least upper bound of $\mathcal{S}$ implies that there exists $y \in \mathcal{S}$ such that $m(\angle XPQ) < y < s$. Therefore ray $\overrightarrow{PY}$ intersects line l at point Y'. But ray $\overrightarrow{PX}$ is

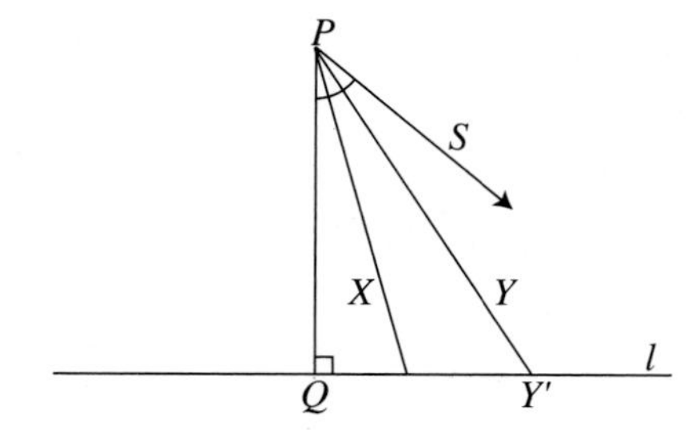

Figure 8.13

between rays $\overrightarrow{PQ}$ and $\overrightarrow{PY'}$ and hence intersects line segment $\overline{QY'}$, by the crossbar theorem. Therefore ray $\overrightarrow{PX}$ intersects l, as we claimed.

Since line $\overleftrightarrow{PQ}$ divides the plane in two sides, we consider for each side of $\overleftrightarrow{PQ}$ the only ray emanating from P that forms with ray $\overrightarrow{PQ}$ an angle of measure s. We call such rays $\overrightarrow{PR}$ and $\overrightarrow{PS}$, and we have proved the theorem. $\square$

The measure s of the angle $\angle QPS$ is called the *angle of parallelism* of l at P, and the rays $\overrightarrow{PS}$ and $\overrightarrow{PR}$ are *limiting parallel rays* for P and l.

Observe that the proof of Theorem 8.4.2 implies that $s \leq 90°$. In fact, if $m(\angle QPX) = 90°$, then $\overleftrightarrow{PX} \parallel l$ by the alternate angle theorem, which implies that $90 \notin \mathcal{S}$. Since s is the least upper bound, $s \leq 90$. We also point out that we did not use the hyperbolic parallel postulate in the proof of Theorem 8.4.2. Therefore the angle of parallelism can be defined in neutral geometry. If we assume Euclid's fifth postulate, then the angle of parallelism is constant and equal to $90°$.

Proposition 8.4.3 *Let l and l' be two distinct lines, $P \notin l$ and $P' \notin l'$. Let $Q \in l$ and $Q' \in l'$ be such that $\overleftrightarrow{PQ} \perp l$ and $\overleftrightarrow{PQ'} \perp l'$. If $PQ = P'Q'$, then the angle of parallelism of l at P is equal to the angle of parallelism of l' at P'.*

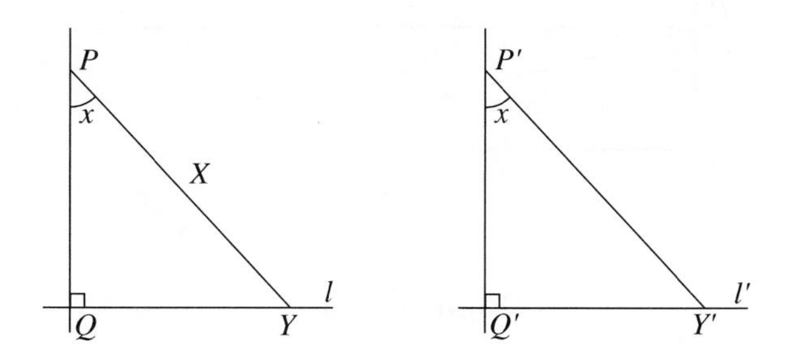

Figure 8.14

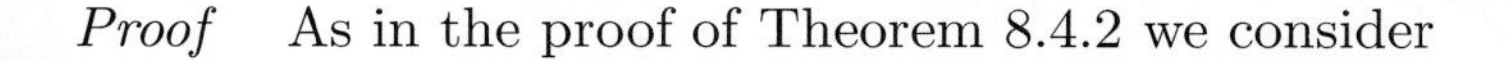

Proof As in the proof of Theorem 8.4.2 we consider

$$\mathcal{S} = \{x \in [0, 180] \mid \overrightarrow{PX} \text{ intersects } l\}$$

and

$$\mathcal{S}' = \{x \in [0, 180] \mid \overrightarrow{P'X'} \text{ intersects } l'\}.$$

If $x \in \mathcal{S}$, let Y denote the point on l where $\overrightarrow{PX}$ intersects l. Consider point Y' on l' such that $QY = Q'Y'$. Then $\triangle PQY \simeq \triangle P'Q'Y'$ by SAS, and then $m(\angle Y'P'Q') = x$, which implies that $x \in \mathcal{S}'$, that is, $\mathcal{S} \subset \mathcal{S}'$. Similarly, one proves that $\mathcal{S}' \subset \mathcal{S}$. Therefore $\mathcal{S}' = \mathcal{S}$, and hence they have the same least upper bound. $\square$

Definition 8.4.4 *Let P be a point not lying on a given line l. We define the distance from P to l as PQ, where Q is the foot of the perpendicular dropped from P to l.*

Using Theorem 8.4.2, we assign to each P on one ray that is perpendicular to l and emanates from Q a real number which is the angle of parallelism of l at P. Notice that Proposition 8.4.3 is saying that measure of this angle does not depend on the point and the ray but on the distance from the point to the line. In other words, for each positive t we assign $t \overset{\varphi}{\mapsto} s$. The next lemma gives one of the properties of the function φ.

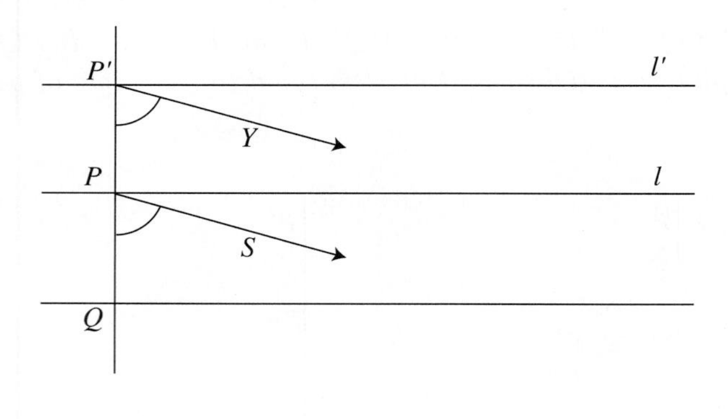

Figure 8.15

Lemma 8.4.5 *The function $s = \varphi(t)$ is monotonically decreasing; that is, if $t' > t$, then $\varphi(t') = s' \leq s = \varphi(t)$.*

Proof Let P, P', and Q be points on a line m such that $PQ = t$ and $P'Q = t'$, with $t < t'$. Let l and l' be lines perpendicular to m through P and P', respectively. Let us consider rays $\overrightarrow{PS}$ and $\overrightarrow{P'Y}$ (see Figure 8.15) such that $m(\angle QPS) = m(\angle QP'Y) = s$, where s denotes the angle of parallelism at P. It follows from the alternate interior angle theorem that lines $\overleftrightarrow{PS}$ and $\overleftrightarrow{P'Y}$ are parallel, and hence all points of $\overleftrightarrow{P'Y}$ lie in the half-plane bounded by $\overleftrightarrow{PS}$ containing P'. Since $\overleftrightarrow{PS} \parallel l$ and P' and Q are on opposite sides of $\overleftrightarrow{PS}$, we have that all points of l are in the half-plane bounded by $\overleftrightarrow{PS}$ containing Q. This means that $\overrightarrow{P'Y}$ does not intersect l. Now let

$$\mathcal{S}' = \{x \in [0, 180] \mid \overrightarrow{P'X} \text{ intersects } l\},$$

and $s' = \sup\mathcal{S}'$. From Theorem 8.4.2 and from the fact that $\overrightarrow{P'Y}$ does not intersect l, we conclude that $s' \leq s$, as we claimed. □

Proposition 8.4.6 *If $\varphi(t_0) < 90$ for some $t_0 > 0$, then $\varphi(t) < 90$ for every $t > 0$.*

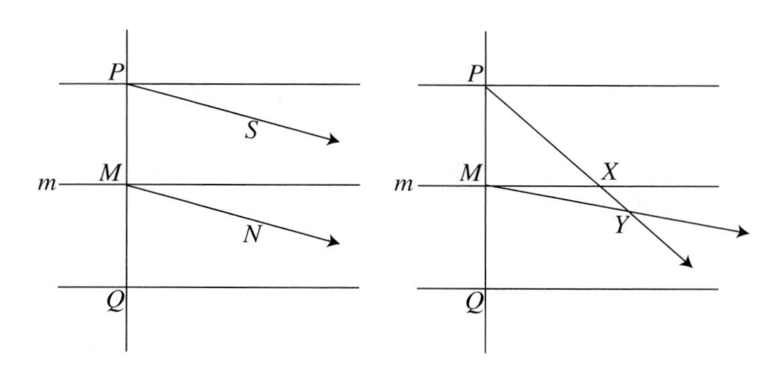

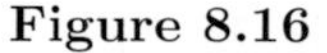
Figure 8.16

Proof From Lemma 8.4.5 we get that $\varphi(t) < 90$ for every $t > t_0$. Then we will consider $t < t_0$. We start by considering $t_1 = t_0/2$. Let P and Q be such that $\overline{PQ} \perp l$, $Q \in l$, and $PQ = t_0$. Let M be the midpoint of $\overline{PQ}$ and m a line through M that is perpendicular to $\overline{PQ}$ (see Figure 8.16). Let $\overrightarrow{PS}$ be a limiting ray and N be a point on the same side of $\overleftrightarrow{PQ}$ as S and such that $\angle NMQ \simeq \angle SPQ$. If ray $\overrightarrow{PS}$ does not intersect m, then the angle of parallelism of m at P is at most $\varphi(t_0)$. Since Proposition 8.4.3 implies that the angle of parallelism l at M is equal to the angle of parallelism of m at P, we conclude that $\varphi(t) < 90$.

Now we assume that ray $\overrightarrow{PS}$ does intersect m at a point X. Let $Y \in \overrightarrow{PS}$ such that X is between P and Y. We have $m(\angle YMQ) < 90$. Therefore, if we prove that ray $\overrightarrow{MY}$ does not intersect l, we conclude that $\varphi(t_1) < 90$. Then we suppose that it does intersect at a point that we call Z. Then we have $\triangle MQZ$ and line $\overleftrightarrow{PS}$ intersecting side $\overline{MZ}$. Therefore $\overleftrightarrow{PS}$ intersects $\overleftrightarrow{PQ}$ or $\overleftrightarrow{QZ}$ by Pasch's theorem. In both cases we have a contradiction.

Now we repeat this procedure, using the midpoint of $\overline{QM}$; after repeating this procedure n times we have

$$t_n = \frac{t_0}{2^n} \quad \text{and} \quad \varphi(t_i) < 90, \ \forall \ i \leq n.$$

From the Archimedean property of the real numbers we know that, given any number $0 < t < t_0$, there exists N such that

$$t_N = \frac{t_0}{2^N} < t.$$

Since $\varphi(t_N) < 90$, Proposition 8.4.5 implies that $\varphi(t) < 90$. □

Recall that in Euclidean geometry the function $\varphi(t) = 90$, $\forall t > 0$. Therefore Proposition 8.4.6 implies that in hyperbolic geometry $\varphi(t) < 90$, $\forall\, t > 0$. In fact, in the next section we will prove that in hyperbolic geometry $\varphi(t)$ is onto $(0, 90)$ (see Theorem 8.5.2).

Lemma 8.4.7 *Given line l and point $P \notin l$, let m denote the line through P that contains one of the limiting rays $\overrightarrow{PS}$ for P and l. Then for every $A \in m$, m contains the limiting ray for A and l in the direction of $\overrightarrow{PS}$.*

Proof We first consider a point $A \in m$ that is also on ray $\overrightarrow{PS}$ (see Figure 8.17). Let B be the foot of the perpendicular dropped from A to l, $C \in m$, such that A is between P and C, and D is a point on the half-plane bounded by $\overleftrightarrow{AB}$ containing C. We want to show that if $m(\angle DAB) < m(\angle CAB)$, then ray $\overrightarrow{AD}$ intersects l. But if $m(\angle DAB) < m(\angle CAB)$, then ray $\overrightarrow{PD}$ is between rays $\overrightarrow{PC}$ and $\overrightarrow{PQ}$ (why?) and then it intersects l by Theorem 8.4.2. Let E denote the point of intersection.

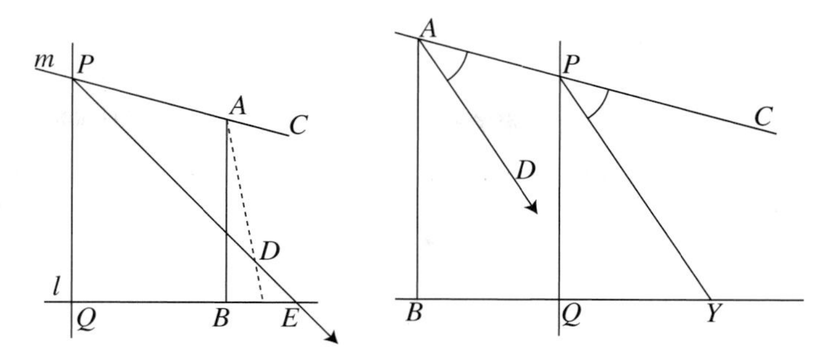

Figure 8.17

Then line $\overleftrightarrow{AD}$ intersects side $\overline{PE}$ of triangle $\triangle PQE$, and then it must intersect side $\overline{QE}$ by Pasch's theorem. Therefore $\overrightarrow{AD}$ intersects l.

Now suppose that point A is on m but on the ray that is opposite to $\overrightarrow{PS}$. Let B be as before and D such that $m(\angle DAB) < m(\angle PAB)$. Consider $\triangle APB$. Since ray $\overrightarrow{AD}$ is between rays $\overrightarrow{AP}$ and $\overrightarrow{AB}$, it intersects side $\overline{PB}$ by the crossbar theorem. Now we consider $\triangle PBY$, where Y is a point on l such that $\angle YPC \simeq \angle DAP$, that is, $\overleftrightarrow{AD} \parallel \overleftrightarrow{PY}$. Then Pasch's theorem implies that $\overrightarrow{AD}$ intersects side $\overline{BY}$. □

If a line m is as defined in Lemma 8.4.7, then we say that m is *critically parallel* to l. Notice that all limiting rays contained in m are in the same direction. Such a direction is called the *direction of parallelism.*

Proposition 8.4.8 *If m is critically parallel to l, then l is critically parallel to m.*

Proof Let $P \in m$ and $Q \in l$ such that $\overleftrightarrow{PQ} \perp l$. Let $T \in m$ such that $\overleftrightarrow{QT} \perp m$ (see Figure 8.18). First, we claim that if m contains the limiting ray $\overrightarrow{PS}$, then T and S are on the same side of line $\overleftrightarrow{PQ}$. If not, $m(\angle TPQ) > 90°$, for it would be the supplement of the angle of parallelism, which measures at most $90°$. Thus $\triangle TPQ$ would contain one right angle and one obtuse angle, which is absurd. Let $R \in l$ on the same side of $\overleftrightarrow{PQ}$ as T.

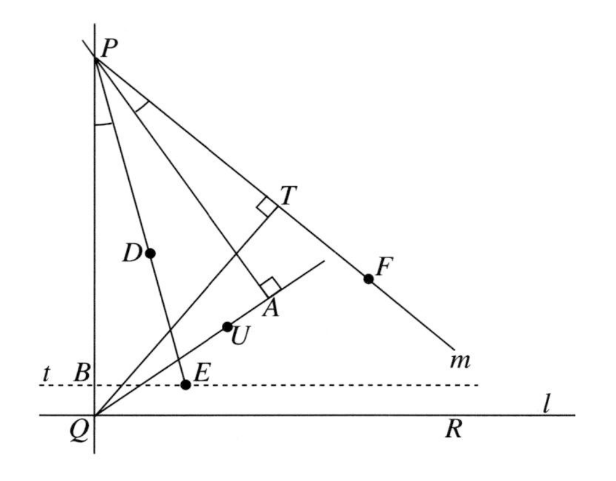

Figure 8.18

In order to prove the proposition we need to show that if ray $\overrightarrow{QU}$ is between rays $\overrightarrow{QT}$ and $\overrightarrow{QR}$, then $\overrightarrow{QU}$ intersects m. For that, let $A \in \overleftrightarrow{QU}$ such that $\overleftrightarrow{PA} \perp \overleftrightarrow{QU}$. We have that $PQ > PA$. Then, consider $B \in \overline{PQ}$ such that $PB = PA$ and line $t \perp \overleftrightarrow{PQ}$ through B. We also consider ray $\overrightarrow{PD}$ such that $\angle BPD \simeq \angle APT$. Since $m(\angle BPD) < m(\angle BPT)$, we conclude that ray $\overrightarrow{PD}$ intersects l at a point denoted by C. This implies that $\overrightarrow{PD}$ also intersects t. In fact, if $\overleftrightarrow{PD}$ did not intersect t, all points of $\overleftrightarrow{PD}$ would be on the same side of t as P. Since $t \parallel l$, all points of l are on the same side of t as Q. Recall that P and Q are on opposite sides of t; therefore we would have a contradiction, namely, that $\overleftrightarrow{PD}$ does not intersect l. Let E denote the point where $\overrightarrow{PD}$ also intersects t. Let $F \in m$ such that T is between P and F and $PF = PE$. We then have $\triangle PBE \simeq \triangle PAF$. It follows from this congruence that $\angle PAF \simeq \angle PBE$, and this implies that $\angle PAF$ is a right angle; that is, $\overrightarrow{AU} = \overrightarrow{AF}$, showing that $\overrightarrow{QU}$ intersects m at point F. $\square$

Proposition 8.4.9 *Let lines l, m, and n be such that l and m are both critically parallel to n in the same direction. Then l and m are critically parallel with the same direction of parallelism.*

Proof Let $P \in m$ and $Q \in l$ such that $\overleftrightarrow{PQ} \perp l$. Notice that the hypothesis of the proposition implies that l and m do not intersect each

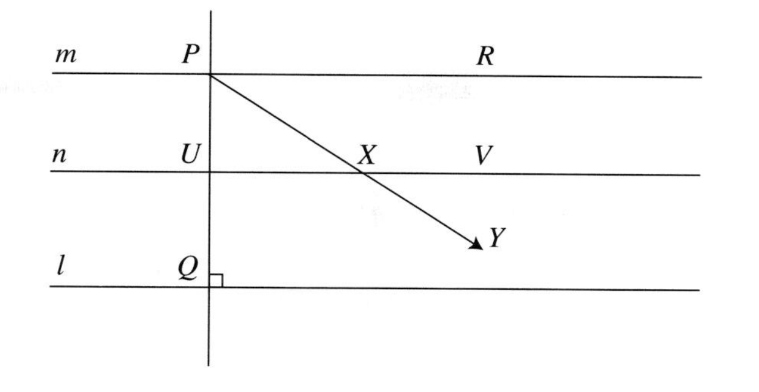

Figure 8.19

other. To see this, suppose that they intersected at a point A. Then we would have triangle $\triangle PQA$ with the property that line n intersects side $\overline{PQ}$. This implies, by Pasch's theorem, that n would intersect $\overline{PA}$ or $\overline{QA}$, which contradicts the hypothesis.

We consider first the case that l and m are on opposite sides of n (see Figure 8.19). Then, since P and Q are on opposite sides of n, ray $\overleftrightarrow{PQ}$ intersects n at point U. Let $R \in m$ be on the side of $\overleftrightarrow{PQ}$ that contains the direction of parallelism. Consider ray $\overrightarrow{PT}$ between rays $\overrightarrow{PR}$ and $\overrightarrow{PQ}$. Since m is critically parallel to n, ray $\overrightarrow{PT}$ intersects line n at point X. Let $Y \in \overrightarrow{PT}$ be such that X is between P and Y and $V \in n$ such that X is between U and V. From Proposition 8.4.8 we know that n is critically parallel to l, for we are assuming that l is critically parallel to n. Therefore ray $\overrightarrow{XY}$ intersects line l, since it is between rays $\overrightarrow{XQ}$ and $\overrightarrow{XV}$. Then ray $\overrightarrow{PT}$ intersects l, and m is critically parallel to l.

Now we suppose that l and n are on opposite sides of m. We choose a point $P \in l$ and consider line l' through P that is critically parallel to m in the same direction of parallelism. Now we have l' and n both critically parallel to m and on opposite sides of m. The first part of this proof implies that l' and n are critically parallel. Therefore $l' = l$, since, given one direction of parallelism, there is only one line through P that is critically parallel to n. □

Definition 8.4.10 *Let l and l' be parallel lines and A and B points on l. Let $A', B' \in l'$ such that $\overleftrightarrow{AA'} \perp l'$ and $\overleftrightarrow{BB'} \perp l'$. We say that points*

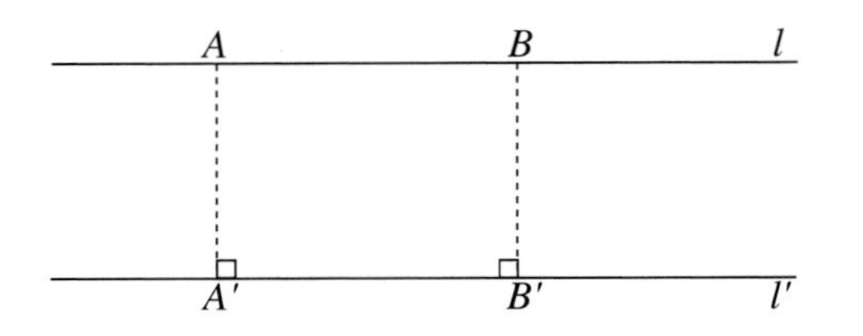

Figure 8.20

A and B are equidistant *from l' if $AA' = BB'$* (see Figure 8.20).

Lemma 8.4.11 *Let l and l' be distinct parallel lines. Let A and B be points on l equidistant from l' and $C \in l, C \neq A, B$. Then $CC' \neq AA' = BB'$, where C' denotes the foot of the perpendicular from C to l'.*

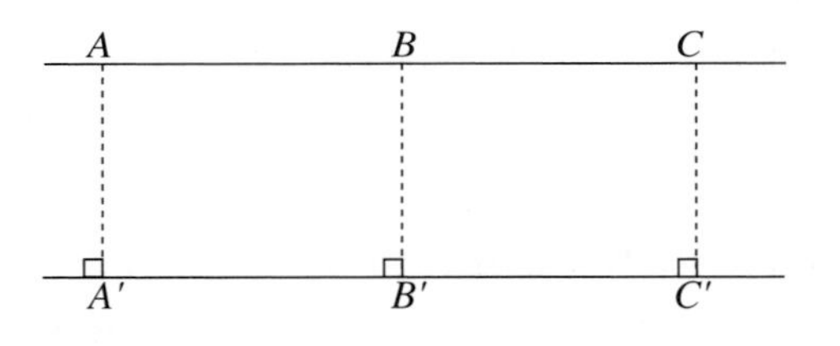

Figure 8.21

Proof Let us assume that $AA' = BB' = CC'$ and, without losing generality, that B is between A and C. Notice that $\square ABB'A', \square ACC'A'$, and $\square BCC'B'$ are Saccheri quadrilaterals and hence their summit angles are congruent, by Exercise 2. This implies that $\angle ABB' \simeq \angle CBB'$, and, since they are supplementary angles, we conclude that $\square BCC'B'$ is a rectangle. But this is a contradiction in hyperbolic geometry. Therefore A, B, and C cannot be equidistant from l'. $\square$

Theorem 8.4.12 *Let l and l' be distinct parallel lines such that the points A and B on l are equidistant from l'. Then l and l' have a common perpendicular segment which is the shortest segment between l and l'.*

Proof As before, let A', B' be on l' such that $AA' = BB'$. Let M be the midpoint of $\overline{AB}$ and M' the midpoint of $\overline{A'B'}$ (see Figure

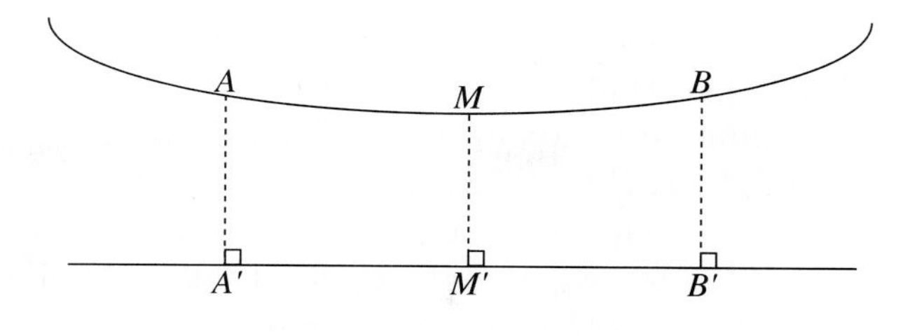

Figure 8.22

8.22). First observe that $\square A'B'BA$ is a Saccheri quadrilateral and hence $\triangle AMA' \simeq \triangle BMB'$, by SAS. Therefore $A'M = B'M$, which in turn implies $\triangle MA'M' \simeq \triangle MB'M'$ by SSS. It follows from the congruence of these triangles that $\angle A'M'M \simeq \angle B'M'M$, and since they are supplementary angles, each of them is right angle. Moreover,

$$\angle AMA' \simeq \angle BMB' \quad \text{and} \quad \angle A'MM' \simeq \angle B'MM'.$$

Therefore $\angle AMM'$ is also a right angle. We have then that $\overline{MM'}$ is perpendicular to the base and to the summit of $\square A'B'BA$. This implies that $\square A'M'MA$ is a Lambert quadrilateral. It follows from Exercise 6 that $AA' > MM'$. Now let $\overline{CD}$ be any other segment such that $C \in l$ and $D \in l'$. Let D' denote the foot of the perpendicular line dropped from C to l'. Then $\square MM'D'C$ is a Lambert quadrilateral and hence $MM' < CD'$. Since $CD' < CD$, by the exterior angle theorem, we have proved the theorem. $\square$

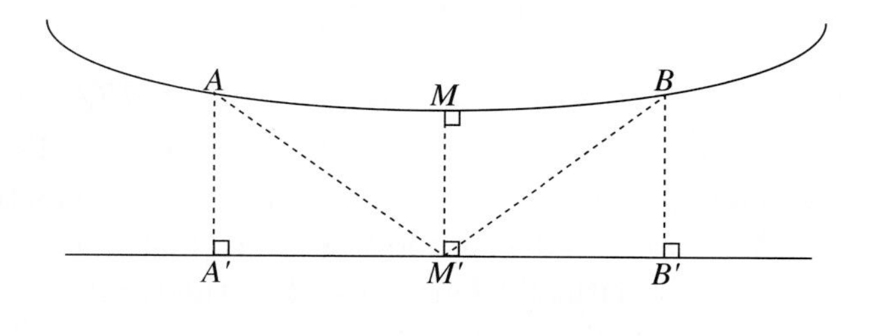

Figure 8.23

Theorem 8.4.13 *If lines l and l' have a common perpendicular line segment $\overline{MM'}$, then $l \parallel l'$. Moreover, if A and B are points on l such that $AM = MB$, then A and B are equidistant from l'* (see Figure 8.23).

Proof The first assertion is just the alternate interior angle theorem. Now let $A, B \in l$ such that M is the midpoint of $\overline{AB}$ and $A', B' \in l'$ obtained by dropping perpendicular lines from A and B to l'. We have that $\triangle AMM' \simeq \triangle BMM'$ by SAS. This implies that $AM' = BM'$ and $\angle AM'M \simeq \angle BM'M$. The congruence of these angles combined with the fact that $\angle A'M'M$ is a right angle implies that $\angle AM'A' \simeq \angle BM'B$. Therefore $\triangle AM'A' \simeq \triangle BM'B'$ by AAS, which gives that $AA' = BB'$. □

Theorem 8.4.14 *Let l and l' be two critically parallel lines. Let $P \in l'$ and φ the angle of parallelism of l at P. If $\varphi < 90°$, then given $\epsilon > 0$ there exists $X \in l'$ such that $d(X, l) = \epsilon$. Moreover, ϵ decreases as P moves in the direction of parallelism, and it increases as P moves in the opposite direction.*

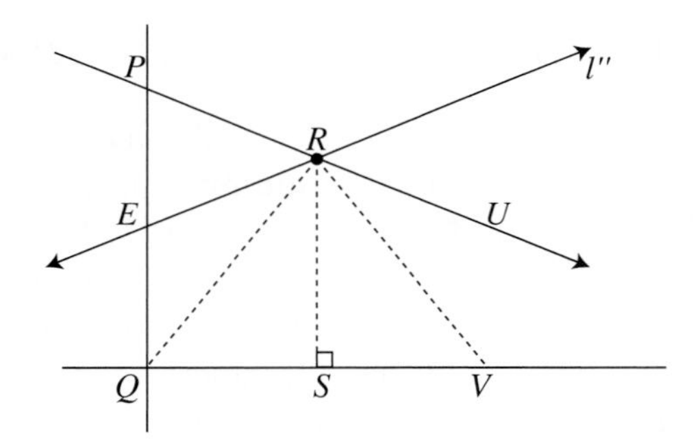

Figure 8.24

Proof Let $Q \in l$ such that $\overline{PQ} \perp l$. Given $\epsilon > 0$, if $PQ = \epsilon$, then P is the desired point. If not, let $E \in \overrightarrow{QP}$ such that $QE = \epsilon$. Let l'' be the line through E that is critically parallel to l, but in the opposite direction of parallelism. We claim that l'' intersects l'. In fact, consider also the line m through E and critically parallel to l in the same direction of l'. Since the angle of parallelism of l at E is less than 90, l'' and m are two distinct lines intersecting at E. Proposition 8.4.9 implies that m and l' are critically parallel, and if l'' is parallel to l' we would conclude that l'' is parallel to m, which is a contradiction.

Let $R = l'' \cap l'$ and $S \in l$ the foot of the perpendicular dropped from R to l. Now we consider points $U \in l'$ and $V \in l$ such that $RU = RE$

and $SV = SQ$. Then we have $\triangle RSV \simeq \triangle RSQ$, by SAS, which implies $RV = RQ$. It also implies that $\angle VRS \simeq \angle QRS$, which, combined with the fact that the angles of parallelism are equal, implies $\angle URV \simeq \angle ERQ$. Then we conclude that U is in the direction of parallelism (since $U \in l'$), $\angle UVQ$ is a right angle, and $UV = EQ = \epsilon$. Therefore U is the desired point.

Now for the second part suppose that we have P and P' on l such that $\overrightarrow{PP'}$ is in the direction of parallelism. Let Q and Q' be on l obtained by dropping perpendicular lines. Observe that $\angle QPP'$ is acute and $\angle PP'Q'$ is obtuse. Then from Exercise 5 we obtain $PQ > P'Q'$. □

It follows from Theorem 8.4.14 that if two lines are critically parallel, then we cannot find a pair of points on one that are equidistant from the other. Therefore, they cannot have a common perpendicular segment, by Theorem 8.4.13.

Theorem 8.4.15 *Let l and l' be distinct parallel lines that are not critically parallel. Then there exists a common perpendicular line to l and l'.*

Proof The proof of this theorem is sketched in Exercise 9. □

Exercises

1. Consider $\triangle ABC$, M and N the midpoints of $\overline{AB}$ and $\overline{AC}$, respectively. Show that $MN < BC$.

2. Prove that the summit angles of a *Saccheri* quadrilateral are congruent.

3. Prove that the summit angles of a Saccheri quadrilateral are right angles if and only if the angle sum of a triangle is 180°.

4. Let $\square ABCD$ be a quadrilateral such that the base angles $\angle A$ and $\angle B$ are right angles. Show that $AD < BC$ if and only if $m(\angle C) < m(\angle D)$.
   Hint: Use the exterior angle theorem.

5. (a) Show that the lines that contain the midpoints of opposite sides of a Saccheri quadrilateral are perpendicular.

(b) Show that the line that contains the midpoints of the congruent sides of a Saccheri quadrilateral bisects its diagonals.

6. Let $\square ABCD$ be a *Lambert* quadrilateral such that A, B, and C denote the right angles.
(a) Show that $m(\angle D) \leq 90°$.
(b) Show that $m(\angle D) < 90°$ if and only if $AD > BC$.
(c) Show that $m(\angle D) = 90°$ if and only if the angle sum of a triangle is $180°$.

7. Show that $\varphi(t) = 90°$ for some t implies the Euclidean parallel postulate.

8. Let m be critically parallel to l, and consider $P \in m$, $Q \in l$ such that $\overline{PQ} \perp l$. Let $S \in \overline{PQ}$ such that $QS < PQ$ and $n \perp \overline{PQ}$ and through S. Show that m intersects n.

9. Prove Theorem 8.4.15.
Hint: Find two points on l' that are equidistant from l; then use Theorem 8.4.12.
To find such points, choose any two points A and B on l and points $A', B' \in l'$ such that $\overleftrightarrow{AA'} \perp l'$ and $\overleftrightarrow{BB'} \perp l'$. Suppose that $AA' > BB'$ and consider $C \in \overline{AA'}$ such that $CA' = BB'$. Let ray $\overrightarrow{CD}$ be such that $\angle DCA' \simeq \angle FBB'$ where $F \in l$ such that B is between A and F, and such that D and F are on the same side of $\overleftrightarrow{AA'}$. Let $\overrightarrow{A'P}$ and $\overrightarrow{A'Q}$ be limiting parallel rays, in the same direction, to lines $\overleftrightarrow{CD}$, l respectively; let $\overrightarrow{B'R}$ be the limiting parallel ray to l, also in the same direction as $\overrightarrow{A'P}$. Justify the following steps, pointing out where the assumption that l and l' are not critically parallel is used:
(1) $\overrightarrow{A'Q}$ and $\overrightarrow{B'R}$ are not subsets of line l.
(2) $\angle CA'P \simeq \angle BB'R$.
(3) $\overrightarrow{B'R}$ is a limiting parallel ray to $\overleftrightarrow{A'Q}$.
(4) $m(\angle QA'S) < m(\angle RB'S)$, where $S \in l'$, such that B' is between A' and S.
(5) $\overrightarrow{A'P}$ is between $\overrightarrow{A'A}$ and $\overrightarrow{A'Q}$.
(6) $\overrightarrow{A'P}$ intersects line l at a point E.

(7) E and A' are on the same side of $\overleftrightarrow{CD}$.
(8) $\overleftrightarrow{AE} = l$ intersects $\overrightarrow{CD}$ at a point G.
(9) Let $H \in l$ such that $CG = BH$ and G is between B and H. Then $\triangle CA'G \simeq \triangle BB'H$.

(10) Let G' and H' be such that $\overleftrightarrow{GG'} \perp l'$ and $\overleftrightarrow{HH'} \perp l'$. Then $GG' = HH'$.

10. Let l and m be two lines concurrent at points O and $P \in m$. Show the following:
(a) $d(P, l)$ increases as P moves away from O; in addition, given $\epsilon > 0$, there exists $X \in m$ such that $d(X, l) = \epsilon$.
(b) $d(P, l)$ decreases as P moves toward O and, given $\epsilon > 0$, there exists $X \in m$ such that $d(X, l) = \epsilon$.

11. Let m and n be two parallel lines but not critically parallel. Let $M \in m$ and $N \in n$ and $\overline{MN}$ the common perpendicular segment. Let $P \in m$. Show that $d(P, n)$ increases as P moves away from M and decreases as P moves toward M. Moreover, for every $\epsilon > 0$, there exists $X \in m$ such that $d(X, n) = \epsilon$.

8.5 The Angle of Parallelism

In this section we prove more properties of the angle of parallelism function φ defined in the previous section.

Proposition 8.5.1 *The function $s = \varphi(t)$ is strictly decreasing; that is, if $t' > t$, then $s' < s$.*

Proof Let P, P', and Q be as in the proof of Lemma 8.4.5. From the midpoint M of $\overline{P'P}$ drop a line that is perpendicular to l (see Figure 8.25). Let F denote the foot of such a perpendicular. Let $E \in l'$ such that E and F are on opposite sides of line $\overleftrightarrow{P'P}$ and $P'E = PF$. If $s = s'$, then $\triangle EP'M \simeq \triangle FPM$ and hence $\angle MEP'$ is a right angle. Further, we have that $\angle EMP' \simeq \angle FMP$, which implies E, M, and F are collinear points. Therefore $\overline{EF}$ is a common perpendicular segment to l and l'. But l and l' are critically parallel by Proposition 8.4.9 and therefore cannot have a common perpendicular. Since $s' \leq s$ by Lemma 8.4.5 and $s' = s$ gives a contradiction, we obtain that $s' < s$. $\square$

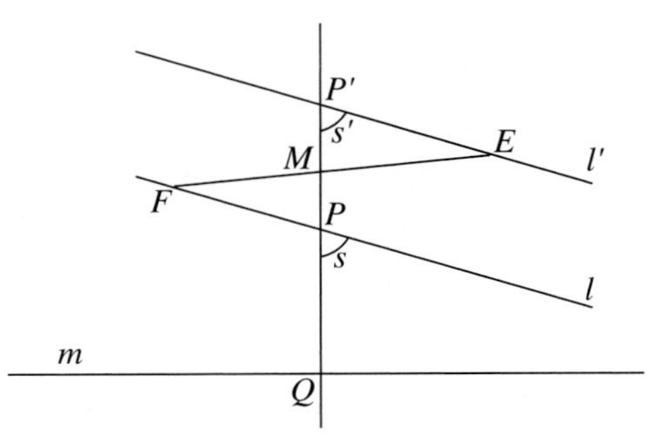

Figure 8.25

Theorem 8.5.2 *The function $s = \varphi(t)$ is onto the interval $(0, 90)$; that is, for each α such that $0 < \alpha < 90$ there exist a point A and a line l such that the angle of parallelism of l at A is α.*

Proof We start by considering an angle $\angle BAC$ of measure α. We claim that there exists a line perpendicular to $\overleftrightarrow{AC}$ that does not intersect $\overrightarrow{AB}$. In fact, suppose that all lines perpendicular to $\overleftrightarrow{AC}$ intersect $\overrightarrow{AB}$. Let $\overleftrightarrow{C_1B_1}$ be one of them (see Figure 8.26). Then we have a triangle whose angle sum is less than 180°; let us write its sum as $180 - \epsilon$. Now we consider on $\overrightarrow{AC}$ a point C_2 such that $AC_1 = C_1C_2$ and let B_2 be the point where the line perpendicular to $\overleftrightarrow{AC}$ intersects $\overrightarrow{AB}$. Then $\triangle AB_1C_1 \simeq \triangle C_2B_1C_1$, by SAS. The reader can verify that this congruence implies that angle sum of $\triangle AB_2C_2$ is $180 - 2\epsilon$. This argument can be repeated several times, and after n times we have a triangle of angle sum $180 - n\epsilon$. Since for every positive ϵ there exists a natural number N such that

$$\frac{1}{N} < \frac{\epsilon}{180},$$

if we repeat the argument above N times we will have a triangle of negative angle sum, which is absurd. This proves our claim.

Let t denote the coordinate of a point $T \in \overrightarrow{AC}$ and m_t the line through T which is perpendicular to $\overleftrightarrow{AC}$. We define two sets of real numbers.

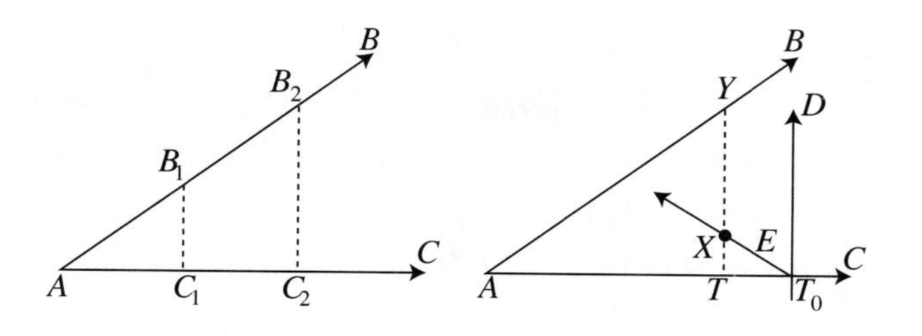

Figure 8.26

$$\mathcal{S} = \{t \mid m_t \text{ intersects } \overrightarrow{AB}\},$$
$$\mathcal{I} = \{t \mid m_t \text{ does not intersect } \overrightarrow{AB}\}.$$

First, notice that $\mathcal{S}$ is nonempty; to see this, pick any point on $\overrightarrow{AB}$ and drop a perpendicular to $\overleftrightarrow{AC}$; since angle $\angle BAC$ is acute, the foot of the perpendicular is on $\overrightarrow{AC}$. The set $\mathcal{S}$ is also bounded above. In fact, from the previous paragraph we know that there exists t_1 such that m_{t_1} does not intersect $\overrightarrow{AB}$. This implies that if $t_2 > t_1$, then m_{t_2} does not intersect $\overrightarrow{AB}$ either. (Why? Use Pasch's theorem.) Therefore if $t \in \mathcal{S}$, then $t < t_1$.

The set $\mathcal{I}$ is also nonempty and bounded below, for the arguments in the previous paragraph imply that each $t \in \mathcal{S}$ is a lower bound of $\mathcal{I}$. Then let us consider

$$s = \sup \mathcal{S} \quad \text{and} \quad i = \inf \mathcal{I}.$$

Since each $t \in \mathcal{S}$ is a lower bound of $\mathcal{I}$ and each $t \in \mathcal{I}$ is an upper bound of $\mathcal{S}$, we have $s \leq t$. The reader can easily see now that $s < t$ gives a contradiction. Therefore $s = t$, and let us call this number t_0.

Let T_0 be the point on ray $\overrightarrow{AC}$ of coordinate t_0. Let l be the line through T_0 that is perpendicular to $\overleftrightarrow{AC}$. We claim now that the angle of parallelism of l at A is α.

Consider a ray $\overrightarrow{T_0E}$ between $\overrightarrow{T_0D}$ ($D \in l$ and on the same side of $\overleftrightarrow{AC}$ as B) and $\overrightarrow{T_0A}$. We will show that $\overrightarrow{T_0E}$ intersects $\overrightarrow{AB}$. Consider a point $X \in \overrightarrow{T_0E}$ and the line through X that is perpendicular to $\overleftrightarrow{AC}$. The coordinate t of the foot T of this perpendicular satisfies $t < t_0$, and hence m_t intersects $\overrightarrow{AB}$, say at point Y. Then we have line $\overleftrightarrow{T_0E}$ intersecting side $\overline{TY}$ of triangle $\triangle AYT$ at point X between T and Y. Applying Pasch's theorem, we conclude that $\overrightarrow{T_0E}$ intersects $\overrightarrow{AB}$. Since $\overrightarrow{T_0D}$ does not intersect $\overrightarrow{AB}$ but every ray between $\overrightarrow{T_0D}$ and $\overrightarrow{T_0A}$ does intersect $\overrightarrow{AB}$, we conclude that $\overrightarrow{T_0D}$ is a limiting parallel ray for $\overleftrightarrow{AB}$ and T_0. It follows that $\overleftrightarrow{AB}$ and l are critically parallel, by Lemma 8.4.7 and Proposition 8.4.8. Thus, $\overrightarrow{AB}$ is a limiting parallel ray for l at A. □

Exercises

1. Let l and m be critically parallel. Consider $A, C \in l$ and $B, D \in m$ such that $\overrightarrow{AC}$ and $\overrightarrow{BD}$ are in the direction of the parallelism. The figure $\overline{AB} \cup \overrightarrow{AC} \cup \overrightarrow{BD}$ is a *generalized triangle* (sometimes it is called an *open triangle* or an *ideal triangle*) (see Figure 8.27). Prove the following:

   (a) If a line passes through vertex A, then it intersects $\overrightarrow{BD}$.
   (b) If a line crosses one side of a generalized triangle at a point between the vertices A and B, then it intersects only one of the other two sides. This statement is known as the *generalized Pasch's theorem.*

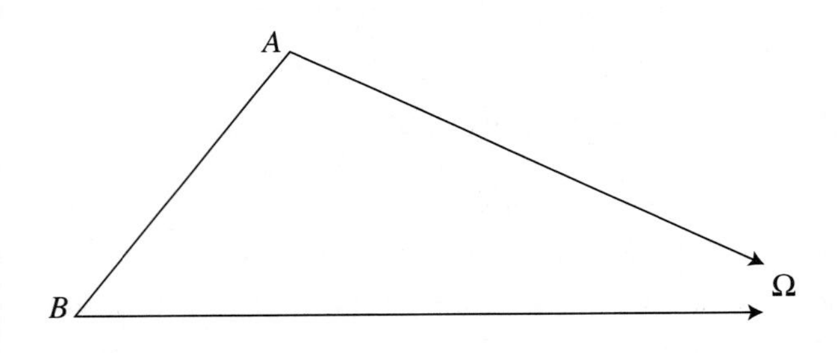

Figure 8.27 A generalized triangle

2. Prove that the measure of an exterior angle of a generalized triangle is greater than the measure of its remote angle. This statement is known as the *generalized exterior angle theorem.*
   Hint: Follow the steps of the proof of Proposition 8.5.1.

8.6 Defect of Hyperbolic Triangles

Definition 8.6.1 *The* defect *of a triangle $\triangle ABC$, denoted by $\delta(\triangle ABC)$, is defined to be*

$$\delta(\triangle ABC) = 180^\circ - m(\angle A) - m(\angle B) - m(\angle C).$$

Lemma 8.6.2 *(i) Let D be a point between B and C. Then*

$$\delta(\triangle ABC) = \delta(\triangle ADB) + \delta(\triangle ADC).$$

(ii) $\delta(\triangle ABC) = 0$ if and only if $\delta(\triangle ADB) = 0$ and $\delta(\triangle ADC) = 0$.

Proof Its proof is similar to that of Lemma 8.2.1. □

The lemma above implies an important property of the angle of parallelism function φ. We already know that it is strictly decreasing and hence $\varphi(t) \to 0$ as $t \to \infty$. The next proposition describes precisely the situation.

Proposition 8.6.3 *Let $\varphi(t)$ be the angle of parallelism function. Then*

$$\inf\{\varphi(t),\, t > 0\} = 0.$$

Proof If zero is not the greatest lower bound, then let us suppose that $0 < \alpha \leq \varphi(t)$, $\forall t > 0$. Then we consider line $l, Q \in l$ and points $P_1, \ldots, P_n, \ldots$ such that $P_iP_{i+1} = P_{i+1}P_{i+2}$ (see Figure 8.28). On one side of line $\overleftrightarrow{QP_1}$ we draw rays emanating from $P_1, \ldots, P_n, \ldots$ and forming an angle of measure α. If the angle of parallelism of l at P_i is α, then at P_{i+1} it will be less than α, and this contradicts $\alpha \leq \varphi(t)$. Then the angle of parallelism at P_i is less than α for all $i = 1, \ldots, n, \ldots$. Thus the rays emanating from them intersect line l at points that we call $R_1, \ldots, R_n, \ldots$. Now, from Q and each P_i we drop a perpendicular to line $\overleftrightarrow{P_iR_i}$ and let $S_1, \ldots, S_n, \ldots$ denote the feet of these lines. We have that

$$\triangle QS_1P_1 \simeq \triangle P_1S_2P_2 \simeq \cdots \simeq \triangle P_nS_{n+1}P_{n+2} \simeq \cdots,$$

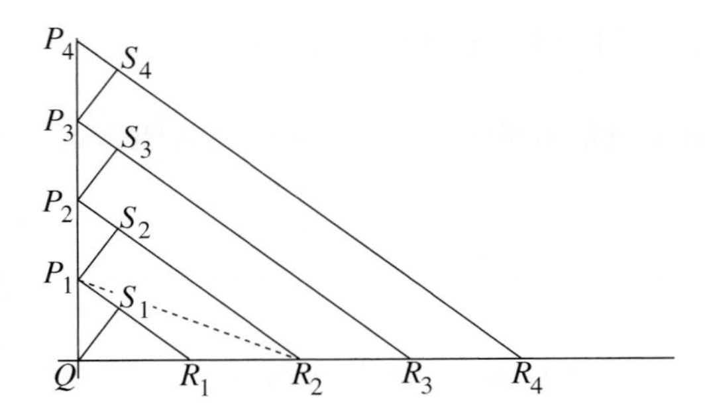

Figure 8.28

and hence they all have the same defect, which will be denoted by d_0. Let d_i denote $\delta(\triangle P_iQR_i)$. Using Lemma 8.6.2, we obtain

$$\begin{aligned} d_1 &= d_0 + \delta(\triangle S_1QR_1), \\ d_2 &= \delta(\triangle P_2P_1R_2) + \delta(\triangle P_1QR_2) \\ &= d_0 + \delta(\triangle S_2P_1R_2) + \delta(\triangle R_1P_1R_2) + d_1, \end{aligned}$$

and in general

$$\delta(\triangle P_{n+1}QR_{n+1}) = d_0 + \delta(\triangle S_{n+1}P_nR_{n+1}) + \delta(\triangle R_{n+1}P_nR_n) + d_n.$$

Since in hyperbolic geometry all triangles have positive defect, we conclude that

$$d_1 > d_0, \quad d_2 > d_1 + d_0, \quad d_3 > d_2 + d_1 > d_1 + 2d_0,$$

and after n steps,

$$d_n > d_1 + (n-1)d_0.$$

Since the Archimedean property of real numbers implies that there exists N such that $(N-1)d_0 > 180$, we would have a triangle $\triangle P_nQR_n$ of defect greater than 180, which is a contradiction. Therefore, there exists t such that $\varphi(t) < \alpha$. This implies that no positive number is a lower bound of the set $\{\varphi(t), t > 0\}$. □

Proposition 8.6.4 *For every real number x such that $0 < x < 180$ there exists a triangle $\triangle$ such that $\delta(\triangle) > x$.*

Proof Let $\triangle PQR$ be an isosceles right triangle with the right angle at the vertex Q. Let t denote the length of the leg and $\beta(t)$ the base angle of the right triangle. Then $\beta(t) < \varphi(t)$. On line $\overleftrightarrow{QR}$ we consider point S such that Q is between S and R and $SQ = RQ$. Then $\triangle PSR$ is also an isosceles triangle of angle sum $4\beta(t)$. Therefore $\delta\triangle PSR = 180 - 4\beta(t)$.

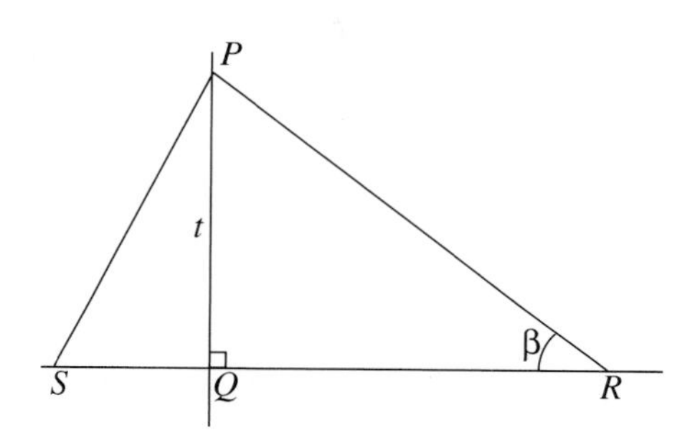

Figure 8.29

It follows that

$$\delta\triangle PSR > 180 - 4\varphi(t).$$

For $x < 180$, let $\alpha < (180 - x)/4$. From Proposition 8.6.3 we know that there exists t such that $\varphi(t) < \alpha$. Therefore, there exists a triangle of defect greater than x. □

The defect of triangles is also associated with the area of a hyperbolic triangle. We will not study here an area theory in hyperbolic geometry. Instead, we define hyperbolic area in the models of the hyperbolic plane that we will study in the next chapter. Then we will prove, with that definition, that the area of a triangle is equal to its defect. The reader can find a nice presentation of hyperbolic area theory in the book by Moise.[2]

Exercises

1. Prove Proposition 8.6.2.

[2]E. E. Moise, *Elementary Geometry from an Advanced Standpoint*, Addison-Wesley Publishing Co., New York, 1974.

2. *For readers who have studied analysis*: Prove that a decreasing function from $(0, \infty)$ onto an interval of real numbers is continuous. Conclude that the angle of parallelism function φ is continuous.

Chapter 9

MODELS FOR PLANE GEOMETRIES

9.1 Introduction

In the last section of Chapter 3 we coordinatized the plane using the axioms of Euclidean plane geometry, i.e., we showed that we can set a one-to-one correspondence between points of the plane and $\mathbf{R}^2$ that enable us to find formulæ for lengths of line segments, equations of lines, and formulæ for rigid motions.

Conversely, given $\mathbf{R}^2$, we can define a *line* as the set of points (x, y) that satisfies an equation of type $y = ax + b$, where a and b are constants. A point is then said to be on the line, or *incident* with the line, if its coordinates (x_0, y_0) satisfy the equation of the line. We just gave a definition for an undefined object, a line, and a definition for an undefined relation, the incidence of a point with a line. Notice that with these definitions *we can prove* that, given two distinct points, there exists only one line through them. Then we say that these definitions satisfy Incidence Axiom $\mathbf{I_1}$. Actually, we can prove that all axioms of Euclidean geometry are verified by these definitions, and then we say that the plane $\mathbf{R}^2$ is a *model* for Euclidean geometry.

A model for a geometry built on a set of postulates can be thought of as a concrete realization of a geometry. We first choose a set, whose elements we call points. Then we give a definition (or interpretation) for each undefined object and relation of the axiom system. If all axioms hold, we then have a model for that geometry. When we have a model for a geometry, all theorems proved using the axioms of that geometry hold in the model.

Example 1

Consider $\mathcal{S} = \{1, 2, 3\}$. We say that a subset of $\mathcal{S}$ is a "line" if it contains exactly two elements of $\mathcal{S}$. A point will be incident with a line if it is an element of the subset that determines the line. It is easy to show that, with these definitions, all incidence axioms hold. Therefore $\mathcal{S}$ with this definition of line constitutes a model for incidence geometry.

Example 2

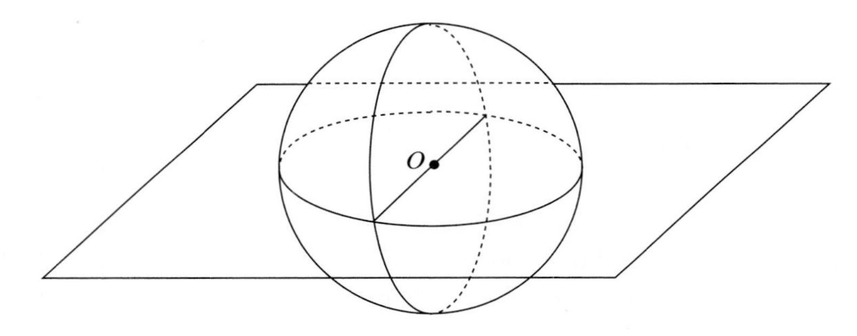

Figure 9.1

Let $\mathcal{S}$ be a sphere and define its "lines" to be the great circles (i.e., circles obtained by cutting the sphere by planes through its center). Observe that there are infinitely many "lines" through the north and south poles. Therefore, with this interpretation, $\mathcal{S}$ is not a model for incidence geometry. Further, two great circles always intersect each other in two points, and therefore there are no parallel lines. To see this, notice first that the planes containing the great circles must intersect, since they have a common point, namely, the center of the sphere. It follows that the intersection of these two planes is a line, and this line intersects the sphere in two points. Moreover, this line goes through the center, and hence the two intersection points are end points of a diameter of the great circle. We call them *antipodal points*. Therefore, with this interpretation for lines, the sphere is not a model for a geometry with parallel lines.

We are interested in constructing models for the axiomatic geometries studied in this text. We already did this for Euclidean plane geometry in Chapter 3, and in this chapter we will construct models for hyperbolic geometry. The situation for hyperbolic geometry is less intuitive. We have a concrete idea of what a line in the Euclidean plane is, or a circle on a sphere, but so far, a hyperbolic line is a very abstract concept; in the diagrams that we usually draw, parallel lines are equidistant. This is one of the advantages of having a model: We can "visualize" the hyperbolic plane, making drawings that help us to understand the theorems of hyperbolic geometry.

There are several models for hyperbolic geometry, although all of them are *equivalent*, or *isomorphic*, that is to say: If M_1 and M_2 are two models, then there is a one-to-one correspondence $T : M_1 \to M_2$ such that if P lies on line l, then $T(P)$ lies on line $T(l)$.

Exercises

Use the model described in Example 2 for the following set of exercises.

1. We say that two great circles are *perpendicular* if their tangent lines at the point of intersection are perpendicular lines. Give an example to show that the following result of neutral geometry does not hold in spherical geometry: Given a "line" l and a point $P \notin l$, the perpendicular dropped from P to l is unique.

2. In neutral geometry, the existence of parallel lines and the uniqueness of the perpendicular is guaranteed by the *alternate interior angle theorem.* Review its proof and explain why such a theorem cannot be proven in spherical geometry.

3. Give an example to show that the exterior angle theorem does not hold in spherical geometry.

4. Let $\mathcal{S}$ be a sphere and define its "lines" to be the great circles and consider two nonantipodal points on the sphere, A and B. The "line segment" $\overset{\frown}{AB}$ is the minor arc of the great circle through A and B (recall that if A and B are antipodal points, then there are infinitely many of them, which are semicircles). We then say that point $C \neq A, B$ *is between* A and B if $C \in \overset{\frown}{AB}$ (in the case that A

and B are antipodal points, C is one of the semicircles connecting them). Which betweenness axiom(s) of neutral geometry are not satisfied by the betweenness definition given above?

5. Let γ be a great circle of the sphere. We say that two points A and B are on the same side of γ if $\overset{\frown}{AB}$ does not intersect γ.
(a) Show that the definition above satisfies betweenness axiom $\mathbf{B_4}$.
(b) Prove the *plane separation property* for spherical geometry.

6. Let us say that C is between points A and B if C is in one arc of the great circle containing A to B. Which betweenness axiom(s) of neutral geometry are not satisfied by such a definition of betweenness?

7. We say that a set $\mathcal{S}$ of the sphere is convex if for all $A, B \in \mathcal{S}$ there exists an arc of great circle γ from A to B that is contained in $\mathcal{S}$. Give an example showing that if $\mathcal{S}_1$ and $\mathcal{S}_2$ are convex, then $\mathcal{S}_1 \cap \mathcal{S}_2$ is not necessarily convex.

9.2 The Poincaré Models

The French mathematician H. Poincaré (1854–1912) constructed two models for the hyperbolic plane that we study in this section; they were called the *disk model* and the *upper half-plane.* His models and their generalizations for higher dimensions are still used for research in hyperbolic geometry.

In the rest of this text, the hyperbolic plane will be denoted by $\mathbf{H}^2$ and "lines" in this plane will be called *geodesics.* In the next chapter we will define the hyperbolic length and show that geodesics are the curves of minimal length. Since the models we construct will be subsets of the Euclidean plane, in order to avoid confusion we reserve the words line, line segment, and collinear only for the appropriate *Euclidean* notions.

The Disk Model Let S^1 denote a circle of radius 1 and D be the set of points inside S^1. In this model, $\mathbf{H}^2 = D$ as a set and circle S^1 is called *circle at infinity* (this name will be justified in the next chapter). The *geodesics* of this model are diameters of S^1 (without the end points) and arcs of circles in D that meet S^1 orthogonally. Since diameters are perpendicular to tangent lines, they can be thought of as " circles" centered at ∞ that meet S^1 orthogonally. The incidence relation between a point and a geodesic is obviously defined.

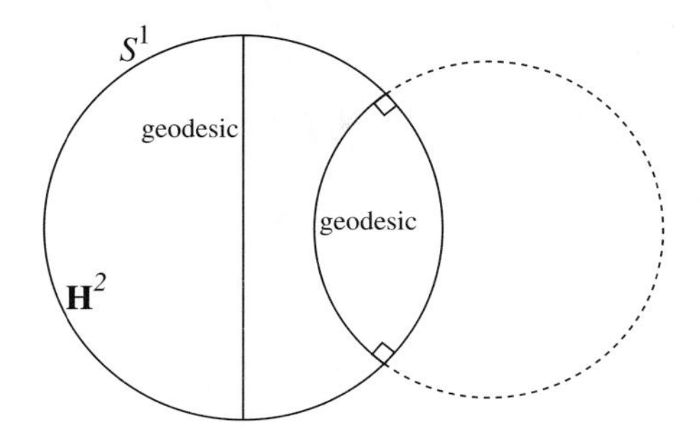

Figure 9.2

Proposition 9.2.1 *Given $A, B \in D$, there exists only one geodesic γ through them.*

Proof Let O denote the center of S^1. If O, A, and B are collinear, then there is a diameter through them. If not, Exercise 5 of Section 3.3 shows that there is a unique circle through P and Q that is orthogonal to S^1. □

Definition 9.2.2 *We say that point $C \neq A, B$ is between A and B if C lies on the arc of geodesic from A to B.*

In Exercises 1 and 2 of this section the reader is asked to show that the notion above satisfies all betweenness axioms of neutral geometry. Consequently, segments, rays, half-planes, convex sets, and so on can be defined exactly as in Chapter 1. In order to discuss congruence we need to define length and angle measure.

In the following, the geodesic segment from A to B will be denoted by γ_{AB}. We denote the ray emanating from A and containing B by $\gamma_{\overrightarrow{AB}}$.

Definition 9.2.3 *An angle $\angle_H BAC$ is the union of (geodesic) rays $\gamma_{\overrightarrow{AB}}$ and $\gamma_{\overrightarrow{AC}}$, which are called the sides of the angle; point A is called the*

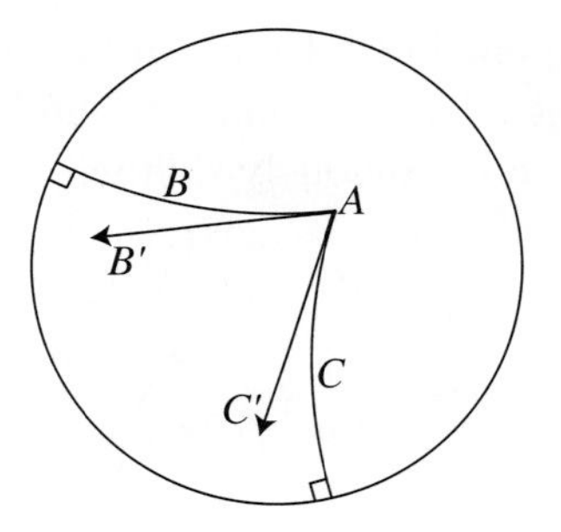

Figure 9.3 Hyperbolic angle

vertex of the angle. Its measure is given by

$$m(\angle_H BAC) = m(\angle B'AC'),$$

where $\overrightarrow{AB'}$ *and* $\overrightarrow{AC'}$ *are vectors tangent to* $\gamma_{\overrightarrow{AB}}$ *and* $\gamma_{\overrightarrow{AC}}$*, respectively, at point* A (see Figure 9.3).

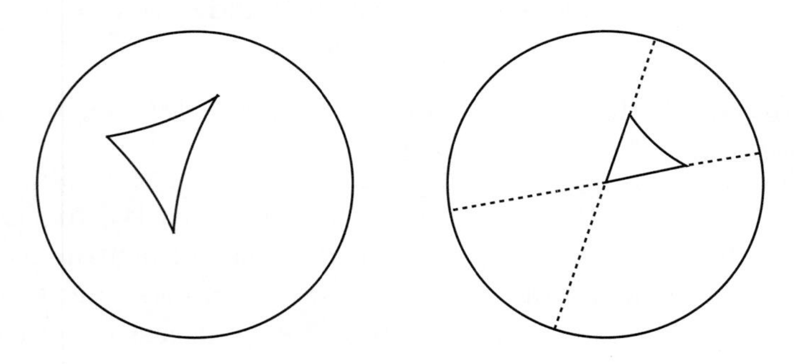

Figure 9.4

Definition 9.2.4 *A* hyperbolic triangle *of vertices* $A, B,$ *and* C *is the geometric figure containing these points, three sides, which are segments of geodesics joining the pairs of vertices, and three angles* (see Figure 9.4).

The concept of length is more complicated. If we used the Euclidean length, the length of all geodesics would be finite. Since geodesics represent "lines," in order to have Axiom $\mathbf{S}_2$ of Chapter 1 we need a one-to-one correspondence f between a geodesic γ and the set of real numbers $\mathbf{R}$ such that for $A, B \in \gamma$ we have

$$\text{Length}(\gamma_{AB}) = |f(A) - f(B)|.$$

Therefore, hyperbolic length has to be measured in a different way from the Euclidean measurement. We start by defining the transformations that will preserve the hyperbolic length.

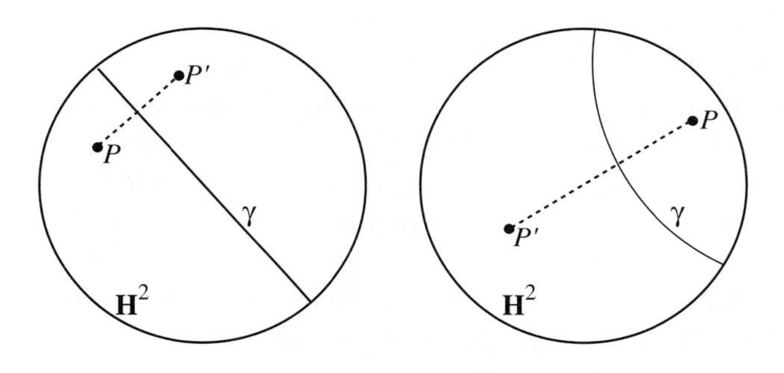

Figure 9.5

Definition 9.2.5 *Let γ be a geodesic of $D = \mathbf{H}^2$* (see Figure 9.5).
(i)If γ is a diameter, then a reflection *of the hyperbolic plane is a Euclidean reflection in γ.*
(ii)If γ is not a diameter, then a reflection *of the hyperbolic plane is an inversion in γ.*

In Exercise 3 of Section 3.3 you proved that if P is inside S^1 and P' is its image by an inversion in a circle orthogonal to S^1, then P' is also inside S^1. Then a hyperbolic reflection is well defined. We also saw that inversions share some of the properties of the reflections; for instance, if f is an inversion, $f^{-1} = f$. Recall that a Euclidean isometry is a composition of reflections. Analogously, we make the following definition.

Definition 9.2.6 *An* isometry *of the hyperbolic plane is any composition of hyperbolic reflections. An isometry is said to be* orientation-preserving *if it is the composition of an even number of reflections. Otherwise, we say that the isometry is* orientation-reversing.

Notice that a rotation about the center of S^1 through an angle θ is an isometry, for it is the composition of two successive reflections in diameters. Observe also that

$$G = \{g \mid g \text{ is an isometry of } \mathbf{H}^2\}$$

is a group where the group multiplication is composition of isometries.

The subgroup of orientation-preserving isometries is usually denoted by $\text{Aut}(\mathbf{H}^2)$. In the next chapter we will write explicitly all orientation-preserving isometries of $\mathbf{H}^2$ and show that such a group is isomorphic to a group of matrices. It will be shown that if two geodesics γ_1 and γ_2 meet at point A, making an angle θ, then $g(\gamma_1)$ and $g(\gamma_2)$ are geodesics and the angle they make at $g(A)$ also measures θ. Then we will show the existence of the map described below.

Theorem 9.2.7 *Given a geodesic γ, there exists a one-to-one correspondence $f : \gamma \to \mathbf{R}$ with the following properties:*
(i) $\text{Length}(\gamma_{AB}) = |f(A) - f(B)|$.
(ii) If g is an isometry of D, then $\text{Length}(\gamma_{AB}) = \text{Length}(g \circ \gamma_{AB})$.
(iii) If σ is any other curve from A to B, then $\text{Length}(\gamma_{AB}) \leq \text{Length}(\sigma)$.

With these notions of angle and length we define congruence of angles, segments, and triangles as in Chapter 1. These notions also allow us to show that this model satisfies the SAS axiom for congruence of triangles. For simplicity, $\text{Length}(\gamma_{AB})$ will be denoted by AB.

Proposition 9.2.8 (SAS): *Consider two triangles $\triangle ABC$ and $\triangle DEF$ such that $AB = DE$, $AC = DF$, and $\angle A \simeq \angle D$. Then $\triangle ABC \simeq \triangle DEF$.*

Proof First we claim that there is an isometry g that maps γ_{DE} to γ_{AB}. In fact, let g_1 be an isometry that maps D to the center O (see Exercise 7.3.3). Let $E' = g_1(E)$. Similarly, there is an isometry g_2 that maps A to O; let us denote $B' = g_2(B)$. Since $AB = DF$, a rotation g_3

centered at O carries $\overline{OE'}$ to $\overline{OB'}$. Therefore $g = g_2^{-1} \circ g_3 \circ g_1$ maps γ_{DE} to γ_{AB}. Let vector v be tangent to γ_{DF} at D. Since $\angle A \simeq \angle D$, $g(\gamma_{DF})$ is a segment of geodesic starting at A, and g has the angle-preserving property, from Exercise 3 we conclude that $g(\gamma_{DF}) = \gamma_{AC}$.

Then we have $g(A) = D, g(B) = E$, and $g(C) = F$. Since the geodesic through B and C is unique and g maps geodesics to geodesics, we get $g(\gamma_{EF}) = \gamma_{BC}$; the congruence of angles $\angle B$ and $\angle E$, and between $\angle C$ and $\angle F$, follows from the fact that g preserves angles. □

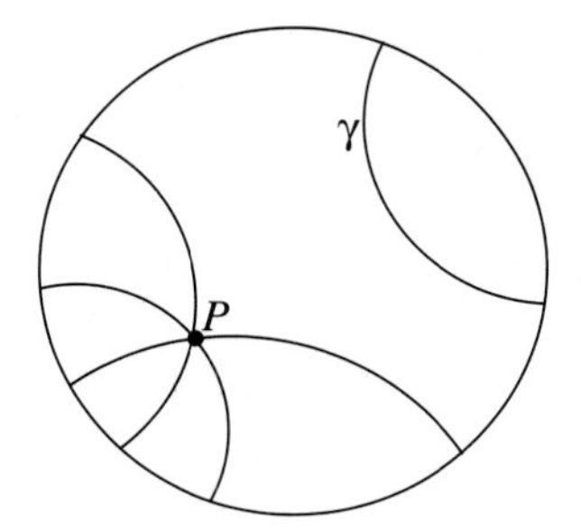

Figure 9.6 The hyperbolic parallel postulate

We then conclude that all results whose proofs depended upon the axioms of neutral geometry hold in this model. Having in mind that parallel lines are geodesics with no common point, we can show the hyperbolic parallel postulate.

Proposition 9.2.9 *There exits a geodesic γ such that for one point P not on l at least two geodesics parallel to γ pass through P.*

The upper half-plane model

The hyperbolic plane $\mathbf{H}^2$ in this model is the set

$$H = \{(x, y) \in \mathbf{R}^2 \mid y > 0\}.$$

The geodesics are curves of two types (see Figure 9.7):
(i) Rays emanating from points on the x-axis and perpendicular to the x-axis.

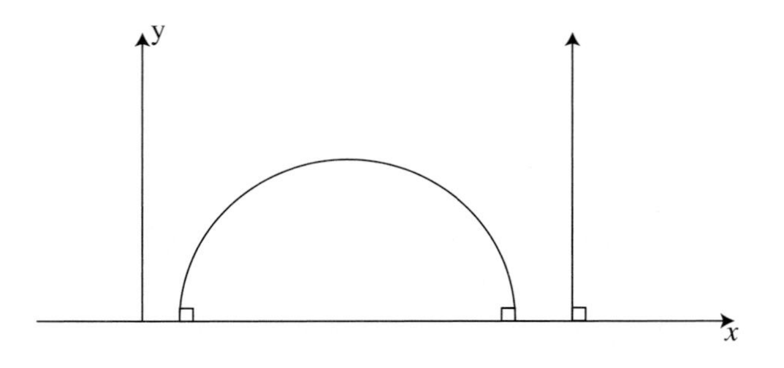

Figure 9.7 The upper half-plane model

(ii) Arcs of circles meeting the x-axis in right angles.

Incidence and betweenness are defined in the usual way. Angles are measured by their tangent vectors and, in the next chapter we will define length of geodesics. Also in the next chapter we will show that the disk and the half-plane models are equivalent.

Exercises

1. Show that Definition 9.2.2 verifies all betweenness axioms of neutral geometry.

2. Define *half-plane* in the disk model and show the plane separation property.

3. Show that, given a vector v in D with initial point A, there is only one geodesic γ through A such that v is tangent to γ at A.
   Hint: First consider the case that the line through A and in the direction of v intersects S^1 at two antipodal points. If not, let A' denote the image of A by the inversion in circle S^1. Then use Propositions 1.6.4 and 1.6.6 of Chapter 1.

4. Show that the measurement of hyperbolic angles satisfies Axiom $\mathbf{A}_2$ of Chapter 1, i.e., given a geodesic γ, a point $P \in \gamma$, and a number $0 < x < 180$, there is only one ray $\sigma_{\overrightarrow{PQ}}$, with Q in one side of γ, such that the measure of the angle between γ and $\sigma_{\overrightarrow{PQ}}$ is x.
   Hint: Use Exercise 3.

5. Prove Proposition 9.2.9.

6. Show that, given P in D, there exists an isometry g of $\mathbf{H}^2$ such that $g(P) = O$, the center of S^1.
Hint: Use Exercise 1 of Section 3.3 to show that there exists an inversion that maps P to the center O of S^1 and leaves S^1 invariant. Such an inversion is an isometry of $D = \mathbf{H}^2$.

7. Let γ_{PQ} be a geodesic through points P and Q. Show that there is an isometry g of $\mathbf{H}^2$ that maps γ onto a segment of the diameter of S^1 that goes through $Q' = g(Q)$.
Hint: Consider the isometry g of Exercise 6 and let $Q' = g(Q)$. Since isometries map geodesics onto geodesics, the geodesic through P and Q is mapped to the only geodesic through O and Q'.

8. Let g be a rotation of D about its center O. Show that g is an orientation-preserving isometry of $\mathbf{H}^2$.

9. Show that given P and Q in D there exists an isometry g of $\mathbf{H}^2$ such that $g(P) = Q$.
Hint: Use Exercise 6 to conclude that there exist isometries g_1 and g_2 such that $g_1(P) = O$ and $g_2(Q) = O$.

10. Let l and m be lines in the hyperbolic plane that are critically parallel. Show that they meet in the disk model at a point on S^1. This point is called an *ideal point.* We then define a *generalized triangle* as a triangle with the property that one of its vertices is an ideal point.

11. Let $\triangle PQR$ be a generalized triangle in the disk model D, with P being an ideal point. Show that there exists a circle γ through P that meets S^1 and γ_{QR} orthogonally. The arc of γ inside S^1 is a geodesic in D, and it is called an *altitude* of the generalized triangle $\triangle PQR$.
Hint: Let P' be the image of P by an inversion in the circle through Q and R, and P'' the image of P' by an inversion in circle S^1. The only circle through P, P', and P'' is the desired circle. Justify the procedure.

12. Draw a figure in the disk model that illustrates:
(a) two limiting parallel rays.

(b) two parallel lines (geodesics) with a common perpendicular.
(c) a Lambert quadrilateral.
(d) a Saccheri quadrilateral.
(e) a generalized triangle.

13. Draw a figure in the disk model that illustrates:
    (a) Generalized Pasch's theorem.
    (b) Generalized exterior angle theorem.

14. Draw a figure in the upper half-plane model that illustrates:
    (a) two limiting parallel rays.
    (b) a generalized triangle.

15. Draw a figure in the upper half-plane model that illustrates:
    (a) Generalized Pasch's theorem.
    (b) Generalized exterior angle theorem.

9.3 A Note on Elliptic Geometry

Spherical Geometry is a "plane geometry" in which the plane is a sphere and lines are interpreted as the great circles. Example 2 of Section 9.1 showed us that such a geometry does not satisfy the incidence axioms of neutral geometry. Further, we saw that two great circles intersect each other in two antipodal points. In Exercises 1, 2, and 3 we point out some of the important results of neutral geometry that do not hold in spherical geometry. A reason for this is that for any pair of antipodal points there are infinitely many great circles through them. Because of this, F. Klein thought of a model for a "plane geometry" in which *the plane is the set of pairs of antipodal points of a sphere* and lines are "great circles," with their antipodal points identified. Such a "plane" cannot be seen in 3-dimensional space, and actually, one can prove that its notion of length is not induced by Euclidean length of curves in 3-space. This geometry is called *elliptic geometry.*

The elliptic plane satisfies the incidence axioms; lengths and angles can be measured, and hence the congruence axioms of neutral geometry can be stated for it. Of course, it still does not contain parallel lines; in fact, since two great circles intersect in two antipodal points, the "great circles" of the elliptic plane still have one common point. Therefore, the alternate interior angle theorem does not hold in this geometry. This is because it is impossible to define a relation of betweenness that satisfies the betweenness axioms. However, the Klein model satisfies axioms of another type, usually called *separation axioms*, which replace the betweenness axioms of neutral geometry. These axioms can be found in Appendix A of Greenberg.[1]

An axiom system for elliptic geometry contains the incidence, congruence, and separation axioms as well as the following postulate:

> **Elliptic Parallel Postulate** There exists a line l such that for one point $P \notin l$ there is no line through P parallel to l.

We finish this chapter by pointing out that spherical and elliptic geometry are two distinct geometries. We single out a crucial difference: Although some of the betweenness axioms are not verified in spherical geometry, one can still talk about the sides of a great circle and prove the plane separation property (Exercise 5) of Section 9.1. However, this is not true in elliptic geometry: When we identify antipodal points, we can no longer talk about sides of a line, for if γ is a great circle and A and B are on opposite sides of it, B', the antipodal of B, is on the same side of γ as A. But B and B' are the same point on the elliptic plane.

[1]M. J. Greenberg, *Euclidean and Non-Euclidean Geometries, Development and History*, W. H. Freeman and Co., New York, 1996.

Chapter 10

THE HYPERBOLIC METRIC

10.1 Introduction

In this chapter we show that a metric can be defined on the hyperbolic plane for which the geodesics are the curves of minimal length. We will also show the equivalence of the two Poincaré models described in Chapter 9. Such results appear naturally if we consider the disk and the upper half-plane as subsets of the complex plane and interpret the isometries as conformal maps of the plane.

In order to make this chapter self-contained, in the next three sections we review some notions and results of complex variable functions.

10.2 The Complex Plane

Let us consider $\mathbf{R}^2$, the set of ordered pairs (a, b) of real numbers with the usual vector addition

$$(a, b) + (c, d) = (a + c, b + d)$$

and scalar multiplication

$$c(a, b) = (ca, cb).$$

Definition 10.2.1 *The set* $\mathbf{C}$ *of the* complex numbers *is the vector space* $\mathbf{R}^2$*, where a multiplication is defined by*

$$(a, b)(c, d) = (ac - bd, bc + ad).$$

Notice that for numbers of the type $(c, 0)$, the multiplication defined above is just the scalar multiplication. So, we will write c for the complex number $(c, 0)$. Let i denote the complex number $(0, 1)$. We then have

$$(0,1)(0,1) = (-1,0) = -1.$$

Moreover,

$$a + ib = (a,0) + (0,1)(b,0) = (a,0) + (0,b) = (a,b).$$

From now on we use the notation $z = a + ib$ for a complex number. The numbers a and b are called the *real part* and *imaginary part* of z, respectively. Denoting $a = \mathrm{Re}(z)$ and $b = \mathrm{Im}(z)$ we can write

$$z = \mathrm{Re}(z) + i\mathrm{Im}(z).$$

A complex number of the type $z = ib$, $b \in \mathbf{R}$ is called *purely imaginary*.

Definition 10.2.2 *If $z = a + ib$ is a complex number, its* conjugate *is defined as $\bar{z} = a - ib$. The* absolute value *of z is defined by $|z| = \left[z\,\bar{z}\right]^{1/2} = \sqrt{a^2 + b^2}$.*

It follows from the definition above that if $z = a + ib$ and $w = c + id$ are in the complex plane, then

$$|z - w| = [(z - w)\,\overline{(z - w)}]^{1/2} = \sqrt{(c-a)^2 + (d-b)^2} = ||z - w||,$$

which is the distance between z and w in the Euclidean metric.

Proposition 10.2.3 *Let z and w be complex numbers. Then:*
(a) $\overline{z + w} = \bar{z} + \bar{w}$
(b) $\overline{z\,w} = \bar{z}\,\bar{w}$
(c) $|z| = |\bar{z}|$
(d) $|z\,w| = |z|\,|w|$
(e) $|z + w|^2 = |z|^2 + 2\mathrm{Re}(z\bar{w}) + |w|^2$
(f) $|z - w|^2 = |z|^2 - 2\mathrm{Re}(z\bar{w}) + |w|^2$
(g) $|z + w| \le |z| + |w|$.

Proof We prove only (g); the other parts are left to the reader. For any complex number z we have

$$-|z| \le \mathrm{Re}(z) \le |z|$$

and then

$$\mathrm{Re}(z\bar{w}) \le |z\bar{w}| = |z||\bar{w}| = |z||w|.$$

Substituting in (e), we obtain

$$|z+w|^2 \le |z|^2 + 2|z||w| + |w|^2 = (|z|+|w|)^2,$$

which implies (g). □

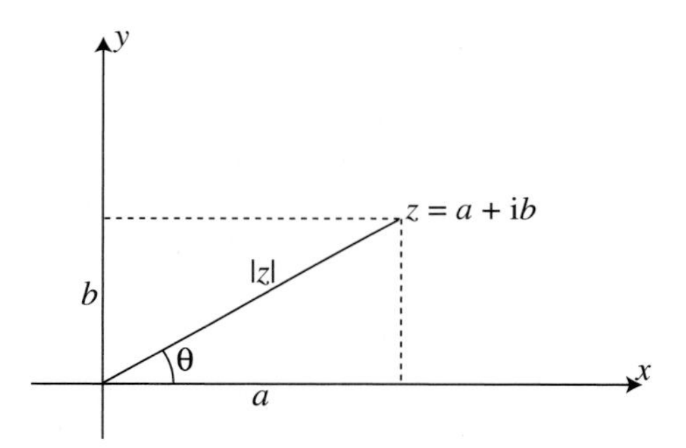

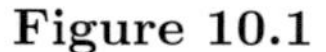

Figure 10.1

Recall that a point (x, y) in $\mathbf{R}^2$ has polar coordinates (r, θ) given by

$$x = r\cos\theta \quad \text{and} \quad y = r\sin\theta.$$

If $z = x+iy$, then $|z| = r$ and θ is the measure of the angle (measured in radians) between the positive real axis and the segment from the origin to z; θ is called an *argument* of z and denoted $\theta = \arg(z)$. Then we write

$$z = x + iy = |z|(\cos\theta + i\sin\theta).$$

Notice that the argument of 0 cannot be defined, since for any value of θ we can write $0 = 0(\cos\theta + i\sin\theta)$.

Proposition 10.2.4 *Let* $z_1 = |z_1|(\cos\theta_1 + i\sin\theta_1)$ *and* $z_2 = |z_2|(\cos\theta_2 + i\sin\theta_2)$ *be complex numbers. Then:*
(a) $1/z_1 = (1/|z_1|)(\cos\theta_1 - i\sin\theta_1)$.
(b) $z_1z_2 = |z_1||z_2|(\cos(\theta_1+\theta_2) + i\sin(\theta_1+\theta_2))$.
(c) $z_1/z_2 = |z_1|/|z_2|(\cos(\theta_1-\theta_2) + i\sin(\theta_1-\theta_2))$.
(d) if $z_1, z_2 \ne 0$ *then* $\arg(z_1z_2) = \arg(z_1) + \arg(z_2)$ *and* $\arg(z_1/z_2) = \arg(z_1) - \arg(z_2)$.

Proof We write $\bar{z_1} = |z_1|(\cos\theta_1 - i\sin\theta_1)$. Since $z_1\bar{z_1} = |z_1|^2$, we obtain

$$\frac{\bar{z_1}}{|z_1|^2} = \frac{1}{z_1},$$

which substituted above, gives (a).

We leave part (b) as an exercise. Now, for (c), first we apply (a) to z_2. Thus

$$\frac{1}{z_2} = \frac{1}{|z_2|}(\cos\theta_2 - i\sin\theta_2) = \frac{1}{|z_2|}[\cos(-\theta_2) + i\sin(-\theta_2)].$$

Then, applying (b) to the product $z_1(1/z_2)$, we obtain (c).

Part (d) follows from (b) and (c). We point out here that if one of them is 0, then no result can be stated on the argument of $0w$, since the argument of 0 is not defined. □

We finish this section with a lemma that will be an essential tool in Section 10.2. The proof is straightforward.

Lemma 10.2.5 *Let $J : \mathbf{R}^2 \to \mathbf{R}^2$ be the linear map given by*

$$J(x, y) = (-y, x),$$

that is, J is the counterclockwise rotation about the origin through the angle $\pi/2$. Then
(a) $J(z) = iz$, where $z = x + iy$.
(b) For any two complex numbers $z = a + bi$ and $w = c + id$ we have

$$z\,\bar{w} = z \cdot w + iz \cdot J(w),$$

where "$\cdot$" denotes the dot product of $\mathbf{R}^2$.

The map J defined in Lemma 10.2.5 is called the *complex structure* of $\mathbf{R}^2$.

Exercises

1. Write the complex numbers below as points of $\mathbf{R}^2$.
   (a) $3(1 - i) + 2(2 - 5i)$.
   (b) $(1 - i)(3 - 2i) + (2 + i)$.

2. Write the complex numbers below in polar coordinates:
   (a) $\sqrt{2} - \sqrt{2}i$.
   (b) $(3 - i)(3 + i) - 4i$.

3. Express the fractions below in both, cartesian and polar form.
   (a) $\dfrac{2}{1-i}$.
   (b) $\dfrac{3-4i}{3-2i}$.
   Hint: Rationalize the denominator for the Cartesian form.

10.3 Analytic Functions

Definition 10.3.1 *An* open ball *of center z_0 and radius $r \in \mathbf{R}, r > 0$, is the set*

$$B_r(z_0) = \{z \in \mathbf{C} \mid |z - z_0| < \epsilon\}.$$

In this section Ω will denote a subset of the plane $\mathbf{C}$ with the property that for $z_0 \in \Omega$ there exists a positive real number ϵ such that the *open ball* $B_\epsilon(z_0) \subset \Omega$. Subsets with this property are called *open sets.*

Definition 10.3.2 *Let Ω be as above. Then a* complex function *defined on Ω is a function which for each $z \in \Omega$ assigns a complex number w.*

In the rest of this chapter, to indicate a complex function of a complex variable we use the notation $w = f(z)$. Observe that if $z = x + iy$, then $w = u(x, y) + iv(x, y)$, where u and v are real functions.

Example 1

Let $f(z) = z^2$. Then

$$f(z) = (x + iy)^2 = x^2 - y^2 \ + \ i2xy$$

and hence

$$u(x, y) = x^2 - y^2 \quad \text{and} \quad v(x, y) = 2xy.$$

Definition 10.3.3 *Given a complex $f : \Omega \subset \mathbf{C} \to \mathbf{C}$ and $z_0 \in \Omega$, we say that*

$$\lim_{z \to z_0} f(z) = w_0$$

if for every positive real number ϵ there exists a positive real number δ such that

$$|f(z) - w_0| < \epsilon \quad \text{whenever} \quad |z - z_0| < \delta.$$

Lemma 10.3.4 *Let $z_0 = x_0 + iy_0$, $f(z) = u(x, y) + iv(x, y)$, and $w_0 = u_0 + iv_0$. Then*

$$\lim_{z \to z_0} f(z) = w_0$$

if and only if

$$\lim_{(x,y) \to (x_0, y_0)} u((x, y)) = u_0 \quad \text{and} \quad \lim_{(x,y) \to (x_0, y_0)} v((x, y)) = v_0.$$

Proof Recall that

$$\begin{aligned} |f(z) - w_0| &= ||f(z) - w_0|| \\ &= ||\Big(u(x, y), v(x, y)\Big) - (u_0, v_0)|| \\ &= \Big((u(x, y) - u_0)^2 + (v(x, y) - v_0)^2\Big)^{1/2}, \end{aligned}$$

and thus

$$|f(z) - w_0| \geq |u(x, y) - u_0| \quad \text{and} \quad |f(z) - w_0| \geq |v(x, y) - v_0|.$$

Therefore if $|f(z) - w_0| < \epsilon$, then

$$|u(x, y) - u_0| < \epsilon \quad \text{and} \quad |v(x, y) - v_0| < \epsilon.$$

Since $|z - z_0| = ||(x, y) - (x_0, y_0)||$, we obtain that $\lim_{z \to z_0} f(z) = w_0$ implies

$$\lim_{(x,y) \to (x_0, y_0)} u((x, y)) = u_0 \quad \text{and} \quad \lim_{(x,y) \to (x_0, y_0)} v((x, y)) = v_0.$$

□

We leave the proof of the converse to the reader, since it uses the same arguments.

Definition 10.3.5 *Let Ω be an open set containing point z_0. A function $f : \Omega \subset \mathbf{C} \to \mathbf{C}$ is said to be* continuous *at z_0 if*

$$\lim_{z \to z_0} f(z) = f(z_0),$$

that is, for every $\epsilon > 0$ there exists $\delta > 0$ such that

$$|f(z) - f(z_0)| < \epsilon \quad \text{whenever} \quad |z - z_0| < \delta.$$

Observe that Lemma 10.3.4 implies that the following result.

Proposition 10.3.6 *A function $f : \Omega \subset \mathbf{C} \to \mathbf{C}$ is continuous at $z_0 = x_0 + iy_0$ if and only if the functions*

$$u : \mathbf{R}^2 \to \mathbf{R} \quad \text{and} \quad v : \mathbf{R}^2 \to \mathbf{R}$$

are continuous at (x_0, y_0).

There is, however, a fundamental difference between a complex function $w = f(z)$ and the corresponding functions $u(x, y)$ and $v(x, y)$ with respect to their differentiability. Before we illustrate this point, we define differentiable functions.

Definition 10.3.7 *Let Ω be an open set of $\mathbf{C}$ containing point z_0. We say that $f : \Omega \to \mathbf{C}$ is differentiable at z_0 if*

$$\lim_{h \to 0} \frac{f(z_0 + h) - f(z_0)}{h}$$

exists. The value of the limit is denoted by $f'(z_0)$.

Example 2

Consider the function $f(z) = |z|^2$. Then $f(z_0 + h) - f(z_0) = |z_0 + h|^2 - |z_0|^2$ is a real number. Therefore, if $h \in \mathbf{R}$, the quotient

$$\frac{f(z_0 + h) - f(z_0)}{h}$$

is real. If the limit exists, and as $h \to 0$ through real numbers, the limit must be a real number. Now, if $h = ik, k \in \mathbf{R}$, the quotient

$$\frac{f(z_0 + h) - f(z_0)}{h}$$

is purely imaginary, and as $k \to 0$, the limit must be purely imaginary. Therefore, the only possible value for $f'(z_0) = 0$. Computing the quotient above for the function $f(z) = |z|^2$, we have

$$\begin{aligned} \frac{f(z_0 + h) - f(z_0)}{h} &= \frac{(z_0 + h)\,(\overline{z_0 + h}) - z_0 \bar{z}_0}{h} \\ &= z_0 + \bar{h} + z_0 \frac{\bar{h}}{h}. \end{aligned}$$

Notice that the limit will be zero if and only if $z_0 = 0$. Therefore, the function $f(z) = |z|^2$ is differentiable only at $z_0 = 0$. However, if we write $f(z) = |z|^2 = x^2 + y^2$, we have

$$u(x, y) = x^2 + y^2 \quad \text{and} \quad v(x, y) = 0,$$

which are differentiable at all points of $\mathbf{R}^2$.

In Example 2 we used the particular function $|z|^2$ as an example of a more general fact. Notice that the same reasoning implies that a real function of a complex variable, that is, $f(z) \in \mathbf{R}$, $\forall z \in \Omega$, either has the derivative zero, or the derivative does not exist.

The case of a complex function of a real variable is simpler, and in fact one can prove the following result.

Proposition 10.3.8 *Let (a, b) be an open interval of real numbers and $z : (a, b) \to \mathbf{C}$ a curve that we write as $z(t) = x(t) + iy(t)$. Then z is differentiable if and only if x and y are differentiable. Further,*

$$z'(t) = x'(t) + iy'(t).$$

Proof If z is differentiable, then

$$\begin{aligned} z'(t) &= \lim_{h \mapsto 0} \frac{z(t+h) - z(t)}{h} \\ &= \lim_{h \mapsto 0} \frac{x(t+h) + iy(t+h) - x(t) - iy(t)}{h}. \end{aligned}$$

It follows from the definition of limit that, for any $\epsilon > 0$, there exists $\delta > 0$ such that, if $|h| < \delta$, then

$$|z'(t) - \frac{x(t+h) + iy(t+h) - x(t) - iy(t)}{h}| < \epsilon,$$

which is equivalent to

$$||z'(t) - \frac{(x(t+h) - x(t), y(t+h) - y(t))}{h}|| < \epsilon.$$

Therefore, repeating the arguments used in the proof of Lemma 10.3.4, we have

$$||\mathrm{Re}(z'(t)) - \frac{x(t+h) - x(t)}{h}|| < \epsilon,$$

$$||\mathrm{Im}(z'(t)) - \frac{y(t+h) - y(t)}{h}|| < \epsilon,$$

implying

$$\mathrm{Re}(z'(t)) = \lim_{h \to 0} \frac{x(t+h) - x(t)}{h} = x'(t),$$

$$\mathrm{Im}(z'(t)) = \lim_{h \to 0} \frac{y(t) + h) - y(z_0)}{h} = y'(t).$$

Therefore, $z'(t) = \mathrm{Re}\,(z'(t)) + i\,\mathrm{Im}\,(z'(t)) = x'(t) + iy'(t)$. □

The study of the general case, that is, complex functions of complex variables, leads us to an important class of functions that are fundamental in complex analysis and mathematics. They are the *analytic functions.* Before we define them, we observe that the differentiabilty of a real function f does not imply that the function f' is continuous; the reader can verify this fact using, for example, the function $f(x) = x^2 \sin(1/x)$. However, a remarkable result of complex variable functions is that *every differentiable function is infinitely differentiable and has a power-series expansion about each point of its domain.* We will not give a proof of this result here, since it goes beyond the scope of the text. We quoted it only to justify the use of the word *analytic* for differentiable functions.

Definition 10.3.9 *A function $f : \Omega \subset \mathbf{C} \to \mathbf{C}$ is said to be* analytic *on Ω if it is differentiable at all points $z \in \Omega$.*

Proposition 10.3.10 *Let $f : \Omega \subset \mathbf{C} \to \mathbf{C}$ be an analytic function. Then for all $(x, y) \in \Omega$ we have*

$$\frac{\partial u}{\partial x} = \frac{\partial v}{\partial y} \quad \text{and} \quad \frac{\partial u}{\partial y} = -\frac{\partial v}{\partial x}$$

and

$$f'(z) = \frac{\partial u}{\partial x} + i\frac{\partial v}{\partial x} = \frac{\partial v}{\partial y} - i\frac{\partial u}{\partial y}.$$

Proof Consider the limit

$$\lim_{h \to 0} \frac{f(z_0 + h) - f(z_0)}{h}.$$

As $h \to 0$ through real numbers, we have

$$\lim_{h\to 0}\frac{f(x+h+iy)-f(x+iy)}{h} = \lim_{h\to 0}\Big[\frac{u(x+h,y)-u(x,y)}{h} + i\frac{v(x+h,y)-v(x,y)}{h}\Big],$$

which gives

$$f'(z) = \frac{\partial u}{\partial x}(x,y) + i\frac{\partial v}{\partial x}(x,y).$$

Now, as $h \to 0$, $h = ik$, $k \in \mathbf{R}$, that is, through purely imaginary numbers, we have

$$\lim_{k\to 0}\frac{f(x+ik+iy)-f(x+iy)}{ik} = \lim_{k\to 0}\frac{1}{i}\Big[\frac{u(x,k+y)-u(x,y)}{k} + i\frac{v(x,k+y)-v(x,y)}{k}\Big],$$

which implies

$$f'(z) = -i\frac{\partial u}{\partial y}(x,y) + \frac{\partial v}{\partial y}(x,y).$$

It follows that

$$\mathrm{Re}(f'(z)) = \frac{\partial u}{\partial x} = \frac{\partial v}{\partial y},$$

and

$$\mathrm{Im}(f'(z)) = -\frac{\partial u}{\partial y} = \frac{\partial v}{\partial x}.$$

□

The equations above involving the partial derivatives of u and v are called the *Cauchy-Riemann* equations. Notice that we proved that they are *necessary* conditions for the existence of $f'(z)$. For instance, for the function of Example 2 we have

$$u(x,y) = x^2 + y^2 \quad \text{and} \quad v(x,y) = 0.$$

Therefore

$$\frac{\partial u}{\partial x} = 2x, \quad \frac{\partial v}{\partial y} = 0, \quad \frac{\partial v}{\partial x} = 0, \quad \frac{\partial u}{\partial y} = 2y.$$

Observe that u and v satisfy the Cauchy-Riemann equations only when $x = 0$ and $y = 0$. It follows that f cannot be differentiable at points

$z \neq 0$, a fact that we already knew from the definition of derivative. However, Proposition 10.3.10 does not guarantee that f is differentiable at $z = 0$.

Actually, the converse of Proposition 10.3.10 is also true. This fact is stated in the next proposition.

Proposition 10.3.11 *Let Ω be an open set of the complex plane and u and v real functions defined on Ω. If u and v satisfy the Cauchy Riemann equations for all $z \in \Omega$, then $f(z) = u(x,y) + iv(x,y)$, where $z = x + iy$, is* analytic *on Ω.*

Example 3

Consider $f(z) = z^2$. Then we have

$$f(z) = (x + iy)\,(x + iy) = x^2 - y^2 + i2xy,$$

that is, $u(x,y) = x^2 - y^2$ and $v(x,y) = 2xy$, and thus

$$\frac{\partial u}{\partial x} = 2x = \frac{\partial v}{\partial y}, \quad \frac{\partial u}{\partial y} = -2y = -\frac{\partial v}{\partial x}.$$

Let us now consider a smooth curve in **C**, that is, $z(t) = x(t) + iy(t),\ \ t \in I$, where I is an open interval of real numbers. Let $w(t) = f(z(t))$, where f is an analytic function. It follows from the fact that analytic functions are infinitely differentiable that $w(t)$ is also a smooth curve. We then prove the following lemma.

Lemma 10.3.12 *Let $w(t)$ be the smooth curve given by $w(t) = f(z(t))$, where f is analytic and $z(t)$ is smooth. Then*

$$w'(t) = f'(z(t))\, z'(t).$$

Proof We write $f(z) = u(x,y) + iv(x,y)$. Then $w(t) = u(x(t), y(t)) + iv(x(t), y(t))$. Using Proposition 10.3.8, we have

$$w'(t) = \frac{d(u(x(t), y(t)))}{dt} + i\frac{d(v(x(t), y(t)))}{dt}.$$

From the chain rule for real functions we obtain

$$\frac{d(u(x(t), y(t)))}{dt} = \frac{\partial u}{\partial x} x'(t) + \frac{\partial u}{\partial y} y'(t),$$

$$\frac{d(v(x(t), y(t)))}{dt} = \frac{\partial v}{\partial x}x'(t) + \frac{\partial v}{\partial y}y'(t).$$

On the other hand, using Lemma 10.2.5(b) we have

$$f'(z(t))\, z'(t) = f'(z) \cdot \bar{z}'(t) + if'(z) \cdot J(\bar{z}'(t)),$$

and since f is analytic,

$$f'(z) = \frac{\partial u}{\partial x} + i\frac{\partial v}{\partial x},$$

yielding

$$f'(z(t))\, z'(t) = \frac{\partial u}{\partial x}x'(t) - \frac{\partial v}{\partial x}y'(t) + i[\frac{\partial u}{\partial x}y'(t) + \frac{\partial v}{\partial x}x'(t)].$$

Now, the Cauchy-Riemann equations imply

$$f'(z(t))\, z'(t) = \frac{\partial u}{\partial x}x'(t) + \frac{\partial u}{\partial y}y'(t) + i[\frac{\partial v}{\partial y}y'(t) + \frac{\partial v}{\partial x}x'(t)] = w'(t).$$

□

Using the lemma above, we prove the following geometric result for analytic functions.

Proposition 10.3.13 *Let $z_1(t)$ and $z_2(t)$ be two smooth curves in* **C** *which intersect at point $z_0 = z_1(t_1) = z_2(t_2)$. Let θ denote the measure of the angle between $z_1'(t_1)$ and $z_2'(t_2)$. Let $w = f(z)$ be an analytic function defined on an open ball of* **C** *containing z_0 such that $f'(z_0) \neq 0$. Then the measure of the angle between $w_1'(t_1)$ and $w_2'(t_2)$ is θ.*

Proof Here we look at the tangent vectors as complex numbers. Writing $z_1'(t_1)$ and $z_2'(t_2)$ in polar coordinates, we have

$$\theta = |\arg(z_1'(t_1)) - \arg(z_2'(t_2))|.$$

From Lemma 10.3.12 and from the fact that $f'(z_0) \neq 0$ we conclude that ϕ, the measure of the angle between $w_1'(t_1)$ and $w_2'(t_2)$, is given by

$$\phi = |\arg(f'(z_0)z_1'(t_1)) - \arg(f'(z_0)z_2'(t_2))|.$$

Applying Proposition 10.2.4(d), we obtain

$$\phi = |\arg(f'(z_0)) + \arg(z_1'(t_1)) - \arg(f'(z_0)) - \arg(z_2'(t_2))| = \theta.$$

□

Corollary 10.3.14 *Let Ω be an open set of the complex plane* $\mathbf{C}$. *Then an analytic function $f : \Omega \to \mathbf{C}$ such that $f'(z) \neq 0$ for all $z \in \Omega$ is a conformal map.*

Exercises

1. Show that each function below is analytic on $\mathbf{C}$.
   (a) $f(z) = 3x + y + (3y - x)i$.
   (b) $f(z) = e^{-y}(\cos x + i \sin x)$.
   (c) $f(z) = (z^2 - 2)e^{-x}(\cos y - i \sin y)$.

2. Explain why the functions below are not analytic at any point of the complex plane.
   (a) $f(z) = xy + iy$.
   (b) $f(z) = e^{y}(\cos x + i \sin x)$.

10.4 Geometric Transformations

In this section we study some complex variable functions that are transformations of the plane.

Proposition 10.4.1 *A complex linear function $f(z) = az + b$, a, $b \in$ $\mathbf{C}$, maps circles onto circles, lines onto lines, and preserves angle measure.*

Proof Let $z = |z|(\cos\theta + i\sin\theta)$ and $a = |a|(\cos\varphi + i\sin\varphi)$. Then

$$az = |z||a|(\cos(\theta + \varphi) + i\sin(\theta + \varphi)),$$

which is a rotation about $z = 0$ through the angle φ followed by the dilation (or contraction) of ratio $|a|$. Therefore the map $z \mapsto az$ is a similarity and the map $z \mapsto az + b$ is a similarity composed with a translation. The result then follows. □

Proposition 10.4.2 *Let S^1 denote the circle centered at $z = 0$ and radius 1. The function $f(z) = z^{-1}$ is an inversion in S^1 composed with a reflection in the real line. Therefore it maps circles (not passing through the origin) onto circles and lines onto punctured circles. Further, f has the angle-preserving property for any $z \neq 0$.*

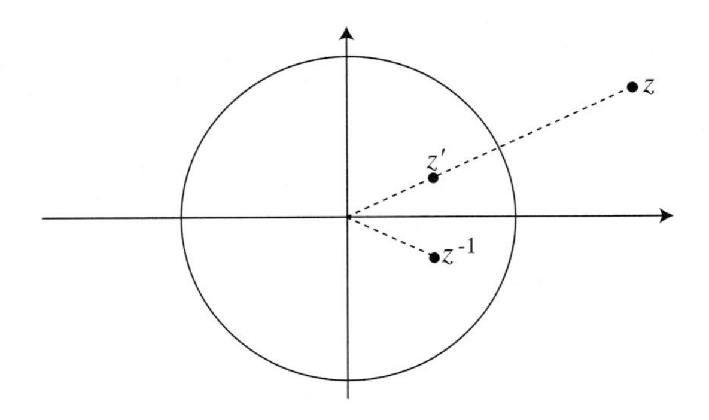

Figure 10.2

Proof Let $z = |z|(\cos\theta + i\sin\theta)$ and consider $z' = |z|^{-1}(\cos\theta + i\sin\theta)$. Then

$$|z'|^2 = z'\bar{z'} = |z|^{-1}(\cos\theta + i\sin\theta)|z|^{-1}(\cos\theta - i\sin\theta) = \frac{1}{|z|^2}.$$

Therefore, $|z'|^2|z|^2 = 1$, that is, $|z'|\,|z| = 1$, and the map $z \mapsto z'$ is an inversion in S^1. Since $f(z) = \bar{z'}$, we conclude that f is the inversion followed by reflection in the real line.

For the second part of the proposition, we write $f(z)$ in the coordinates x, y. We have

$$u(x,y) \;+\; iv(x,y) = \frac{1}{x+iy} = \frac{x-iy}{x^2+y^2},$$

which implies

$$u(x,y) = \frac{x}{x^2+y^2} \quad \text{and} \quad v(x,y) = \frac{-y}{x^2+y^2}.$$

We easily see that u and v verify the Cauchy-Riemann equations for $z \neq 0$. Therefore f is an analytic function on the region $z \neq 0$. Moreover,

$$f'(z) = \frac{y^2-x^2}{(x^2+y^2)^2} + i\frac{2xy}{(x^2+y^2)^2} \neq 0, \quad \text{for } z \neq 0,$$

and hence preserves angle measure by Proposition 10.3.13. □

The proof above has shown that the map $z \mapsto 1/\bar{z}$ describes the inversion in circle S^1. The same arguments show that an inversion in a circle of radius μ and centered at $(0,0)$ is given by $z \mapsto \mu^2/\bar{z}$.

We now introduce an important class of complex functions; they will be used to study the isometries of the Poincaré models for the hyperbolic plane.

First, we consider the *extended complex plane*, which is $\mathbf{C} \cup \{\infty\} = \mathbf{C}_\infty$. Formally, this point ∞ is just the preimage of $w = 0$ by the transformation $w = 1/z$. A concrete way to see it is the following: identify the plane $\mathbf{C}$ with the sphere $S^2 - \{(0,0,1)\}$ in $\mathbf{R}^3$, using the stereographic projection, defined in Chapter 7; then identify the point ∞ with the north pole $(0,0,1)$. The extended plane $\mathbf{C}_\infty$ is represented then as the sphere S^2. The function $w = 1/z$ is defined for any $z \in \mathbf{C}_\infty$.

Definition 10.4.3 *A* Möbius transformation *is a complex function given by*

$$f(z) = \frac{az+b}{cz+d}, \quad a,b,c,d \in \mathbf{C}, \quad ad - bc \neq 0.$$

Remarks:

(i) The condition $ad - bc \neq 0$ is necessary to guarantee that f is not a constant function. In fact, suppose $ad = bc$:

(a) if $a = 0$: then $bc = 0$ and hence $b = 0$ or $c = 0$; the case $b = 0$ implies $f(z) = 0$ while $c = 0$ gives $f(z) = b/d$;

(b) if $a \neq 0$: then $d = bc/a$. We then have

$$f(z) = \frac{az+b}{cz+(bc/a)} = \frac{a(az+b)}{acz+bc} = \frac{a(az+b)}{c(az+b)} = \frac{a}{c}.$$

(ii) The condition $ad - bc \neq 0$ can be restated as $ad - bc = 1$. To see this, first notice that if α is a nonzero complex number, then

$$f(z) = \frac{az+b}{cz+d} = \frac{\alpha(\ az+\alpha b)}{\alpha\ (cz+\alpha d)}.$$

Therefore, if $\alpha = (ad - bc)^{-1}$, we write

$$f(z) = \frac{\tilde{a}z+\tilde{b}}{\tilde{c}z+\tilde{d}}, \quad \tilde{a}\tilde{d} - \tilde{b}\tilde{c} = 1,$$

where $\alpha a = \tilde{a}, \alpha b = \tilde{b}, \alpha c = \tilde{c}, \alpha d = \tilde{d}$.
(iii) f is invertible and if $w = f(z)$,

$$f^{-1}(w) = \frac{dw - b}{-cw + a}.$$

(iv) The composition of Möbius transformations is another Möbius transformation (the reader should verify this). This fact combined with (iii) implies that the set of Möbius transformations is a group whose operation is composition.
(v) A Möbius transformation is defined on the extended plane $\mathbf{C}_\infty$. In fact, consider

$$f(z) = \frac{az + b}{cz + d}.$$

If $c \neq 0$, f maps ∞ to a/c, that is,

$$f\Big(\frac{1}{z}\Big) = \frac{a(1/z) + b}{c(1/z) + d} = \frac{a + bz}{c + dz},$$

which tends to a/c as $z \to 0$. Likewise we show that f^{-1} maps ∞ to $-d/c$ and hence f maps $-d/c$ to ∞. The case $c = 0$ can be seen as the map $f : \mathbf{C}_\infty \to \mathbf{C}_\infty$ given by $f(z) = az + b$. It maps ∞ to ∞. To see this, just notice that

$$f\Big(\frac{1}{z}\Big) = \frac{a + bz}{z},$$

which approaches ∞ as $z \to 0$.

A *Möbius transformation* can be expressed using matrix notation. For that we write $z = z_1/z_2$, and if $A = \begin{pmatrix} a & b \\ c & d \end{pmatrix}$, we have

$$\begin{pmatrix} w_1 \\ w_2 \end{pmatrix} = \begin{pmatrix} a & b \\ c & d \end{pmatrix}\begin{pmatrix} z_1 \\ z_2 \end{pmatrix},$$

yielding

$$w_1 = az_1 + bz_2,$$

$$w_2 = cz_1 + dz_2.$$

Therefore

$$w = \frac{w_1}{w_2} = \frac{a\ z_1/z_2\ +\ b}{c\ z_1/z_2\ +\ d} = \frac{az + b}{cz + d} = f(z).$$

This notation will be used in the next section to identify the isometries of the hyperbolic plane with some types of matrices.

Proposition 10.4.4 *A Möbius transformation is the composition of rotations, dilations or contractions, translations, and at most one inversion.*

Proof If $c = 0$, then f is just a complex linear function. If $c \neq 0$, we write

$$w = \frac{az+b}{cz+d} = \frac{a(z+(d/c)) + b - (ad/c)}{c(z+(d/c))} = \frac{a}{c} + \frac{b-(ad/c)}{cz+d}.$$

Setting $z' = cz + d$ and $z'' = 1/z'$, we have

$$w = \frac{a}{c} + \frac{b-(ad/c)}{cz+d} = \frac{a}{c} + \frac{bc-ad}{cz'} = \frac{a}{c} + \frac{bc-ad}{c}z''.$$

Therefore the transformation f can be seen as the following composition

$$z \overset{f_1}{\longmapsto} z' = cz + d \overset{f_2}{\longmapsto} z'' = \frac{1}{z'} \overset{f_3}{\longmapsto} \frac{a}{c} + \frac{bc-ad}{c}z''.$$

Notice that f_1 and f_2 are linear complex functions and f_2 is an inversion composed with a reflection. □

Corollary 10.4.5 *Let $\Omega = \mathbf{C} - \{-d/c\}$. Then the Möbius transformation $f : \Omega \to \mathbf{C}$ given by*

$$f(z) = \frac{az+b}{cz+d}$$

is a conformal map on Ω.

Proof The proof of Proposition 10.4.4 shows that if $z \neq -d/c$, then f is the composition of analytic maps that have nonzero derivatives on Ω. The result now follows from Corollary 10.3.14. □

Now we look for a Möbius transformation that shows the equivalence between the two Poincaré models for the hyperbolic plane. We want the real line $y = 0$ to be mapped onto the circle at infinity S^1 and the set

$$H = \{(x, y) \in \mathbf{R}^2 \mid y > 0\} = \{z \in \mathbf{C} \mid \mathrm{Im}(z) > 0\}$$

to be carried onto the disk

$$D = \{(x, y) \in \mathbf{R}^2 \mid x^2 + y^2 < 1\} = \{z \in \mathbf{C} \mid |z| < 1\}.$$

Recall that inversions map lines onto punctured circles, i. e., circles that pass through the center of the inversion. This is because the center of the inversion (considering the circle of the inversion in $\mathbf{C}_\infty$) is mapped to ∞. This implies that if a Möbius transformation carries a line onto a circle, ∞ must be sent to a point on the circle (see Exercise 7).

Proposition 10.4.6 *Let z_1 be a complex number such that* $\mathrm{Im}(z_1) > 0$. *Then the Möbius transformation*

$$f(z) = \frac{z - z_1}{z - \bar{z_1}}$$

carries the upper half-plane $\{x + iy, y > 0\}$ to the interior of the circle S^1 and maps the real line $y = 0$ to S^1.

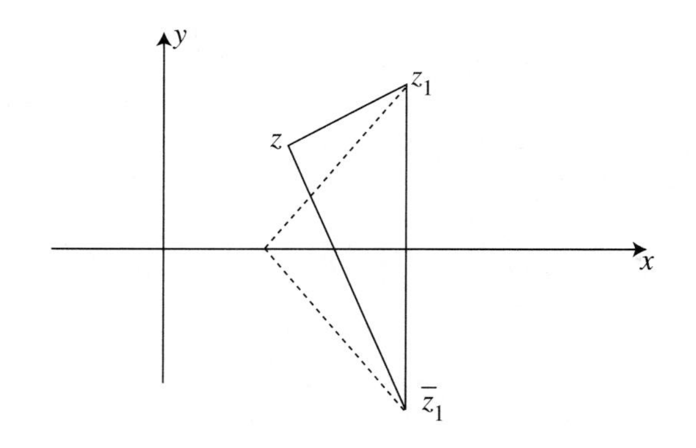

Figure 10.3

Proof Notice that $f(z_1) = 0$. Further, if $z \in \{x + yi, y \geq 0\}$, then $|z - z_1| \leq |z - \bar{z_1}|$ with equality if and only if z is on the real line. This fact implies:
(i) $f(H) \subset D$,
(ii) $f(\{x + yi \mid y = 0\}) \subset S^1$.
Let us consider the inverse f^{-1} of f. We have

$$f^{-1}(w) = \frac{-\bar{z_1}w + z_1}{-w + 1} = \frac{\bar{z_1}w - z_1}{w - 1}.$$

Observe that f^{-1} is defined for every $z \neq 1$. Therefore $f(H) = D$; that is, every $w \in D$ is the image of $z = f^{-1}(w)$ in the upper half-plane by f. We also conclude that every $w \neq 1$ such that $|w| = 1$ is the image of $z = f^{-1}(w)$ for z on the real line. We show now that $w = 1$ is the image of ∞. In fact,

$$f\left(\frac{1}{z}\right) = \frac{1/z - z_1}{1/z - \bar{z_1}} = \frac{1 - z_1 z}{1 - \bar{z_1} z},$$

which goes to 1 as $z \mapsto 0$. Therefore $f(\{x + yi \mid y = 0\}) = S^1$. □

Corollary 10.4.7 *Let f be as in the above proposition, and let us consider*

$$g : \{x + iy, y \geq 0\} \to S^1.$$

given by the restriction of f to the upper half-plane. Then g has the angle-preserving property.

Proof Notice that f is analytic for $z \neq \bar{z_1}$. Since $\bar{z_1} \notin \{x + yi, y \geq 0\}$, we conclude that g is analytic. Moreover, the proof of Proposition 10.4.4 shows that f is the composition of functions f_1, f_2, and f_3 such that f_1 and f_3 have the angle-preserving property for all points of the plane, while f_2 has the angle-preserving property for points $z \neq \bar{z_1}$. Since $\bar{z_1}$ is not in the domain of g, we conclude the desired property for g. □

Exercises

1. Show that a Möbius transformation f can have at most two fixed points in $\mathbf{C}_\infty$, unless f is the identity map.
   Hint: Note that a fixed point satisfies the equation $cz^2 + (d - a)z - b = 0$.

2. Show that Möbius transformation is uniquely determined by its action on any three given points in $\mathbf{C}_\infty$.

3. Find the image of ∞ by the functions below:
   (a) $f(z) = 1/(z - 1)$.
   (b) $f(z) = (z - i)/(iz + i)$.

4. If $f(z) = (az + b)/(cz + d)$, show that $f(\mathbf{R} \cup \infty) = \mathbf{R} \cup \infty$ if and only if a, b, c, d can be chosen to be real numbers.

5. Given any three distinct complex numbers z_1, z_2, and z_3, show that there is a Möbius transformation f such that

$$f(z_1) = 1, \quad f(z_2) = 0, \quad \text{and} \quad f(z_3) = \infty.$$

Hint: Consider

$$f(z) = \Big(\frac{z - z_2}{z - z_3}\Big)\Big(\frac{z_1 - z_2}{z_1 - z_3}\Big)^{-1}.$$

6. Show that there is a unique Möbius transformation f that maps any three distinct complex numbers z_1, z_2, and z_3 to three other distinct complex numbers w_1, w_2, and w_3.

7. Determine all Möbius transformations that map H onto D and $\mathbf{R} \cup \infty$ onto S^1.
Hint: Let $w = (az + b)/(cz + d)$. Choose $z = 0, z = 1$, and $z = \infty$. From $|w| = 1$ one obtains equations that relate a, b, c, d; then proceed as in the proof of Proposition 10.4.6.

10.5 Classification of Isometries

We use the inverse of the transformation g defined at the end of the previous section to carry all geodesics defined in the Poincaré disk model to the upper half-plane. It follows from Corollary 10.4.7 that the geodesics in this model are vertical rays and arcs of circles meeting the real line in right angles, i.e., semicircles. Therefore each isometry in this model is a composition of reflections through these geodesics. We will identify the set of orientation-preserving isometries of the upper half-model with a special group of matrices. For that we will study two types of orientation-preserving isometries; each type is a composition of two reflections only.

The first type is a composition of two reflections in vertical rays and hence it is a translation parallel to the real axis (see Figure 10.4). This transformation can be written as $z \mapsto z + \lambda$, where λ is a real number. Writing $z = a + ib$, in the matrix notation we have

$$\begin{pmatrix} 1 & \lambda \\ 0 & 1 \end{pmatrix}\begin{pmatrix} z \\ 1 \end{pmatrix} = \begin{pmatrix} z+\lambda \\ 1 \end{pmatrix}.$$

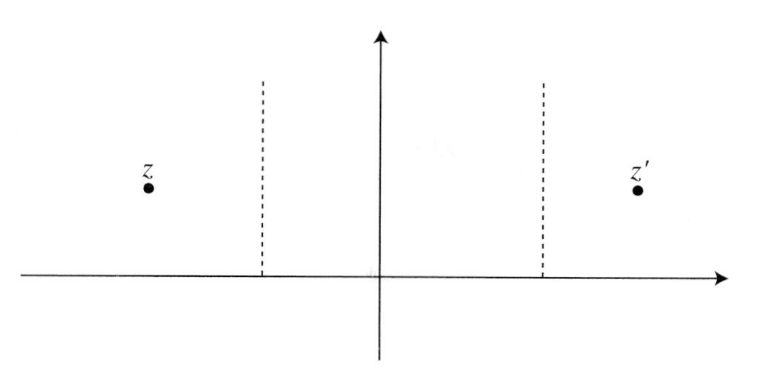

Figure 10.4

The second type we study is an inversion in the circle C of radius μ and centered at $(0,0)$, composed with a reflection in the y-axis. Observe that the map $z \mapsto \mu^2/z$ is an inversion in circle C composed with a reflection in the real line. But the map we are interested in reflects the image of the inversion in the y-axis. Then, to obtain the final image, we rotate μ^2/z about the origin through $180°$, and hence the desired map is given by $z \mapsto -\mu^2/z$, which can be written as $z \mapsto \dfrac{-\mu z}{1/\mu}$.

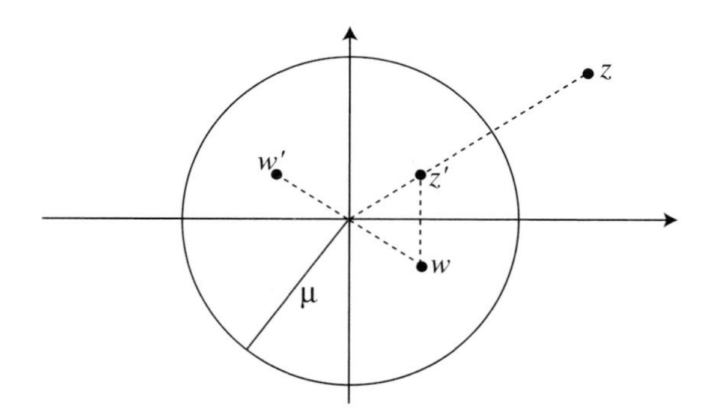

Figure 10.5

In matrix notation this transformation is

$$\begin{pmatrix} -\mu & 0 \\ 0 & 1/\mu \end{pmatrix} \begin{pmatrix} z \\ 1 \end{pmatrix} = \begin{pmatrix} -\mu z \\ 1/\mu \end{pmatrix}.$$

Setting $z = z_1/z_2$, the transformation above can also be seen as

$$\begin{pmatrix} 0 & \mu \\ -1/\mu & 0 \end{pmatrix} \begin{pmatrix} z_1 \\ z_2 \end{pmatrix} = \begin{pmatrix} \mu z_2 \\ -z_1/\mu \end{pmatrix},$$

that is,

$$w = \frac{\mu z_2}{-z_1/\mu} = \frac{-\mu^2}{z}.$$

Now we are ready to study all orientation-preserving isometries. Let us consider a special set of matrices. Let $SL_2(\mathbf{R})$ be the set of all 2×2 real matrices of determinant 1; this group is called *special linear group.* Let $\mathcal{T}$ denote the group of Möbius transformations of the complex plane. We define a map

$$\psi : SL_2(\mathbf{R}) \to \mathcal{T}$$

by

$$A = \begin{pmatrix} p & q \\ r & s \end{pmatrix} \mapsto \psi(A)(z) = \frac{pz+q}{rz+s}.$$

It is a simple exercise of linear algebra to verify that $SL_2(\mathbf{R})$ is a group under matrix multiplication. Observe that ψ is a homomorphism between groups. In fact, if

$$A = \begin{pmatrix} p & q \\ r & s \end{pmatrix}, \qquad B = \begin{pmatrix} a & b \\ c & d \end{pmatrix},$$

then

$$\begin{aligned} \Big(\psi(B) \circ \psi(A)\Big)(z) &= \frac{a\Big((pz+q)/(rz+s)\Big)+b}{c\Big((pz+q)/(rz+s)\Big)+d} \\ &= \frac{(ap+br)z+aq+bs}{(cp+dr)z+cq+ds} \\ &= \psi(BA)(z). \end{aligned}$$

The following lemma shows an important property of Möbius transformations obtained by the homomorphism ψ.

Lemma 10.5.1 *Let $A \in SL_2(\mathbf{R})$. Then $\psi(A)$ leaves the upper half-plane invariant and preserves the real axis.*

Proof Let $A = \begin{pmatrix} p & q \\ r & s \end{pmatrix}$. If z is real, since p, q, r, and s are real numbers, then

$$\frac{pz+q}{rz+s}$$

is on the real line.

Now to prove the second assertion we consider z such that $\mathrm{Im}\,(z) > 0$. We write $pz+q$ and $rz+s$ in polar coordinates. Let $z = a+ib$.

$$p(a+ib)+q = pa+q+ipb = \alpha(\cos\theta + i\,\sin\theta), \quad \text{where} \quad \alpha = |pz+q|.$$

$$r(a+ib)+s = ra+s+irb = \beta(\cos\phi + i\,\sin\phi), \quad \text{where} \quad \beta = |rz+s|.$$

Then we have

$$\cos\theta = \frac{pa+q}{\alpha}, \qquad \cos\phi = \frac{ra+s}{\beta},$$

$$\sin\theta = \frac{pb}{\alpha}, \qquad \sin\phi = \frac{rb}{\beta}.$$

We want to show that $\mathrm{Im}[(pz+q)/rz+s] > 0$. From the above and Proposition 10.2.4(c) we get

$$\frac{pz+q}{rz+s} = \frac{\alpha}{\beta}(\,\cos(\theta-\phi) + i\,\sin(\theta-\phi)).$$

Since $\alpha, \beta > 0$, it is enough to verify that $\sin(\theta-\phi) > 0$. In fact,

$$\begin{aligned} \sin(\theta-\phi) &= \sin\theta\,\cos\phi - \sin\phi\,\cos\theta \\ &= \frac{pb(ra+s) - rb(pa+q)}{\alpha\beta} \\ &= \frac{b(ps-rq)}{\alpha\beta} = \frac{b}{\alpha\beta} > 0. \end{aligned}$$

□

Observe that the homomorphism ψ is not one-to-one, for

$$\begin{pmatrix} -p & -q \\ -r & -s \end{pmatrix}$$

corresponds to the same transformation. Let I denote the identity matrix. We then consider

$$PSL_2(\mathbf{R}) = SL_2(\mathbf{R})/\{\pm I\} = \{(A, -A)\}.$$

This is called the *projective special linear group.* In $PSL_2(\mathbf{R})$ the matrices A and $-A$ are identified and become only one element. Now the map

$$\bar{\psi} : PSL_2(\mathbf{R}) \to \mathcal{T}$$

is injective. From the first isomorphism theorem of group theory (see for instance, Fraleigh[1]) we conclude that $\bar{\psi}(PSL_2(\mathbf{R}))$ is a subgroup of $\mathcal{T}$ which is isomorphic to $PSL_2(\mathbf{R})$ (notice that the kernel of ψ is $\{I, -I\}$). We will show that $\bar{\psi}(PSL_2(\mathbf{R}))$ is the subgroup of all orientation-preserving isometries of the hyperbolic plane in the upper half-plane model. This result will be proved in the next theorem. But before that, we prove a preliminary lemma.

Lemma 10.5.2 *The group $SL_2(\mathbf{R})$ is generated by matrices of the following two forms:*

$$A_1 = \begin{pmatrix} 1 & \lambda \\ 0 & 1 \end{pmatrix}, \qquad A_2 = \begin{pmatrix} 0 & \mu \\ -1/\mu & 0 \end{pmatrix},$$

where $\lambda, \mu \in \mathbf{R}$.

Proof We need to show that any matrix $A \in SL_2(\mathbf{R})$ is the multiplication of matrices of types A_1 and A_2. In fact, if $c \neq 0$, then

$$A = \begin{pmatrix} a & b \\ c & d \end{pmatrix} = \begin{pmatrix} 1 & a/c \\ 0 & 1 \end{pmatrix} \begin{pmatrix} 0 & -1/c \\ c & 0 \end{pmatrix} \begin{pmatrix} 1 & d/c \\ 0 & 1 \end{pmatrix}.$$

If $c = 0$, since $\det(A) = 1$, we have $ad = 1$, which implies $d = a^{-1}$. Therefore

$$A = \begin{pmatrix} a & b \\ 0 & 1/a \end{pmatrix} = \begin{pmatrix} a & 0 \\ 0 & 1/a \end{pmatrix} \begin{pmatrix} 1 & b/a \\ 0 & 1 \end{pmatrix},$$

[1]J. B. Fraleigh, *A First Course on Abstract Algebra*, Addison-Wesley Publishing Co., New York, 1982.

and

$$\begin{pmatrix} a & 0 \\ 0 & 1/a \end{pmatrix} = \begin{pmatrix} 1 & a^2 - a \\ 0 & 1 \end{pmatrix} \begin{pmatrix} 1 & 0 \\ 1/a & 1 \end{pmatrix} \begin{pmatrix} 1 & 1 - a \\ 0 & 1 \end{pmatrix} \begin{pmatrix} 1 & 0 \\ -1 & 1 \end{pmatrix}.$$

To complete the proof, notice that the second and fourth matrix on the right-hand side are of the type described below, which is the multiplication of matrices of type A_1 and A_2. In fact,

$$\begin{pmatrix} 1 & 0 \\ \alpha & 1 \end{pmatrix} = \begin{pmatrix} 0 & -1 \\ 1 & 0 \end{pmatrix} \begin{pmatrix} 1 & -\alpha \\ 0 & 1 \end{pmatrix} \begin{pmatrix} 0 & -1 \\ 1 & 0 \end{pmatrix} \begin{pmatrix} -1 & 0 \\ 0 & -1 \end{pmatrix}$$

and

$$\begin{pmatrix} -1 & 0 \\ 0 & -1 \end{pmatrix} = \begin{pmatrix} 0 & -1 \\ 1 & 0 \end{pmatrix} \begin{pmatrix} 0 & -1 \\ 1 & 0 \end{pmatrix}.$$

□

Theorem 10.5.3 *The group of all orientation-preserving isometries of the hyperbolic plane is isomorphic to $PSL_2(\mathbf{R})$.*

Proof First we have to show that if $A \in PSL_2(\mathbf{R})$, then $\bar{\psi}(A)$ is an orientation-preserving isometry of the hyperbolic plane. Consider all matrices of the following two forms:

$$A_1 = \begin{pmatrix} 1 & \lambda \\ 0 & 1 \end{pmatrix} \qquad A_2 = \begin{pmatrix} 0 & \mu \\ -1/\mu & 0 \end{pmatrix}.$$

We have already shown that $g_1 = \bar{\psi}(A_1)$ and $g_2 = \bar{\psi}(A_2)$ are both orientation-preserving isometries of the hyperbolic plane, since they are compositions of two reflections. It follows from Lemma 10.5.2 that any matrix $A \in PSL_2(\mathbf{R})$ is the multiplication of matrices of types A_1 and A_2. Since $\bar{\psi}$ is a homomorphism, we obtain that $\bar{\psi}(A)$ is the composition of isometries of types g_1 and g_2 and hence it is an orientation-preserving isometry.

Conversely, let g be any orientation-preserving isometry of the hyperbolic plane. We will show that $g = \bar{\psi}(A)$, for some $A \in PSL_2(\mathbf{R})$, i.e., that $\bar{\psi}$ is onto the group of all orientation-preserving isometries of the hyperbolic plane, which in turn implies the desired isomorphism.

Recall that g is the composition of an even number of reflections in geodesics. We sketch below how we show that the composition of any two reflections in geodesics is identified with an element of $PSL_2(\mathbf{R})$. You are asked to verify the details and finish the proof in Exercise 1.
Case 1: composition of two reflections in vertical rays. This is of type $g_1 = \bar{\psi}(A_1)$ studied before.
Case 2: composition of two reflections in semicircles orthogonal to the real line. We have then that each of these transformations is an inversion

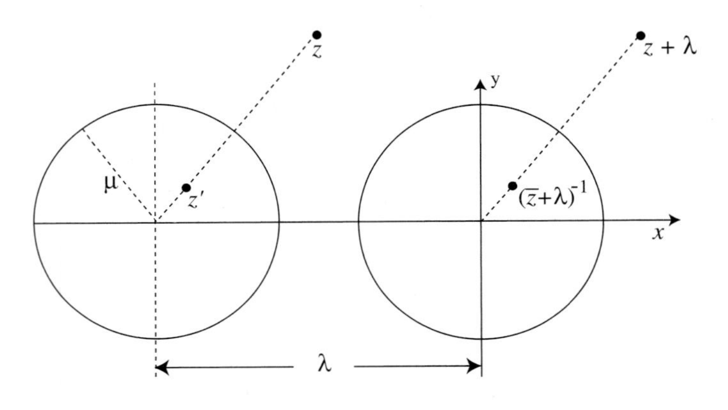

Figure 10.6

in a semicircle. We will use a method similar to the one used in Chapter 5, when we studied rigid motions of Euclidean spaces. We first translate the semicircle of the first inversion so that its center is the origin. Such a translation is given by $z \mapsto z + \lambda$. From the proof of Proposition 10.4.2 we obtain that the image of $z + \lambda$ by an inversion in a circle centered at the origin is $\mu^2/(\overline{z + \lambda}) = \mu^2/(\bar{z} + \lambda)$. To translate back to the original semicircle we need to subtract λ, and therefore the first inversion is obtained by composing the following steps:

$$z \mapsto z + \lambda \mapsto \frac{\mu^2}{\bar{z} + \lambda} \mapsto \frac{\mu^2}{\bar{z} + \lambda} - \lambda = \frac{\mu^2 - \lambda(\bar{z} + \lambda)}{\bar{z} + \lambda}.$$

An expression of the same type is obtained for the second inversion. Composing them, it is straightforward to verify that the composition is of the form

$$z \mapsto \frac{pz+q}{rz+s} = \psi\Big(\begin{pmatrix} p & q \\ r & s \end{pmatrix}\Big)(z), \quad \text{for} \quad p,\ q,\ r,\ s \in \mathbf{R}.$$

Case 3: composition of one reflection in a semicircle with a reflection in a vertical ray.

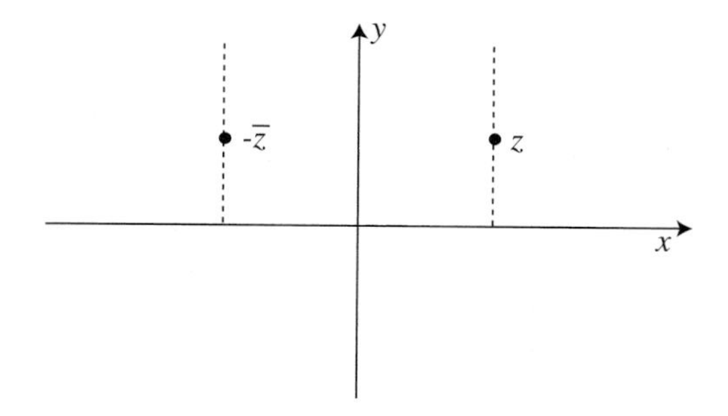

Figure 10.7

For this case, notice first that reflection in the y-axis is given by $z \mapsto -\bar{z}$. Therefore, if the vertical ray is the y-axis, composing, using the expression obtained in case 2 for the inversion, we have the following composition:

$$z \mapsto -\overline{\frac{\mu^2 - \lambda(\bar{z}+\lambda)}{\bar{z}+\lambda}} = \frac{\lambda z + (\lambda^2 - \mu^2)}{z+\lambda} = \frac{(\lambda z + (\lambda^2 - \mu^2))/\mu}{(z+\lambda)/\mu},$$

which is of the form $\bar{\psi}(A)$ for $A \in PSL_2(R)$, since $\lambda, \mu \in \mathbf{R}$. In the case that the vertical ray is not the y-axis, we then compose with suitable translations. □

Now we want to identify the matrices of $PSL_2(\mathbf{R})$ with isometries of the hyperbolic plane in the disk model. Recall that the Möbius transformation

$$f(z) = \frac{z - z_1}{z - \bar{z}_1}$$

carries that upper half-plane to the disk model. We choose $z_1 = i$ and write f in the matrix notation

$$T = \begin{pmatrix} 1 & -i \\ 1 & i \end{pmatrix}, \qquad T^{-1} = \frac{1}{2i}\begin{pmatrix} i & i \\ -1 & 1 \end{pmatrix}.$$

Given $A \in PSL_2(\mathbf{R})$, we define $B = T \circ A \circ T^{-1}$ such that the diagram below commutes.

$$\begin{array}{ccccc} & z & \stackrel{A}{\longmapsto} & z' & \\ T & \downarrow & & \downarrow & T. \\ & w & \stackrel{B}{\longmapsto} & w' & \end{array}$$

Therefore

$$\begin{aligned} B &= \frac{1}{2i}\begin{pmatrix} 1 & -i \\ 1 & i \end{pmatrix}\begin{pmatrix} p & q \\ r & s \end{pmatrix}\begin{pmatrix} i & i \\ -1 & 1 \end{pmatrix} \\ &= \frac{1}{2}\begin{pmatrix} p+s+i^{-1}(r-q) & p-s+i^{-1}(r+q) \\ p-s-i^{-1}(r+q) & p+s-i^{-1}(r-q) \end{pmatrix}. \end{aligned}$$

We know that if we multiply by a constant we obtain the same Möbius transformation. Multiplying by $i^2 = -1$, we still have determinant 1, and we get

$$B = \frac{1}{2}\begin{pmatrix} -(p+s)+i(r-q) & -(p-s)+i(r+q) \\ -(p-s)-i(r+q) & -(p+s)-i(r-q) \end{pmatrix},$$

which is a Möbius transformation of the type

$$z \mapsto \frac{az+\bar{c}}{cz+\bar{a}}, \quad \text{where } |a|^2 - |c|^2 = 1.$$

Therefore we can state:

Theorem 10.5.4 *The group*

$$G = \{g : D^2 \to D^2 \mid g(z) = \frac{az+\bar{c}}{cz+\bar{a}}, \ |a|^2 - |c|^2 = 1\}$$

describes all orientation-preserving isometries of the hyperbolic plane in the disk model. It follows that G is isomorphic to $PSL_2(\mathbf{R})$.

Exercises

1. Complete the proof of Theorem 10.5.3, filling in the details.
2. Show that if an orientation-preserving isometry in the disk model has three fixed points on the circle at infinity, then it is the identity.
3. Show that $g : D \to D$ given by

$$g(z) = \frac{az+\bar{c}}{cz+\bar{a}}, \quad 0 \neq |a|^2 - |c|^2 < 1,$$

is an element of $\mathrm{Aut}(\mathbf{H}^2)$.

10.6 The Hyperbolic Metric

In this section we use the Poincaré disk model to study some properties of hyperbolic geometry. Let g be an orientation-preserving isometry of $\mathbf{H}^2$, that is, an element of the group

$$\mathrm{Aut}(\mathbf{H}^2) = \{g : D \to D \mid g(z) = \frac{az + \bar{c}}{cz + \bar{a}},\ |a|^2 - |c|^2 = 1\}.$$

Proposition 10.6.1 *Let $g \in \mathrm{Aut}(\mathbf{H}^2)$. Then g satisfies the following identity:*

$$g'(z) = \frac{1 - |g(z)|^2}{1 - |z|^2}.$$

Proof Differentiating

$$g(z) = \frac{az + \bar{c}}{cz + \bar{a}}$$

we have

$$g'(z) = \frac{(cz + \bar{a})a - (az + \bar{c})c}{(cz + \bar{a})^2} = \frac{a\bar{a} - c\bar{c}}{(cz + \bar{a})^2} = \frac{1}{(cz + \bar{a})^2}.$$

Therefore

$$g'(z)\overline{g'(z)} = \frac{1}{(cz + \bar{a})^2\,(\overline{cz} + a)^2},$$

which implies

$$|g'(z)| = \sqrt{(g'(z)\overline{g'(z)}} = \frac{1}{|c|^2|z|^2 + acz + \overline{acz} + |a|^2}.$$

We leave it to the reader to verify that

$$\frac{1 - |g(z)|^2}{1 - |z|^2} = \frac{1}{|c|^2|z|^2 + acz + \overline{acz} + |a|^2}.$$

Let $\gamma : [a, b] \to D$ be a smooth curve parametrized by $\gamma(t) = (x(t), y(t))$. Recall that the *Euclidean* length of γ is given by

$$\int_a^b |\gamma'(t)|\, dt = \int_a^b \sqrt{x'(t)^2 + y'(t)^2}\, dt.$$

Writing γ as $z(t) = x(t) + iy(t)$, we get that $dz = [x'(t) + iy'(t)]\, dt$ and hence $|dz| = \sqrt{x'(t)^2 + y'(t)^2}\, dt$. Therefore

$$\int_a^b \sqrt{x'(t)^2 + y'(t)^2}\, dt = \int_\gamma |dz|.$$

□

Definition 10.6.2 *Let $l_D : D \to (0, \infty)$ be given by*

$$l_D(z) = \frac{2}{1 - |z|^2}.$$

The hyperbolic *length is defined by*

$$L(\gamma) = \int_\gamma l_D(z)|dz|.$$

Proposition 10.6.3 *If $g \in \mathrm{Aut}(\mathbf{H}^2)$, then $L(g(\gamma)) = L(\gamma)$; i.e., g preserves length.*

Proof We write $g(\gamma(t)) = w(t) = u(t) + iv(t)$. Therefore

$$L(g(\gamma)) = \int_a^b \frac{2}{1 - |w(t)|^2} |w'(t)|\, dt.$$

Recall that $w'(t) = g'(z(t))z'(t)$ and thus

$$L(g(\gamma)) = \int_a^b \frac{2}{1 - |g(z(t))|^2} |g'(z(t)||z'(t)|\, dt.$$

Using Proposition 10.6.1, we have

$$\frac{2}{1 - |g(z(t))|^2} |g'(z(t)| = \frac{2}{1 - |z(t)|^2},$$

which implies

$$\begin{aligned} L(g(\gamma)) &= \int_a^b \frac{2}{1 - |z(t)|^2} |z'(t)|\, dt \\ &= \int_\gamma \frac{2}{1 - |z|^2\, |dz|} = L(\gamma). \end{aligned}$$

□

Definition 10.6.4 *Let Γ be the set of all smooth curves γ in D connecting the points $z, w \in D$. The* hyperbolic distance *between them is defined by*

$$\rho(z,w) = \inf\{L(\gamma) \mid \gamma \in \Gamma\}.$$

One can show that ρ satisfies the definition of metric, and this is Exercise 1. It follows from Proposition 10.6.3 that

$$\rho(z,w) = \rho(g(z), g(w)) \quad \forall g \in \mathrm{Aut}(H^2),$$

which shows that g is an isometry of ρ.

Proposition 10.6.5 *For each $z \in D$,*

$$\rho(0,z) = \ln\left(\frac{1+|z|}{1-|z|}\right).$$

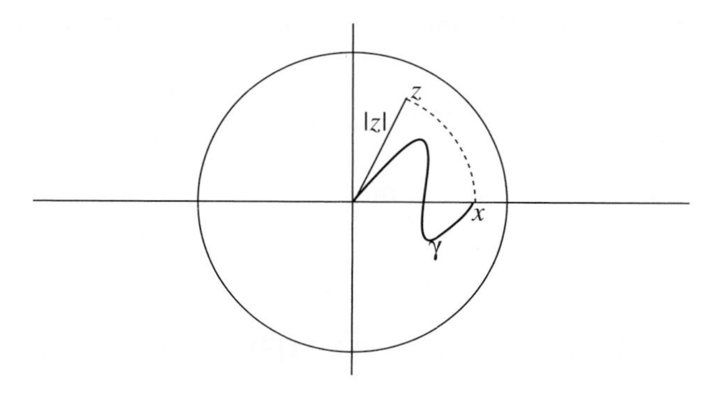

Figure 10.8

Proof Let g be a hyperbolic rotation that maps z to $|z|$ (see Figure 10.8). Then $\rho(0,z) = \rho(0,|z|)$, and it suffices to prove the proposition for the case that

$$z = x, \quad 0 < x < 1.$$

Let $\gamma(t) = a(t) + ib(t)$ be a smooth curve from 0 to x. Thus $a(0) = 0$ and $a(1) = x$. We compute the length of γ.

$$L(\gamma) = \int_\gamma \frac{2}{1-|z|^2}\,|dz| = \int_0^1 \frac{2\sqrt{a'(t)^2 + b'(t)^2}}{1-(a(t)^2+b(t)^2)}\,dt$$

$$
\begin{aligned}
&\geq \int_0^1 \frac{2\sqrt{a'(t)^2}}{1-(a(t))^2}\,dt \geq \int_0^1 \frac{2a'(t)}{1-(a(t))^2}\,dt \\
&= \int_0^x \frac{2}{1-a^2}\,da \\
&= \ln\frac{1+x}{1-x}\,.
\end{aligned}
$$

Therefore

$$\rho(0,z) = \ln\Big(\frac{1+|z|}{1-|z|}\Big).$$

□

Observe that $\rho(0,z) \to \infty$ as $|z| \to 1$, which justifies our calling S^1 the circle at infinity. Since the segment that connects the origin to z is an arc of geodesic, we can say that this line segment can be extended beyond any prescribed length (compare with Euclid's second postulate).

Proposition 10.6.6 *Let z and w be two arbitrary points in D^2. Then*

$$\rho(z,w) = \ln\Big(\frac{|1-z\bar{w}|+|z-w|}{|1-z\bar{w}|-|z-w|}\Big).$$

Proof Consider the conformal map

$$g(z) = \frac{z-w}{1-z\bar{w}}.$$

It follows from Exercise 3 that $g \in \mathrm{Aut}(\mathbf{H}^2)$. Moreover, $g(w) = 0$. Thus

$$
\begin{aligned}
\rho(w,z) &= \rho(0,g(z)) = \ln\Big(\frac{1+|g(z)|}{1-|g(z)|}\Big) \\
&= \ln\Big(\frac{|1-z\bar{w}|+|z-w|}{1-z\bar{w}|-|z-w|}\Big).
\end{aligned}
$$

□

Theorem 10.6.7 *Let γ be a geodesic from z to w in D. Then $\rho(z,w) = L(\gamma)$.*

Proof Given z and w, consider $g \in \mathrm{Aut}(\mathbf{H}^2)$ that takes z to the origin 0 and w to w'. The proof of Proposition 10.6.5 implies that the line segment $\overline{0w'}$ is the arc of shortest length connecting these two points and therefore by the definition of ρ, $\rho(0, w') = L(\overline{0w'})$. Since $\overline{0w'}$ is a geodesic connecting 0 to w' and g takes geodesics to geodesics, we conclude that $g(\overline{0w'})$ is the unique geodesic connecting z to w, that is, $g(\overline{0w'}) = \gamma$. In addition, g preserves length and distance. Therefore

$$\rho(z, w) = \rho(0, w') = L(\overline{0w'}) = L(\gamma).$$

Now we consider the map $f(z) = (z - i)/z + i$, the map that carries $H = \{x + iy \mid y > 0\}$ onto D. We want to define length in the upper half-plane model for the hyperbolic plane. We define a function

$$l_H : H \to (0, \infty)$$

by

$$l_H(z) = l_D(f(z))|f'(z)|.$$

If $\alpha(t)$ is a parametrized curve in H, we consider $\gamma(t) = f(\alpha(t))$. We then define

$$L(\alpha) = \int_\alpha l_H(z)|dz|.$$

□

Lemma 10.6.8 *Let $w = f(z)$ and let us denote $\gamma(t) = f(\alpha(t))$. Then*

$$L(\alpha) = L(\gamma).$$

Proof

$$L(\gamma) = \int_\gamma \frac{2}{1 - |w|^2}|dw| = \int_0^1 \frac{2}{1 - |w(t)|^2}|w'(t)|\, dt.$$

From Lemma 10.3.12 we obtain that $w'(t) = f'(z)z'(t)$, which, substituted above, gives

$$\begin{aligned} L(\gamma) &= \int_0^1 \frac{2}{1 - |f(z((t)))|^2}|f'(z)||z'(t)|\, dt \\ &= \int_0^1 l_H(z)|z'(t)|\, dt \\ &= \int_\alpha l_H(z)|dz| = L(\alpha). \end{aligned}$$

□

Proposition 10.6.9 *Let $l_H(z)$ be defined as above. If $z = x + iy$, then $l_H(z) = 1/y$.*

Proof This proof is computational, and we leave it to the reader. □

Exercises

1. Show that (D, ρ) is a metric space.
   Hint: See Example 3 of Section 7.2 and use the same approach.

2. Prove Proposition 10.6.9

10.7 Hyperbolic Triangles

Definition 10.7.1 *A hyperbolic triangle in D is a geometric figure containing three points, called* vertices, *three sides, which are arcs of geodesics joining the pairs of vertices, and three angles.*

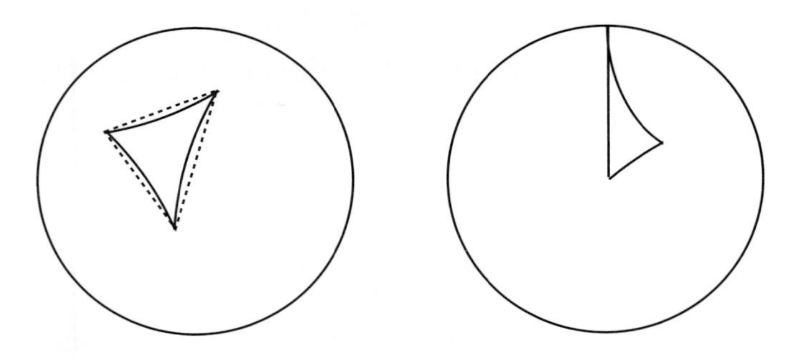

Figure 10.9

Figure 10.9 compares Euclidean and hyperbolic triangles in the disk model of the hyperbolic plane and illustrates the fact that the angle sum of hyperbolic triangles is less than 180°.

Recall that a *generalized triangle* is a triangle that contains one vertex on the circle at infinity. Two sides ending at a such vertex are said to be tangent. We then say that they form an angle of measure 0.

Definition 10.7.2 *The hyperbolic trigonometric functions are defined by*

$$\sinh(x) = \frac{e^x - e^{-x}}{2}, \qquad \cosh(x) = \frac{e^x + e^{-x}}{2}.$$

Lemma 10.7.3

$$\cosh\, \rho(z, w) = \frac{|1 - z\bar{w}|^2 + |z - w|^2}{|1 - z\bar{w}|^2 - |z - w|^2}\,.$$

Proof This proof is left as an exercise for the reader. □

Now consider the triangle with vertices at $0, x$, and iy (see Figure 10.10).

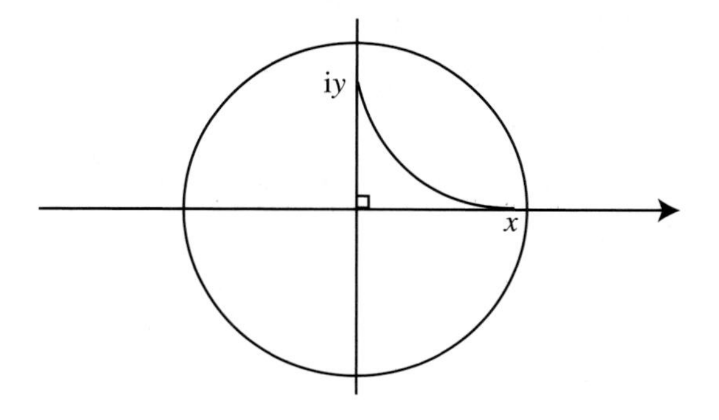

Figure 10.10

Applying Lemma 10.7.3 to points x and iy, we have

$$\begin{aligned}\cosh \rho(x, iy) &= \frac{1 + x^2 + y^2 + x^2y^2}{1 - x^2 - y^2 + x^2y^2} \\ &= \left(\frac{1 + x^2}{1 - x^2}\right)\left(\frac{1 + y^2}{1 - y^2}\right) \\ &= \cosh \rho(0, x) \cosh \rho(0, y).\end{aligned}$$

This gives the following theorem.

Theorem 10.7.4 (The Pythagorean theorem in hyperbolic geometry) *Let a be the hypotenuse of a right triangle with legs b and c. Then*

$$\cosh\, a = \cosh\, b\, \cosh\, c.$$

Proof Choose $g \in \text{Aut}(\mathbf{H}^2)$ that maps the vertex of the right angle to the origin 0. Since g preserves angle measure, $g(\triangle)$ is a right triangle with a vertex at 0 and may be rotated by an isometry fixing the origin so that the other two vertices are on the x-axis and y-axis. From the above we get

$$\cosh a' = \cosh b' \cosh c',$$

where a', b', c' are the sides of $g(\triangle)$. Since g preserves hyperbolic length, we conclude the result. □

Definition 10.7.5 *The* hyperbolic area *of any region $\mathcal{R} \subset D$ is defined as*

$$\text{Area}(\mathcal{R}) = \int\int_{\mathcal{R}} l_D(z)^2 \, dx \, dy = \int\int_{\mathcal{R}} \frac{4}{(1-|z|^2)^2} \, dx \, dy$$

It follows from Lemma 10.6.9 that if $\mathcal{R} \subset H$, then

$$\text{Area}(\mathcal{R}) = \int\int_{\mathcal{R}} \frac{1}{y^2} \, dx \, dy.$$

Theorem 10.7.6 (The Gauss-Bonnet theorem in hyperbolic geometry) *Let $\triangle$ be a triangle with angles measuring α, β, and γ. Then*

$$\text{Area}(\triangle) = \pi - (\alpha + \beta + \gamma).$$

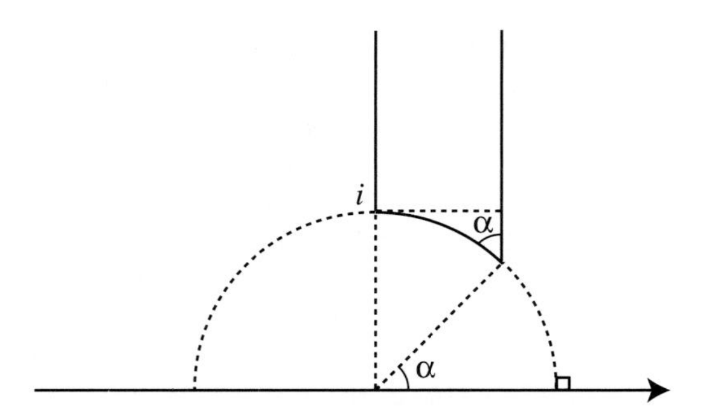

Figure 10.11

Proof We sketch the proof here. The reader is asked to fill in the details in Exercise 2. We start by considering a triangle with angles

$0, \pi/2$, and α. Then, in the disk model, such a triangle must have a vertex $P \in S^1$. Let us denote the other two vertices by R and Q, with the right angle at R. We know that there is a Möbius transformation that maps P to a real number x and the points R and Q to points in the upper half-plane H, say z_1 and z_2. We will compute the area of the triangle of vertices x, z_1, z_2 in the upper half-plane model of the hyperbolic plane.

From Exercise 2 we conclude that this area can be found using the triangle whose vertices are ∞, i, and $\cos\alpha + i\sin\alpha$, with the right angle at i as shown in Figure 10.11. Then we have

$$\begin{aligned}
\text{Area}(\triangle) &= \lim_{a\to\infty}\int_1^a \int_0^{\cos\alpha} \frac{1}{y^2}\,dx\,dy \;+\; \int_0^{\cos\alpha}\int_{\sqrt{1-x^2}}^1 \frac{1}{y^2}\,dy\,dx \\
&= \cos\alpha\Big(\lim_{a\to\infty}\frac{-1}{a}+1\Big) \;+\; \int_0^{\cos\alpha}\Big(-1+\frac{1}{\sqrt{1-x^2}}\Big)\,dx \\
&= \cos\alpha - \cos\alpha - \cos^{-1}(\cos\alpha) + \cos^{-1}0 \\
&= \frac{\pi}{2} - \alpha.
\end{aligned}$$

Now we consider a triangle that has angles $0, \alpha$, and β. In the disk model, such a triangle also has a vertex at $P \in S^1$. Using the same notation for the other two vertices and using Exercise 11 of Section 9.2, we consider the altitude of $\triangle PQR$ that has its foot S on the geodesic segment γ_{QR}. We then have two cases (see Figure 10.12):

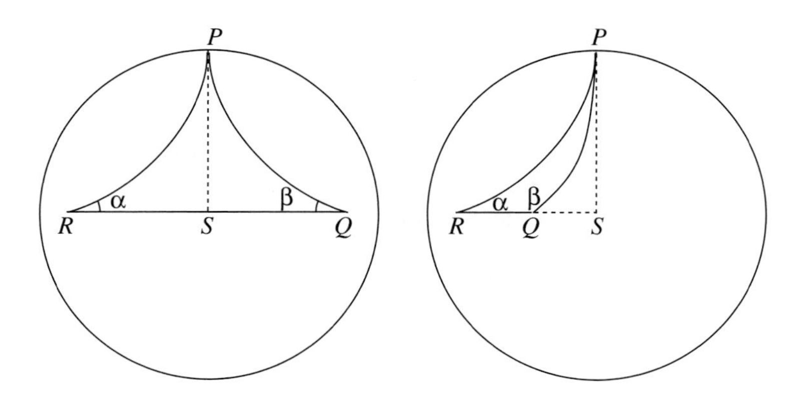

Figure 10.12

(i) S is between Q and R. We have two triangles with angles 0 and $\pi/2$. From the first part of this proof we get

$$\begin{aligned}\text{Area}(\triangle) &= \text{Area}(\triangle PQS) + \text{Area}(\triangle PSR) \\ &= \frac{\pi}{2} - \alpha + \frac{\pi}{2} - \beta \\ &= \pi - (\alpha + \beta).\end{aligned}$$

(ii) S is not between Q and R, and without loss of generality we assume that Q is between R and S. Using again the first part of the proof, we obtain

$$\begin{aligned}\text{Area}(\triangle) &= \text{Area}(\triangle PSR) - \text{Area}(\triangle PSQ) \\ &= \frac{\pi}{2} - \alpha - (\frac{\pi}{2} - (\pi - \beta)) \\ &= \pi - (\alpha + \beta).\end{aligned}$$

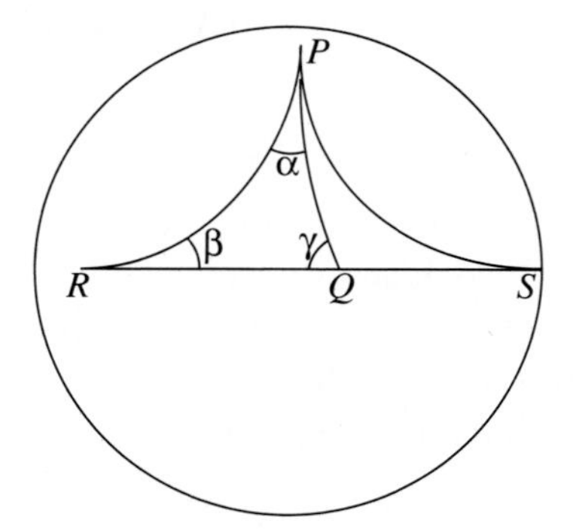

Figure 10.13

We conclude the proof by considering the general case of a triangle $\triangle$ of angles α, β, and γ (see Figure 10.13). Let S denote the point where geodesic γ_{QR} meets the circle S^1. Then we have

$$\begin{aligned}\text{Area}(\triangle) &= \text{Area}(\triangle PSR) - \text{Area}(\triangle PSQ) \\ &= \pi - \alpha'' - \beta - (\pi - \alpha' - \gamma'),\end{aligned}$$

where α' and α'' are the measures of the angles that triangles $\triangle PQS$ and $\triangle PRS$ have, respectively, at vertex P. Notice that

$$\alpha = \alpha'' - \alpha' \quad \text{and} \quad \pi - \gamma' = \gamma,$$

which yields

$$\text{Area}(\triangle) = \pi - (\alpha + \beta + \gamma).$$

□

Exercises

1. Consider the maps given by $g_1(z) = z + \lambda$ and $g_2(z) = \mu^2 z$, where λ and μ are real numbers.
   (a) Show that g_1 and g_2 are orientation-preserving isometries of the hyperbolic plane in the upper half-plane model.
   (b) Show that g_1 and g_2 both map ∞ to ∞.

2. Let $\triangle$ be a hyperbolic triangle in the disk D with angles of measure $0, \pi/2$, and α. In the upper half-plane model, such a triangle has vertices at $x \in \mathbf{R}$, z_1, and z_2 with the right angle at z_1.
   Show that there is an isometry g of the hyperbolic plane that maps the vertices of the triangle to ∞, i, and $\cos\alpha + i\sin\alpha$, with the right angle at i. For that, follow these steps:
   (1) Find $p, q, r, s \in \mathbf{R}$ such that $ps - rq = 1$ and $rx + s = 0$. Consider the isometry

$$z \mapsto h(z) = \frac{pz + q}{rz + s}$$

   and let w_1 and w_2 denote the images of z_1 and z_2, respectively.
   (2) Compose h with isometries of types g_1 and g_2 described in Exercise 1 so that w_1 is mapped to i and w_2 to $\cos\alpha + i\sin\alpha$ (see the figure below). Justify the procedure.

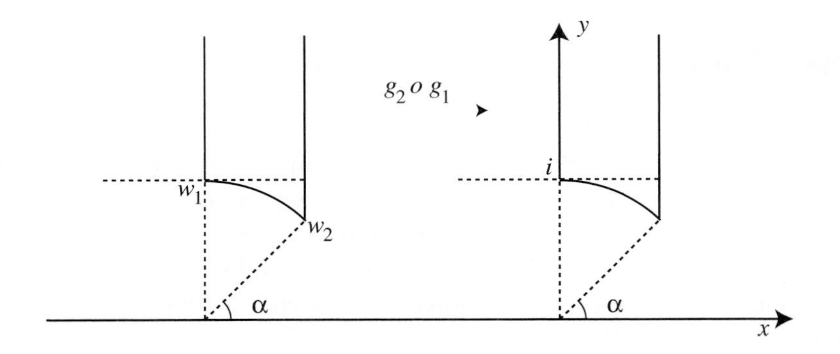

10.8 Hyperbolic Circles

In Chapter 1, using the axioms of neutral geometry, we defined circles and proved some results that did not require the Euclidean parallel postulate. This implies that they hold for both hyperbolic and Euclidean circles. Now in the disk model for the hyperbolic plane we will see an interesting result relating circles in these two geometries. Let us consider the sets:

$$\begin{aligned} C_H &= \{z \in D \mid \rho(0, z) = r\}, \\ C_E &= \{z \in D \mid d(0, z) = R\}, \end{aligned}$$

where ρ and d denote the hyperbolic and the Euclidean metric, respectively. Then, C_H denotes the hyperbolic circle and C_E a Euclidean circle.

Lemma 10.8.1 *Let C_H and C_E be as above. Then $C_H = C_E$ if and only if*

$$r = \ln(1 + R) - \ln(1 - R).$$

Proof If $z \in C_H$, then by Proposition 10.6.5 we have

$$\rho(0, z) = r = \ln\left(\frac{1 + |z|}{1 - |z|}\right).$$

If $z \in C_E$, then $|z| = R$ and hence

$$r = \ln\left(\frac{1 + R}{1 - R}\right).$$

The converse is proved in the same way. □

Proposition 10.8.2 *A set of points in D is a hyperbolic circle if and only if it is a Euclidean circle.*

Proof Let C be a hyperbolic circle centered at w and radius r. Let g be an isometry of (D^2, ρ) that maps w to 0. Then $\rho(g(z), 0) = \rho(z, w) = r$, implying that $g(C)$ is a hyperbolic circle centered at 0. From the previous lemma we get that $g(C)$ is also a Euclidean circle, and since hyperbolic isometries carry Euclidean circles to Euclidean circles, C is also a Euclidean circle.

The converse is proved in a similar manner. □

We point out that, although a hyperbolic circle and a Euclidean circle are the same set of points in the disk model, these two circles do not have the same center, unless the center is the origin. This fact follows from the expression for the hyperbolic metric ρ, i.e.,

$$\{z \mid \|z-w\| = R\} = \left\{z \mid \ln \frac{|1-z\bar{w}|+|z-w|}{|1-z\bar{w}|-|z-w|} = \ln\left(\frac{1+R}{1-R}\right)\right\}$$

if and only if $w = 0$.

Proposition 10.8.3 *The length of a hyperbolic circle of radius r is $2\pi \sinh r$.*

Proof Let R be the Euclidean radius of a circle γ. Its hyperbolic length is given by

$$L(\gamma) = \int_\gamma l_H(z)|dz| = \int_0^{2\pi} \frac{2R}{1-R^2}d\theta = \frac{4\pi R}{1-R^2}.$$

From Lemma 10.8.1 we get that

$$(e^r - e^{-r}) = \frac{4R}{1-R^2}$$

and then

$$L(\gamma) = 2\pi\frac{e^r - e^{-r}}{2} = 2\pi \sinh r.$$

□

The circumference grows rapidly as $|z| \to 1$. Since $e^{-r} \to 0$ as $r \to \infty$, a circle through a point near the boundary S^1 has a very large length.

Proposition 10.8.4 *The area of a hyperbolic circle of radius r is $4\pi \sinh^2(r/2)$.*

Proof This proof is left as an exercise. □

Notice that the area also grows rapidly as $r \to \infty$.

LIST OF AXIOMS

Incidence Axioms

I_1 Given two distinct points P, Q, there exists a unique line incident with P and Q.
I_2 For every line l there exist at least two distinct points incident with l.
I_3 There exist three distinct points with the property that no line is incident with all three of them.

Betweenness Axioms

B_1 If B is between A and C, then A, B, and C are three distinct collinear points and B is between C and A.
B_2 Given two distinct points B and D, let l be a line incident with B and D. Then there exist points A, C, and E such that B is between A and D, C is between B and D, and D is between B and E.
B_3 Given three distinct points lying on the same line, one and only one of the points is between the other two.
B_4 Given a line l and three points A, B, and C not lying on l,
(a) If A and B are on the same side of l and B and C are on the same side of l, then A and C are on the same side of l.
(b) If A and B are on opposite sides of l and B and C are on opposite sides of l, then A and C are on the same side of l.

Measurement Axioms for Line Segments

S_1 To every pair of points A and B there corresponds a real number $x \geq 0$ such that $x = 0$ if and only if $A = B$. The number x will be called the *length* of $\overline{AB}$ and it will be denoted by AB.
S_2 There is a one-to-one correspondence between the points of a line l and the set of real numbers $\mathbf{R}$ such that, if to point A corresponds number a and to point B corresponds b, then $AB = |b - a|$.
S_3 If C is between A and B, then $AC + CB = AB$.

Measurement Axioms for Angles

A_1 To every angle $\angle ABC$ corresponds a unique real number x such that $0 \leq x \leq 180$. Further,
(i) $x = 0$ if and only if $\overrightarrow{BA} = \overrightarrow{BC}$.
(ii) $x = 180$ if and only if $\overrightarrow{BA}$ and $\overrightarrow{BC}$ are opposite rays.
A_2 Let $\overrightarrow{AB}$ be a ray and H one half-plane determined by line through points A and B. Then for every x between 0 and 180 there is only one ray $\overrightarrow{AC}$, with $C \in H$ such that $m(\angle CAB) = x$.
A_3 If ray $\overrightarrow{AD}$ divides angle $\angle BAC$, then $m(\angle BAD) + m(\angle DAC) = m(\angle BAC)$.

Congruence Axiom

SAS: If two sides and the included angle of one triangle are congruent to two sides and the included angle of another triangle, then these two triangles are congruent.

Euclidean Parallel Postulate

For every line l and every point P not lying on l, there exists a *unique* line through P that is parallel to l.

Axioms of Neutral Geometry

Incidence, betweenness, measurement, congruence axioms.

Axioms of Plane Euclidean Geometry

Axioms of neutral geometry and the Euclidean parallel postulate.

Hyperbolic Parallel Postulate

There exists a line l such that for some point P not on l at least two lines parallel to l pass through P.

Axioms of Plane Hyperbolic Geometry

Axioms of neutral geometry and the hyperbolic parallel postulate.

Axioms for Euclidean 3-Space

Sp$_1$ Any three points lie in at least one plane, and any three non-collinear points lie in exactly one plane.
Sp$_2$ There exist four distinct points with the property that no plane contains all four of them.
Sp$_3$ If two points lie in a plane, then the line incident with these points lies in the same plane.
Sp$_4$ The intersection of two planes contains at least two distinct points.

Area Axioms

Ar$_1$: If two triangular regions are congruent, then they have the same area.
Ar$_2$: If the intersection of two polygonal regions does not contain interior points (only edges or vertices), then the area of their union is the sum of their areas.
Ar$_3$: If a square region has edges of length s, then its area is s^2.

Volume Axioms

V$_1$: If $A, B \in \mathcal{V}$ and $A \subset B$, then $V(A) \leq V(B)$.
V$_2$: If $A, B \in \mathcal{V}$ and $V(A \cap B) = 0$, then $V(A \cup B) = V(A) + V(B)$.
V$_3$: The volume of a parallelepiped is the product of the area of a face and its altitude.
V$_4$: *Cavalieri's principle*: Given two solids $\mathcal{S}_1$, $\mathcal{S}_2$ and a plane $\mathcal{P}$. If for every plane that intersects $\mathcal{S}_1$ and $\mathcal{S}_2$ and is parallel to $\mathcal{P}$, the two intersections determine regions that have the same area, then the two solids have the same volume.

Completeness Axiom of the Real Numbers

Let S be a nonempty set of real numbers.
(i) If S is bounded above, then $\sup S$ exists.
(ii) If S is bounded below, then $\inf S$ exists.

Index